“十四五”普通高等院校计算机类专业系列教材

C语言程序设计

（双语版）

赵小蕾　许喜斌　耿江涛◎主　编
何　秀　丁　丰◎副主编

中国铁道出版社有限公司
CHINA RAILWAY PUBLISHING HOUSE CO., LTD.

内 容 简 介

本书是“十四五”普通高等院校计算机类专业系列教材之一，从基础语法到高级应用深入剖析了C语言程序设计的核心知识与实用技巧，详细解读了C语言的基本概念、数据类型、运算符、控制结构、数组、指针、函数、结构体和文件操作等关键内容，并通过实际案例让读者深入理解C语言的编程精髓。

本书系统性强，逻辑严谨，实例丰富，采用循序渐进的方式让读者逐步掌握相关知识，强化编程能力。书中对重点、难点进行了详细解析，并配有丰富的电子教学资源，提供一站式C语言学习体验，引领读者轻松开启编程之旅。中英文对照的编写方式有助于学生提升英语能力，熟悉专业词汇。

本书适合作为普通高等院校软件工程、计算机科学与技术、数字媒体技术、电子信息科学与技术、电气工程及其自动化、智能制造工程、人工智能、大数据等专业平台课程“C语言程序设计”教材，也可作为编程爱好者的参考书。

图书在版编目（CIP）数据

C语言程序设计 : 双语版 / 赵小蕾, 许喜斌, 耿江涛主编. -- 北京 : 中国铁道出版社有限公司, 2024. 10.
(“十四五”普通高等院校计算机类专业系列教材).
ISBN 978-7-113-31341-8

Ⅰ. TP312.8

中国国家版本馆CIP数据核字第2024QW9708号

书　　名：C语言程序设计（双语版）
作　　者：赵小蕾　许喜斌　耿江涛

策　　划：唐　旭　　　编辑部电话：（010）51873090
责任编辑：刘丽丽　徐盼欣
封面设计：穆　丽
封面制作：刘　颖
责任校对：安海燕
责任印制：赵星辰

出版发行：中国铁道出版社有限公司（100054，北京市西城区右安门西街8号）
网　　址：https://www.tdpress.com/51eds
印　　刷：天津嘉恒印务有限公司
版　　次：2024年10月第1版　2024年10月第1次印刷
开　　本：889 mm × 1 194 mm　1/16　印张：18.75　字数：580千
书　　号：ISBN 978-7-113-31341-8
定　　价：59.80元

前　言

随着信息技术的飞速发展，计算机科学与技术已经成为推动社会进步的重要力量。作为这一领域的基石，C语言以其简洁、高效的特点，成为众多编程初学者的首选，高校软件工程、计算机科学与技术、数字媒体技术、电子信息科学与技术、电气工程及其自动化、智能制造工程、人工智能、大数据等专业都开设了C语言程序设计教程。本书旨在满足高校C语言程序设计课程教学要求，为广大编程爱好者提供一套系统、全面且贴近实践的C语言教材及指南，帮助读者掌握C语言的核心知识和编程技巧，培养扎实的编程技能和算法思维，为读者后期学术研究和职业发展提供重要的技能支持。

本书在编写过程中，紧密结合现代教学理论和教学实践，始终坚持“以学生为中心，以实践为导向”的写作思想，理论讲解深入浅出，实践案例讲解详细，具有启发性，可以帮助读者理论联系实践、实践强化理论，形成良性闭环。为满足高校中、英双语教学需求，本书采用中、英双语对照编写方式。

本书具有以下特色：

1. 系统性与实用性结合

本书内容设计系统全面，覆盖C语言的各个重要知识点，可以帮助读者逐步掌握复杂的编程概念和技巧。

2. 注重实践与应用

本书分析了读者在学习过程中的难点和重点，提供了丰富的实践案例，通过详细讲解和案例演示，帮助读者将所学知识应用到实际中，提升编程能力和解决问题的能力。

3. 双语交相呼应

本书关键术语和核心内容以中英文对照的方式呈现，以提升读者的英语应用能力和跨文化交流能力，满足中、英双语教学需求。

4. 灵活多样的学习方式

本书适合作为普通高等院校教材使用，也可作为自学参考书。书中内容结合现代教育技术，提供丰富的电子教学资源和在线学习平台，以方便读者随时随地进行学习和复习；同时也适用多种教学方法，包括讲解、演示、练习、讨论等。

本书共分为九章，内容层层递进。第1章主要论述C语言的基本概念，并带领读者初识C程序，帮助读者建立对C语言的初步认识。第2~4章详细论述C语言的基础语法和程序结构，包括顺序结构、选择结构和循环结构等。这些章节将通过丰富的案例，帮助读者熟练掌握C语言的基础语法和编程技巧。从第5章开始逐步深入C语言的高级特性，论述数组的概念和处理批量数据应用场景。第6章论述指针的概念和用法，促进对内存管理的深入理解。第7章论述函数的定义和调用方法，以及递归编程等高级技巧。第8章论述结构体的概念和用法，帮助读者构建更复杂的数据结构。第9章论述文件操作等实用技能。

为帮助读者更好地学习和掌握本书内容，每一章都设计了详细的课时安排和教学建议。我们建议读者按照章节顺序逐步学习，也鼓励读者根据自己的实际情况和兴趣进行选择性学习或深入研究某一特定领域。在教学过程中，建议教师采用“讲练结合”的方式进行教学。在讲解每个知识点后，可以安排相应的实践环节让学生巩固所学知识并提升编程技能。教师也可以根据学生的学习情况和需求进行灵活调整，优化教学内容，进度安排，详见下表：

周次	章　节	课时	教学活动
1	第 1 章	4	课程导入，C 语言引入，环境安装与运行，算法概念讲解
2~3	第 2 章	8	顺序程序设计基础，数据类型讲解，综合案例演练
4~5	第 3 章	8	选择结构语法与应用，if 和 switch 语句练习，综合案例演练
6~7	第 4 章	8	循环结构语法与应用，while、do...while 和 for 语句练习，综合案例演练
8	第 5 章	4	一维数组、二维数组讲解，字符数组与字符串应用
9	第 6 章	4	指针基础，指针与数组、字符串的关系及应用
10~11	第 7 章	8	函数定义与调用，局部变量与全局变量，递归程序设计
12~13	第 8 章	8	结构体定义与使用，动态内存分配及链表实现
14	第 9 章	4	文件操作基础，文件读写应用
15	综合复习与考试	4	复习，期末考试

课时安排中，每一章内容均配有详细的教学活动安排，具体包括理论讲解、示例演示、编程练习、案例分析等，旨在通过多种教学方式，帮助学生深入理解和掌握C语言编程的各个知识点。

如前所述，本书提供了丰富的学习资源，供教师备课和学生自学使用。读者可扫描书中二维码观看教学视频，在中国铁道出版社教育资源数字化平台（https://www.tdpress.com/51eds）下载配套的教学课件、习题及答案、案例源代码、教学大纲等资源。本书配套的在线慕课课程“计算机程序设计（C语言）”已在智慧树平台上线。

本书的编写工作由一支经验丰富、专业素养高的团队共同完成。团队成员各司其职、通力合作，确保了全书的质量。本书由赵小蕾、许喜斌、耿江涛任主编，由何秀、丁丰任副主编。

在本书的编写过程中，我们还得到了众多同行、审稿专家、技术顾问以及学生的宝贵意见和支持，在此深表感谢。

由于编者的学识和能力有限，书中难免还存在一些不足或疏漏之处，真诚地希望广大读者能够给予指正，您的宝贵意见将成为我们不断改进和完善的动力源泉。同时，我们也期待与您共同探讨、共同进步。愿本书能够成为您学习C语言编程路上的良师益友。

编　者

2024年10月

目　录

Chapter 1 第 1 章 First Introduction to C Programming and Algorithms 初识 C 程序设计与算法 ······ 1

1.1 Introduction to C（C 语言的引入）······ 2

1.1.1 What Is a Computer Language（什么是计算机语言）······ 2

1.1.2 Why to Learn C Language（为什么学习 C 语言）······ 4

1.1.3 How to Learn C Language Well（怎样学好 C 语言）······ 6

1.2 First Look at C Programming（初识 C 程序设计）······ 7

1.2.1 C Standards（C 标准）······ 7

1.2.2 Structure and Execution of C Program（C 程序结构及运行）······ 9

1.2.3 Programming Tasks（程序设计的任务）······ 11

1.3 C Environment Installation and Running（C 环境安装与运行）······ 12

1.3.1 Installation and Startup of RedPanda C++（小熊猫 C++ 的安装与启动）······ 12

1.3.2 RedPanda C++ Tutorial（小熊猫 C++ 的使用）······ 14

1.4 Definition, Characteristics, and Representation of Algorithms（算法定义、特性及表示）······ 16

1.4.1 What Is an Algorithm（什么是算法）······ 16

1.4.2 Features of Algorithms（算法的特性）······ 17

1.4.3 Representation Methods of Algorithms（算法的表示）······ 17

Summary（小结）······ 22

Exercise（习题）······ 23

Chapter 2 第 2 章 Sequential Programming 顺序程序设计 ······ 25

2.1 Sequential Programming Examples and Program Components（顺序程序设计举例及程序组成成分）······ 25

2.1.1 Sequential Programming Example（顺序程序设计举例）······ 26

2.1.2 Program Components（程序组成成分）······ 27

2.2 Basic Data Types（基本数据类型）······ 27

2.2.1 Why Data Types Are Necessary（为什么需要数据类型）······ 28

2.2.2 Basic Data Types（基本数据类型）······ 28

2.3 Constants and Variables（常量与变量）······ 30

2.3.1 Constants（常量）······ 31

2.3.2 Variables （变量） 32

2.3.3 Symbolic Constants （符号常量） 33

2.4 C Operators and Expressions（C 运算符及表达式） 35

2.4.1 Operator Overview （运算符概述） 35

2.4.2 Common Operators and Expressions （常用运算符及表达式） 35

2.4.3 Type Conversions and Casts （类型转换） 38

2.5 C Statements（C 语句） 39

2.5.1 What Are C Statements（什么是 C 语句） 40

2.5.2 Output Statements （输出语句） 41

2.5.3 Input Statements （输入语句） 43

2.5.4 Single-character Input and Output （单个字符的输入和输出） 45

2.6 Comprehensive Cases（综合案例） 47

Summary （小结） 49

Exercise （习题） 50

Chapter 3 第 3 章 Selective Structure Programming 选择结构程序设计 52

3.1 Relational, Logical Operators and Expressions （关系、逻辑运算符及其表达式） 52

3.1.1 Why Do We Need Selection Structures（为什么需要选择结构） 53

3.1.2 Relational Operators and Expressions（关系运算符及其表达式） 53

3.1.3 Logical Operators and Expressions （逻辑运算符及其表达式） 54

3.2 Selection Structures Using if and switch Statements （if 和 switch 语句实现选择结构） 57

3.2.1 Selection Structures Using if Statements（if 语句实现选择结构） 57

3.2.2 Using the switch Statement for Multi-branch Selection Structures （switch 语句实现多分支选择结构） 64

3.3 Comprehensive Cases（综合案例） 66

Summary （小结） 71

Exercise （习题） 72

Chapter 4 第 4 章 Loop Structure Programming 循环结构程序设计 75

4.1 Why Do We Need a Loop Structure（为什么需要循环结构） 75

4.2 The While and do...while Statements（while 和 do...while 语句） 77

4.2.1 The While Statements（while 语句） 77

4.2.2 The do...while Statement（do...while 语句）…… 79
4.2.3 Comparison of while and do...while（while 和 do...while 的比较）…… 81
4.3 The for Statements（for 语句）…… 83
4.3.1 Review and Introduction（回顾与引入）…… 83
4.3.2 The for Statements to Implement Loops（for 语句实现循环）…… 84
4.3.3 Special Forms of the for Statement（for 语句的特殊形式）…… 87
4.3.4 Comparison of Three Loop Statements（三种循环语句的比较）…… 89
4.4 Loop Control Statements（循环控制语句）…… 90
4.4.1 The break Statement（break 语句）…… 90
4.4.2 The continue Statement（continue 语句）…… 93
4.4.3 Difference Between break and continue（break 和 continue 的区别）…… 94
4.5 Nesting of Loops（循环的嵌套）…… 95
4.5.1 What Is Nesting of Loops（什么是循环的嵌套）…… 95
4.5.2 Examples of Loop Nesting（循环嵌套的举例）…… 97
4.6 Comprehensive Cases（综合案例）…… 102
Summary（小结）…… 107
Exercise（习题）…… 107
Chapter 5 第 5 章 Batch Processing of Data Using Arrays 利用数组批量处理数据 …… 110
5.1 Definition and Use of One-dimensional Array（一维数组的定义和使用）…… 110
5.1.1 Source of Array（数组的来源）…… 110
5.1.2 Definition of One-dimensional Array（一维数组的定义）…… 112
5.1.3 Storage of One-dimensional Array（一维数组的存储）…… 112
5.1.4 Reference of One-dimensional Array（一维数组的引用）…… 113
5.1.5 Initialization of One-dimensional Array（一维数组的初始化）…… 114
5.2 Two-dimensional Array（二维数组）…… 116
5.2.1 Definition and Storage Mode of Two-dimensional Array（二维数组定义及存储方式）…… 116
5.2.2 Reference of Two-dimensional Array（二维数组的引用）…… 117
5.2.3 Initialization of Two-dimensional Array（二维数组的初始化）…… 118
5.2.4 Application Example of Two-dimensional Array（二维数组应用举例）…… 119
5.3 Character Arrays and Strings（字符数组及字符串）…… 121
5.3.1 Character Array（字符数组）…… 121
5.3.2 String（字符串）…… 122
5.3.3 String Handling Functions（字符串处理函数）…… 125

5.4 Comprehensive Cases（综合案例）128
Summary（小结）136
Exercise（习题）136

Chapter 6 第 6 章 Pointer 指针 139

6.1 Pointers and Pointer Variables（指针和指针变量）139
6.1.1 What Is a Pointer（什么是指针）140
6.1.2 Definition and Use of Pointer Variables（指针变量的定义与使用）142
6.1.3 Precautions for Using Pointer Variable（指针变量使用注意事项）145
6.2 Pointer Reference Array（指针引用数组）148
6.2.1 Pointer Variable Points to Array（指针变量指向数组）148
6.2.2 Pointer Addition and Subtraction（指针加减运算）149
6.2.3 Pointers Referencing Array Elements（指针引用数组元素）150
6.3 Pointer Reference String（指针引用字符串）153
6.3.1 Storage and Reference of Character Strings（字符串的存储与引用）153
6.3.2 Character Array Reference String（字符数组引用字符串）154
6.3.3 Pointer Variable Reference String（指针变量引用字符串）155
6.3.4 Comparison of Character Array and Character Pointer Variable Reference String Method（字符数组与字符指针变量引用字符串方法的比较）157
6.4 Comprehensive Cases（综合案例）158
Summary（小结）163
Exercise（习题）164

Chapter 7 第 7 章 Function 函数 167

7.1 Definition of Function（函数的定义）168
7.1.1 Why Use Functions（为什么要用函数）168
7.1.2 How to Define a Function（怎样定义函数）170
7.2 Function Declaration and Call（函数的声明和调用）174
7.2.1 What is Function Declaration and Call（什么是函数的声明和调用）174
7.2.2 How Does the Function Work（函数如何工作）177
7.3 Local and Global Variables（局部变量与全局变量）178
7.3.1 Scope and Lifetime（作用域和生存期）178
7.3.2 Local Variables（局部变量）178
7.3.3 Global Variables（全局变量）180

7.4 Storage of Variables（变量的存储方式）…… 182
7.4.1 What Is the Storage of Variables（什么是变量的存储方式）…… 182
7.4.2 Storage Categories of Local Variables（局部变量的存储类别）…… 183
7.4.3 Storage Categories of Global Variables（全局变量的存储类别）…… 186
7.5 Arrays as Function Parameters（数组作为函数参数）…… 188
7.5.1 Array Elements as Function Parameters（数组元素作为函数实参）…… 188
7.5.2 Arrays as Function Parameters（数组作为函数参数）…… 190
7.6 Pointers as Function Parameters（指针作为函数参数）…… 193
7.7 Application of Pointer in Function（指针在函数中的应用）…… 198
7.7.1 Pointer to Function（指向函数的指针）…… 199
7.7.2 Functions That Return Pointer Values（返回指针值的函数）…… 205
7.8 Recursive Programming（递归程序设计）…… 208
7.8.1 What is Recursion（什么是递归）…… 208
7.8.2 How Does Recursion Work（递归如何工作）…… 209
7.8.3 How Is Recursion Designed（递归如何设计）…… 211
7.9 Comprehensive Cases（综合案例）…… 213
Summary（小结）…… 218
Exercise（习题）…… 218

Chapter 8 第 8 章 Structures 结构体 …… 223

8.1 Structure Type and Variable Definition（结构体的类型及变量定义）…… 224
8.1.1 Why Do We Need a Structure（为什么需要结构体）…… 224
8.1.2 What Is a Structure（什么是结构体）…… 225
8.1.3 Structure Variable（结构体变量）…… 227
8.1.4 Storage Space Allocation of Structure Variable（结构体变量的存储空间分配）…… 228
8.1.5 Use of Structure Variables（结构体变量的使用）…… 229
8.2 Structure Array and Structure Pointer（结构体数组及结构体指针）…… 232
8.2.1 Structure Array（结构体数组）…… 232
8.2.2 Structure Pointer（结构体指针）…… 234
8.3 Dynamic Memory Allocation and Linked List Definition（动态内存分配及链表定义）…… 235
8.3.1 Dynamic Memory Allocation（动态内存分配）…… 235
8.3.2 Linked List Definition（链表的定义）…… 238

8.4 Creation and Output of Linked List（链表的创建和输出）…… 239
8.4.1 Creation of Linked List（链表的创建）…… 240
8.4.2 Output of Linked List（链表的输出）…… 243
8.5 Insertion, Search and Deletion of Dynamic Linked List（动态链表的插入、查找和删除）…… 246
8.5.1 Insertion of Linked List（链表的插入）…… 246
8.5.2 Search of Linked List（链表的查找）…… 250
8.5.3 Deletion of Linked List（链表的删除）…… 252
8.6 Comprehensive Cases（综合案例）…… 254
Summary（小结）…… 260
Exercise（习题）…… 260

Chapter 9 第 9 章 File 文件 …… 264

9.1 File Open and Close（文件的打开与关闭）…… 265
9.1.1 What Is a File（什么是文件）…… 265
9.1.2 Classification of Files（文件的分类）…… 265
9.1.3 File Name（文件名）…… 267
9.1.4 File Buffer（文件缓冲区）…… 267
9.1.5 File Open and Close Operations（文件的打开与关闭操作）…… 269
9.2 File Read and Write Operations（文件的读写操作）…… 271
9.2.1 File Read and Write Functions（文件读写函数）…… 272
9.2.2 Application Examples（应用举例）…… 273
9.3 Comprehensive Cases（综合案例）…… 278
Summary（小结）…… 286
Exercise（习题）…… 287

Chapter 1

First Introduction to C Programming and Algorithms

第 1 章

初识C程序设计与算法

Introduction　主要内容

This chapter gives an overview of the basic concepts of computer languages and programs. It explanations into the development history, notable features, and wide applications of the C language. Additionally, it explains the fundamental structure and operating mechanism of C language programs in detail and introduces the installation steps for the C language development environment. Finally, we focus on the concept, characteristics, and representation methods of algorithms, emphasizing two primary ways to represent algorithms: ANSI flowcharts and N-S flowcharts.

本章概述计算机语言与程序的基本概念，深入剖析C语言的发展脉络、显著特点与广泛应用。同时，详细阐释C语言程序的结构与运行机制，以及C语言环境的安装步骤。最后，聚焦于算法的概念、特性及表示方法，重点介绍传统流程图和N-S流程图两种表示算法的方式。

Objectives　学习目标

Students should understand basic concepts of computer languages and programs. They should be familiar with the development, features, and applications of the C language. They should master the structure and operation of C programs, as well as the installation process of the programming environment. Additionally, they should understand the concepts, characteristics, and representation methods of algorithms, and be proficient in the use of ANSI flowcharts and N-S flowcharts to represent algorithms. Meanwhile, emphasis should be placed on developing students' computational thinking, algorithm design skills, and practical operation abilities. Through the combination of theory and practice, team collaboration and communication skills should be enhanced.

学生应理解计算机语言和程序的基本概念，熟悉C语言的发展历程、特点及应用，掌握C语言程序结构与运行、环境安装过程，了解算法的概念、特性及表示方法，并能熟练运用传统流程图和N-S流程图表示算法。同时，注重培养编程思维、算法设计能力与实践操作能力，通过理论与实践结合，提升团队协作和沟通能力。

1.1 Introduction to C
C语言的引入

1.1节视频

Computer technology has brought scientific progress to our country and convenience to people's lives. There have been new breakthroughs in areas like cloud computing, big data, and artificial intelligence. Software and apps make our lives easier. As builders of the new era, we should dare to innovate and challenge, building confidence with our abilities, and using what we've learned to contribute to national development and the well-being of the people. To design programs, we must first learn a computer language. C language is rich in functions, expressive, and flexible to use. It has the advantages of a high-level language while retaining the characteristics of a low-level language. Moreover, C language is easy to learn, making it a good choice for building a solid foundation in programming.

So, what exactly is the C language? Why should we learn it? And how can we learn it well? This chapter will answer these questions one by one, guiding readers to explore the mysteries of the C language together.

计算机技术给国家带来了科技进步，给人们的生活带来了便利，如云计算、大数据、人工智能等有了新的突破，软件和应用便利了人们的生活。作为新时代的建设者，我们要敢于创新、勇于挑战，用实力铸就信心，运用所学为国家发展和人民幸福贡献力量。想要设计程序，必须先学习一种计算机语言，C语言功能丰富、表达力强，使用灵活方便，既具有高级语言的优点，又具有低级语言的特点，且C语言简单易学，学习C语言可以更容易打下编程基础。

那么，什么是C语言？为什么要学习C语言？如何学好C语言？本章主要是对上面的问题进行解答，带领读者一起走进C语言。

1.1.1 What Is a Computer Language（什么是计算机语言）

To answer the question of what C language is, we first need to understand what a computer language is and comprehend its concept and function. With the advancement of information technology and the popularization of computers, computers seem to be omnipotent. However, a computer is still a machine, and all the tasks it completes are controlled by programs designed and implemented by humans.

为了回答什么是C语言这个问题，首先要了解什么是计算机语言，理解计算机语言的概念和作用。随着信息技术的进步、智能计算机的普及和应用，计算机似乎无所不能，但计算机始终是机器，它所完成的任务是程序控制的，而程序是人设计并实现的。

1. Computer Programs（计算机程序）

We are all familiar with computers. A computer consists of two major components: the hardware system and the software system, as shown in Figure 1-1. If a computer only has a hardware system, it is just a non-functional machine. Software gives the computer its "soul". Especially with the development of artificial intelligence, computers seem to be able to think for themselves and are omnipotent. But software is made up of programs. Computers cannot think independently, and all operations are controlled by programs. Without programs, computers cannot accomplish any tasks. So, what is a computer program? A program is a set of instructions that can be recognized and executed by a computer.

计算机是我们再熟悉不过的设备，计算机是由计算机系统组成的，而计算机系统分为硬件系统和软件系统，如图1-1所示。如果一台计算机只有硬件系统它就是一台“死”的机器，软件赋予了计算机灵魂，尤其是随着人工智能的发展，使得计算机看上去可以自己思考、无所不能。而软件是由程序组成的，计算机并不会独立思考，一切都是受程序控制的，离开程序，计算机将无法完成任务。那么，什么是计算机程序？程序是一组计算机能识别和执行的指令集合。

In 1946, von Neumann proposed a revolutionary design principle: to store programs and data together in the computer's internal memory. The core of this principle is that both instructions and data are stored in binary form, achieving a perfect combination of "stored program" and "program control". By integrating programs and data into the same memory and automatically executing under the precise control of the program, von Neumann's idea laid a solid foundation for the development of computers. The von Neumann model is shown in Figure 1-2.

在1946年，冯·诺依曼提出了一个革命性的设计原理，即将程序与数据共同存储在计算机的内部存储器中。这一原理的核心在于指令与数据均以二进制形式存放，实现了“存储程序”与“程序控制”的完美结合。通过将程序与数据整合至内存中，并在程序的精确指导下自动执行，冯·诺依曼的这一思想为计算机的发展奠定了坚实的基础。冯·诺依曼模型如图1-2所示。

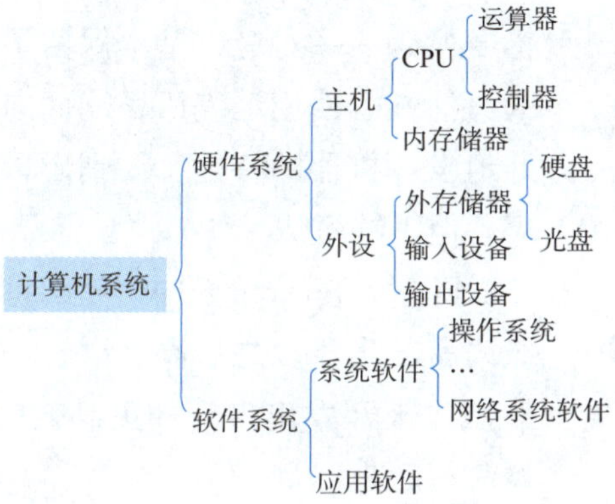

图1-1 计算机系统组成

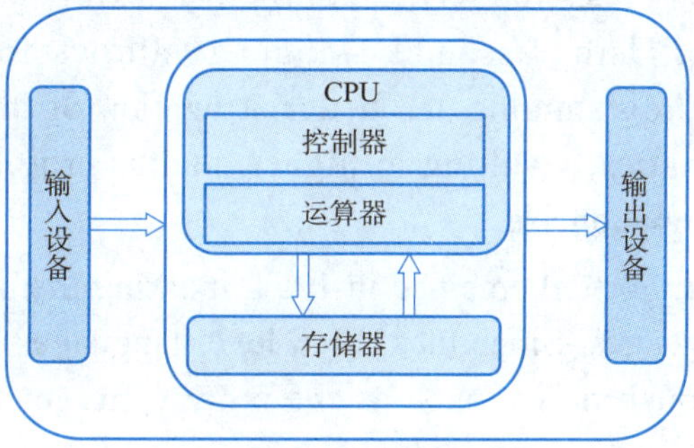

图1-2 冯·诺依曼模型

2. Computer Language and Its Development（计算机语言及发展历程）

So, how are programs written? They are written using computer languages. People and computers interact through a communication medium that both can understand, and this medium is the computer language, as shown in Figure 1-3.

那么，程序是如何编写的呢？程序是使用计算机语言来编写的。人和计算机打交道要通过计算机和人都能识别的沟通媒介，计算机语言就是能实现人与计算机沟通的媒介，如图1-3所示。

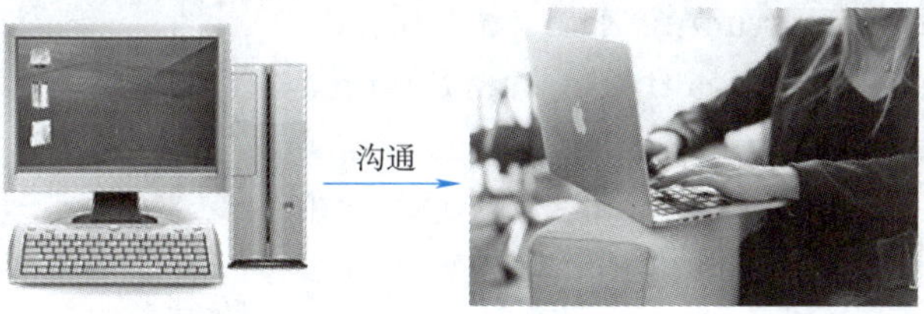

图1-3 计算机语言

The development of computer languages has gone through a process from low-level to high-level. Here, "low-level" or "high-level" doesn't refer to the quality, but to the closeness to the hardware. The earliest and lowest-level language consisted of operation codes made of 0s and 1s, which we call machine language. Each operation code has a corresponding circuit inside the computer to execute it, making it directly recognizable and processable by the machine. Initially, perforated paper tapes were used to control programs, making

计算机语言的发展经历了一个从低级阶段发展到高级阶段的过程，这里的低级或高级并不是水平的高低，而是其距离硬件的远近。最早的语言也是最低级的语言，是由0、1组成的操作代码，称为机器语言，每一个操作代码在计算机内部都有相应的电路来完成，是能够被机器直接识别和处理的语言。最早是用打孔纸带来控

operations extremely complex. Any error required starting the troubleshooting process from scratch. Early programmers were often judged not on their programming skills, but on their ability to correct and proofread. Such a language made it difficult to implement complex programs and was not easy for people to remember.

制程序，操作极为复杂，一旦出错就要从头开始排错。早期的程序员考察的不是程序设计能力，反而是纠错和校对能力。这样的语言不大可能实现复杂的程序，且不便于人们记忆。

To aid memory, mnemonics were used to replace the 0-1 machine codes, such as in assembly language, which simplified the programming process to some extent. However, the machine still recognizes machine code, so people developed an assembler, a program that can automatically generate machine language. Mnemonics correspond one-to-one with instruction codes, largely retaining the flexibility of machine language. Using assembly language allows for machine-oriented programming and better utilization of the machine's characteristics, resulting in higher-quality programs, but its syntax is not concise.

为了便于记忆，人们用助记符代替0-1机器代码，如汇编语言，在一定程度上简化了编程过程。然而机器能识别的仍然是机器代码，故人们开发了一种能够自动生成机器语言的程序——汇编器，助记符与指令代码一一对应，基本保留了机器语言的灵活性。使用汇编语言能面向机器并较好地发挥机器的特性，得到质量较高的程序，但不够简洁。

Later, Fortran emerged in 1954, marking the birth of high-level languages. Since then, high-level languages have sprung up like mushrooms, such as the more efficient C language, which began to gain popularity.

随后，Fortran于1954年问世，至此高级语言产生了。之后高级语言如雨后春笋般依次被开发出来，如更高效的C语言开始流行。

✧Low-level and High-level Language Features（低级语言和高级语言特性）

Low-level languages, like machine language and assembly language, fully rely on specific machine features. In contrast, high-level languages are powerful and not dependent on a specific machine, such as Fortran, and later languages like BASIC, Pascal, C, C++, Java, Python, and more. High-level languages have also gone through different stages of development, mainly from unstructured to structured. Unstructured high-level languages, such as BASIC and Fortran, are early languages with a freer programming style and jumps anywhere in the code, making programs hard to read and maintain. Structured high-level languages, like C, C++, and Java, are built from basic structures and don't allow random jumps, making programs easier to write, read, and maintain.

低级阶段语言完全依赖具体的机器特性，如机器语言、汇编语言；而高级语言功能强、不依赖具体的机器特性，如Fortran，以及随后面世的BASIC、Pascal、C、C++、Java、Python语言等。高级语言同样经历了不同的发展阶段，主要是非结构化到结构化的发展。非结构化高级语言是初期语言，编程风格随意，流程可以随意跳转，程序难以阅读和维护，如BASIC和Fortran语言；而结构化高级语言由基本的结构构成，不允许随意跳转，易于编写、阅读和维护，如C、C++、Java等。

1.1.2 Why to Learn C Language（为什么学习C语言）

With thousands of high-level languages today, why do we still need to learn C? By understanding C's origins, features, and application areas, we can grasp its importance.

高级语言发展到现在已经有上千种，那为什么我们要学习C语言呢？下面通过了解C语言的起源，C语言的特点以及C语言的应用方面，进一步了解C语言。

1. Birth of C Language (C语言的诞生)

Before C, system software was mainly written in assembly language. But assembly code is highly hardware-dependent, making it hard to read and port. Regular high-level languages struggle to directly manipulate hardware. Thus, a new language combining the best of both worlds was needed. Enter C language. Ken Thompson and Dennis Ritchie, two influential scientists in computing history, designed it. Its creation was somewhat accidental. Thompson, a bell labs researcher, wanted to develop a space travel game using computers. He wrote a program for the PDP-7 but needed OS features, so he started developing UNIX for it. In 1970, he modified BCPL, naming it B, and rewrote UNIX in it. Ritchie joined Thompson in 1971, sharing his love for space. In 1972, Ritchie designed a new language based on B, naming it C(taking the second letter of BCPL). By 1973, C was mostly complete. Ritchie eagerly rewrote UNIX in C, forgetting the original game idea and focusing solely on UNIX and C. C's journey is shown in Figure 1-4.

在C语言以前，系统软件主要是由汇编语言编写的，由于汇编语言程序依赖计算机硬件，其可读性和可移植性都极差，一般的高级语言又难以实现对计算机硬件的直接操作，于是人们迫切需要一种兼有汇编语言和高级语言特性的新语言，在这种情况下，C语言应运而生。C语言的设计者是两位对计算机语言发展影响深远的计算机科学家——肯•汤普森（Ken Thompson）和丹尼斯•里奇（Dennis Ritchie）。C语言的诞生可以说是“无心插柳柳成荫”。当时美国实验室的研究员肯•汤姆森一直有遨游宇宙的梦想，又很爱玩游戏，想用计算机开发一款模拟在太阳系航行的电子游戏——Space Travel。他基于机器PDP-7编写程序，但游戏必须使用操作系统的一些功能，于是他着手为PDP-7开发操作系统，这个操作系统就是久负盛名的UNIX。1970年，肯•汤姆森对BCPL进行了修改,并将其命名为B语言，之后他用B语言重写了UNIX操作系统。1971年，同样酷爱Space Travel的丹尼斯•里奇，加入了肯•汤姆森团队开发项目，1972年，他在 B 语言基础上设计出了一种新的语言，他取了BCPL的第二个字母作为这种语言的名字，即 C 语言。1973年，C语言的主体完成，他迫不及待地完全投入到用C语言重写UNIX操作系统中。此时，编程的乐趣使他们已经完全忘记了最初要设计游戏的初衷，全身心投入到了UNIX和C语言的开发中。C语言的发展历程如图1-4所示。

图1-4 C语言的发展

Ken Thompson and Dennis Ritchie, two scientists, won the Turing Award for their groundbreaking work on the UNIX operating system and the C language. The Turing Award, known as the “Nobel Prize of Computing”, was established by the Association for Computing Machinery(ACM) in 1966 to honor individuals who have made significant contributions to the field of computing.

肯•汤姆森和丹尼斯•里奇两位科学家因在UNIX操作系统和C语言方面的开创性工作，获得了计算机界最高奖项——图灵奖[图灵奖由美国计算机协会（ACM）于1966年设立，专门奖励那些对计算机事业做出重要贡献的个人，有着“计算机界诺贝尔奖”之称。]

2. Features of C Language (C语言的特点)

C language has powerful control structures, high execution efficiency, strong portability, and doesn't rely on specific machine features. Its codes are compact, making programs short and precise. However, due to the flexibility and freedom of C language coding, it also increases program risks to some extent. Besides, although C provides rich operators and concise syntax, this might reduce code readability, leading to hard-to-understand codes.

C语言具有强大的控制结构、程序执行效率高、强大的可移植性以及不依赖具体的机器特性等主要特点，C语言的代码较为紧凑、程序更短小，但由于使用C语言编写的代码灵活、自由度大，因此一定程度增加了程序的风险性，丰富的运算符与简洁性往往会降低程序的可读性，可能会编写出含糊难懂的代码。

3. Application Areas of C Language（C语言的应用领域）

C is a powerful, flexible, and versatile programming language with wide and important applications. In system software, because of its efficient execution speed and low-level memory management, C is often used in developing core software like operating systems and compilers. In embedded systems, C also plays a key role, as it can directly access hardware, achieving tight interaction. It's widely applied in smart homes, medical devices, etc. Moreover, C is crucial in game development, graphics processing, and network communication. In short, with its powerful functions and extensive applications, C has become an indispensable part of computer science. Examples of C's application areas are shown in Table 1-1.

C语言是一种功能强大且灵活通用的编程语言，其应用领域广泛且深远。在系统软件领域，C语言因其高效的执行速度和底层的内存管理能力，常被用于操作系统、编译器等核心软件的开发。在嵌入式系统领域，C语言同样占据重要地位，它可以直接访问硬件，实现与硬件的紧密交互，广泛应用于智能家居、医疗设备等领域。此外，C语言还在游戏开发、图形处理、网络通信等领域发挥着重要作用。总之，C语言以其强大的功能和广泛的应用领域，成为计算机科学领域中不可或缺的一部分。C语言的应用领域举例见表1-1。

表 1-1 C 语言的应用领域

应用软件开发	系统软件开发
办公软件：WPS	操作系统：UNIX、Windows、Linux
图形图像多媒体：Photoshop	驱动程序：主板驱动、显卡驱动、摄像头驱动
嵌入式软件开发：单片机	数据库：SQL Server、Oracle、MySQL、DB2
游戏开发：CS 游戏引擎	网络软件：网络路由器和防火墙固件

1.1.3 How to Learn C Language Well（怎样学好C语言）

We now know why we should learn C language. But how can we learn it well? Have you heard of the famous 10,000-hour rule? Psychologist Ericsson found that talent plays a small role, while hard work is the key. This proves Einstein's words: "Genius is 1% inspiration and 99% perspiration." To master C, a complex skill, we need to practice over and over until we reach a critical point—"10,000 hours". Research shows that world-class expertise in any field requires about 10,000 hours of training. So, there's no shortcut to learning C well. We must train our listening, speaking, reading, and writing skills through "practical exercises". "Listening" means listening to lectures carefully. "Speaking" means communicating more. "Reading" refers to reading code. "Writing" is about programming. Write more code, practice more, and constantly train your ability to find, analyze, and solve problems, as well as the logical training process of verifyingresults.

前面阐述了学习C语言的缘由，那怎样才能学好C语言呢？不知大家是否听过著名的一万小时定律？心理学家埃里克森研究发现天赋的作用其实很小，而后天的努力作用其实很大，正印证了爱因斯坦说的那句话：天才是1%的灵感加上99%的汗水。一个人在学习的过程中，要完美掌握某项复杂技能，就要一遍一遍艰苦练习，时长必须达到一个最小临界量“一万小时”。研究发现，任何一个领域的世界级水平都需要大约一万小时的训练。所以要学好C语言没有捷径，“实战练习”做足听说读写。“听”指的是认真听课；“说”指的是多沟通交流；“读”指的是阅读代码；“写”指的是动手编程，多写代码、多实践，不断训练发现问题、分析问题和解决问题以及验证结果的逻辑训练过程。

1.2 First Look at C Programming 初识C程序设计

1.2节视频

After the last section, we learned about computer languages and their roles, and briefly introduced C language. Next, we will initially understand C programming, mainly including C standards, the structure and operation of C programs, and the core tasks of C programming. After a brief understanding of the basic structure and operation steps of C programs, we will clarify the tasks required to write C programs, laying a solid foundation for subsequent learning and practice.

通过上一节我们已经知道了什么是计算机语言，以及计算机语言的作用，并引入了C语言，接下来让我们初步认识C程序设计的相关知识。主要内容有C标准、C程序结构与运行以及C程序设计的任务。在初步认识简单的C程序结构及运行的步骤后，了解编写一个C程序需要完成什么任务，为后续学习与实践奠定基础。

1.2.1 C Standards（C标准）

Any language needs standards for long-term and stable development. To write correct C programs, we must first understand C standards. The development of C language standards has gone through many important stages. K&R C, defined in "The C Programming Language" by Dennis Ritchie and Brian Kernighan in 1978, is an early standard. It includes basic syntax, data types, and operators, laying the foundation for C. Later, in 1989, ANSI C(also known as C89 or C90) was released, marking a rapid promotion and development phase for C. In the 21st century, C standards continued to evolve with versions like C99, C11, and C17, introducing new features like binary literals and thread-local storage, expanding C's functionality and application scope. The latest standard, C17, supplements and corrects C11, further strengthening C's importance in programming. These evolutions not only improve C's performance and stability but also promote its wide application in various fields.

任何一个语言长期稳定的发展均需要制定标准，为了编写正确的C程序，首先要了解C标准。C语言标准的发展经历了多个重要阶段。K&R C是C语言的早期标准，由C语言的创始人丹尼斯·里奇（Dennis Ritchie）和布莱恩·克尼汉（Brian Kernighan）于1978年在其著作*The C Programming Language*中定义。K&R C标准包含了C语言的基本语法、数据类型、运算符等，它奠定了C语言的基础。随后，1989年ANSI C（也称C89或C90）的发布，标志着C语言发展迈入快速推广和发展阶段。进入21世纪，C语言标准继续演进，C99、C11、C17等版本相继问世，引入了二进制字面量、线程局部存储等新特性，不断扩展了C语言的功能和应用范围。最新的C语言标准是C17，它在C11的基础上进行了补充和修正，进一步巩固了C语言在编程领域的重要地位。这些标准的演进不仅提升了C语言的性能和稳定性，也推动了其在各个领域的广泛应用。

This textbook adopts the C99 standard because it introduces a series of new features and improvements, significantly enhancing C's flexibility and functionality, with good stability and compatibility. As an early and widely accepted standard, C99 is supported by many compilers and platforms, with extensive industry applications. Mastering it helps students better adapt to the actual work environment.

本书采用C99标准。使用C99标准的原因在于其引入了一系列新特性和改进，显著增强了C语言的灵活性和功能，且其具有较好的稳定性和兼容性。C99作为较早且广泛接受的标准，许多编译器和平台都支持C99标准，在业界也有广泛的应用，掌握它有助于更好地适应实际工作环境。

A standards-compliant compiler plays a crucial role in programming. So, what is a compiler? C programs are high-level and easy to understand for programmers but hard for computers to directly interpret. Computers can only recognize machine language, or binary code. Therefore, to execute C programs, we need special software to “translate” C into machine code. This software is the compiler.

The compiler’s function is similar to a translator in real life. When people who speak different languages communicate and don’t understand each other’s languages, a translator bridges the gap. Likewise, the compiler converts C into binary code, making it “understandable” for the machine to execute the program. This conversion process is called “compilation”.

During compilation, the compiler first checks if the code follows syntax rules. If there are errors, the compiler usually provides hints to help programmers locate and solve them. Only when the code is syntactically correct does the compiler proceed to the next compilation step. Many C compilers are available for different platforms. For example, Windows often uses Microsoft’s compiler integrated into Visual Studio or Visual C++; Linux commonly uses GCC; Mac uses LLVM/Clang integrated into Xcode.

We also often hear about Integrated Development Environments(IDEs). What are they? IDEs are tool suites combining compilers, code editors, and other development tools, similar to a desktop computer with peripherals. On Windows, common IDEs include Visual Studio, Dev C++, and RedPanda C++. Due to Visual Studio’s large size and complex settings, which can be challenging for beginners, and Dev C++’s debug issues, we chose the small and easily installed RedPanda C++ as our programming environment. Subsequent chapters will introduce it further.

支持标准的编译器在编程中发挥着至关重要的作用。那么，究竟什么是编译器呢？C语言程序虽然对于程序员而言是高级且易于理解的，但对于计算机来说却难以解析。计算机所能识别的，始终是机器语言，即二进制代码。因此，要让计算机执行C语言编写的程序，就需要一个专门的软件来将C语言“翻译”成机器代码，这个软件就是编译器。

编译器的功能与现实生活中的翻译人员颇为相似。当使用不同语言的人进行交流，彼此不懂对方的语言时，翻译人员便成为他们间沟通的桥梁。同样地，计算机只能识别机器代码（二进制代码），而编译器则是将C语言转换成二进制代码的媒介，使得机器能够“理解”并执行，而这个转换或者翻译过程，被称为“编译”过程。

在编译过程中，编译器首先会检查所编写的代码是否符合语法规则。如果代码中存在错误，编译器通常会给出提示信息，帮助程序员定位和提示问题信息。只有当代码语法正确无误时，编译器才会进行下一步的编译过程。C语言的编译器有很多种，不同的平台下有不同的编译器。例如，Windows下常用的微软编译器，被集成在Visual Studio或Visual C++中，Linux下常用GCC，Mac下常用LLVM/Clang，它被继承在Xcode中。

我们也常听说集成开发环境，那么，什么是集成开发环境呢？集成开发环境是编译器、代码提示器等开发工具的组合套装，类似台式机的主机加外围设备，集成项及功能见表1-2。在Windows下，常用的集成开发环境有Visual Studio、Dev C++ 和小熊猫 C++。鉴于VS庞大且不便下载，而Dev C++ 存在调试等问题，本书选择了体积小、安装方便并适合初学者使用的小熊猫C++作为编程环境，后续章节将对其加以介绍。

表 1-2　集成开发环境

集　成　项	功　　能
编译器	编写代码
代码提示器	编译过程中给予代码提示
调试器	观察程序的每一步运行步骤，发现程序的逻辑错误
项目管理工具	对程序涉及的所有资源进行管理
其他工具	界面，各种按钮、面板、菜单、窗口等

1.2.2 Structure and Execution of C Program（C程序结构及运行）

1. The Structure of a C Program（C程序结构）

After understanding the origins of the C language, its standards, and the programming environment, let's further explore the basic structure of a C program. Here's a very simple example of a C program.

已知C语言的由来、标准和编程环境，那么C程序的结构是怎样的呢？下面我们来一探究竟，展示最简单的C程序结构。

Case 1_1（案例 1_1）

Displaying the string "Hello World." on the screen.

在屏幕上输出英文字符串“Hello World.”。

Thinking（解题思路）:

Under a standardized C program structure, we'll use the printf function in the main function to output the string.

在规范的C程序结构下，在主函数中使用标准输出函数printf将英文字符串输出。

Program Code（程序代码）:

Case1_1.c

```
1    #include <stdio.h>                         // 预编译头文件
2    int main()
3    {
4        printf("Hello World.\n");              // 输出语句
5        return 0;
6    }
```

Program Output（运行结果）:

```
Hello World.
```

Case Analysis（案例解析）:

(1)The comment at the end of line 1 explains the code's function. In C, single-line comments use double slashes(//), while multi-line comments use "/*" and "*/". Comments don't affect program compilation or execution. They mainly provide information for programmers, enhancing code readability. Comments can follow a code line or exist independently.

(2)Line 2 defines the main function, where "main" is the function name. The "int" before main indicates that this function returns an integer value. Every C program must have one and only one main function, serving as the starting point for program execution.

(3)Lines 3 to 6 are the main function body, enclosed in curly braces. The function body contains various declarations and statements defining the function's functionality and logic. Detailed explanations of these will be discussed later.

（1）第1行代码后的双斜杠是注释的标识。在C语言中，单行注释采用双斜杠（//）来标记，而多行注释则使用“/*”和“*/”这一对符号来界定。注释在程序中不参与编译和运行过程，其主要作用是为程序员提供信息，帮助理解代码的功能和逻辑。注释可以紧随代码行之后，也可以独立于代码行单独存在，其目的都是使代码更易读易懂。

（2）第2行是主函数的定义，其中main是函数的名称。main前面的int表明该函数的返回类型为整型，意味着函数执行完毕后将返回一个整数值。每个C语言程序都必须包含一个且仅有一个主函数，是程序的执行入口。

（3）第3行和第6行之间的内容被花括号所包围，这构成了主函数的函数体。函数体内部包含了各种声明和语句，它们共同定义了主函数的具体功能和执行逻辑。关于这些声明和语句的详细解释，我们将在后续的学习中深入探讨，此处先不展开说明。

(4) The printf function on line 4 is an output function from the C language standard library, used to display text on the screen. Here, it outputs the string "Hello World". to the console, enclosed in double quotes. The '\n' at the end is a special character for a new line, moving the cursor to the next line after output.

(5) The "return 0;" statement on line 5 has a specific role in C programs. After the main function executes from top to bottom, this statement returns the integer 0 to the calling function(mostly the operating system), indicating the program ended normally. Returning 0 usually means successful execution, while non-zero values might indicate errors or abnormalities.

（4）第4行中的printf函数是C语言标准库中的一个输出函数，用于在屏幕上显示文本。在这里，printf函数将字符串"Hello World"输出到控制台，字符串被双引号（" "）包围起来。字符串末尾的'\n'是一个转义字符，用于实现换行效果，使得输出后光标会移动到下一行。

（5）第5行的"return 0;"语句在C语言程序中具有特定的作用。当主函数自上而下执行完毕后，这条语句会将整数0作为函数的返回值传递给调用函数（在大多数情况下是操作系统），表示程序正常结束。返回0通常表示程序成功执行，而非零值则可能表示出现了某种错误或异常情况。

Common Mistakes and Analyses 常见错误及问题分析

(1) C is case-sensitive, meaning 'A' and 'a' are considered different characters with different ASCII values. When coding in C, attention must be paid to the correct use of uppercase and lowercase.

(2) C source code files consist of a series of C statements, and each statement must end with a semicolon(;) to mark its conclusion. This is part of C language syntax, separating statements.

(3) While multiple statements can be written on one line, for readability and maintainability, it's recommended to write only one statement per line. This clarifies the code, facilitating debugging and modifications.

(4) A complete C source program has one and only one main function. As the entry point, other functions can only be executed through the main function. Thus, when coding in C, ensure the main function's correctness and completeness, avoiding mistakes like writing "mian()" instead of "main()," a common beginner's error when using Chinese input methods.

(5) Aside from comments and string texts, all characters should be entered using an English keyboard to avoid compilation errors.

（1）C语言区分大小写，即C语言对大小写字母非常敏感。例如，'A'和'a'被视为两个不同的字符，它们在ASCII码表中有不同的数值。因此，在编写C语言程序时，必须注意大小写的正确使用。

（2）C源程序文件由一系列C语句组成，每条语句的末尾都必须加分号（;），以标识其结束。这是C语言语法的一部分，用于分隔语句。

（3）虽然一行中可以编写多条语句，但为了提高代码的可读性和可维护性，通常建议每行只写一条语句。这样可以使代码更加清晰明了，方便后续的调试和修改。

（4）一个完整的C源程序有且只有一个main函数。main函数是程序的入口点，其他函数只有通过主函数的调用才能被执行。因此，在编写C语言程序时需要确保主函数的正确性和完整性，更不能将"main()"错误地写为"mian()"，这是使用中文输入法的初学者常犯的一个错误。

（5）除了注释的具体内容、字符串内部文本，其他字符均需要使用英文输入法，否则会导致编译错误。

2. How C Language Works（C语言的运行过程）

After a basic understanding of the structure of a C program, let's explore how it runs. First, we write the source code and save it as a file with a ".c" extension. We call this a source file. Next, we move to the compilation

在初步了解了C程序的结构后，接下来我们探讨它是如何运行的呢。首先，我们需要精心编辑源程序并将其妥善保存，这类文件是以".c"作为扩展名，称为源文件。随后，进入编译阶段，

stage. This is done by the compiler, which turns the source file into an object file. Even though this file is already in binary format, it's not ready to run yet. Its extension is usually ".obj". Then, we need to link it, just like putting together car parts after they're made. In programming, linking includes connecting with library functions. After that, we get an executable file, often with a ".exe" extension on Windows. Now, we can run the program and see the results. The C program's process is shown in Figure 1-5. In a development environment, clicking the right button makes the software do the compiling, linking, and running for us.

这一步由编译器来执行，它将生成目标文件。虽然这个文件已经是二进制格式，但还不能直接执行，其扩展名为".obj"。接着，需要进行连接操作，这类似于汽车零件生产完毕后需要进行的拼装过程。在程序中，连接操作包括与库函数文件的连接，完成后将生成可执行的".exe"文件。至此，运行程序就可以看到程序的运行结果了。C程序运行过程如图1-5所示。在软件上单击对应的功能，程序将自动执行编译、连接和运行过程。

1.2.3 Programming Tasks（程序设计的任务）

To complete a full programming task, we usually need to follow six steps:

(1) Deeply analyze the problem and identify its core issues.

(2) Design solution steps, or algorithms, for these core issues to ensure the program can accurately and efficiently complete tasks.

(3) Based on the designed algorithms, write the program on a programming editor and save the source files.

(4) Verify the program's correctness through compilation, linking, and running.

(5) Check the running results and analyze them in detail. If the results don't meet expectations, promptly revise the program until the desired effect is achieved.

(6) Write corresponding program documentation for others to understand and use.

For beginners, the core goal of programming is to cultivate algorithm design skills. In the next section, we'll further introduce algorithm-related knowledge, laying a solid foundation for subsequent programming practices. The programming task is illustrated in Figure 1-6.

要完成完整的程序设计任务，往往需要进行如下六个步骤：

（1）对问题进行深入的分析，明确问题的核心所在。

（2）针对这些核心问题，设计相应的解决步骤，即算法，以确保程序能够准确高效地完成任务。

（3）根据设计的算法在程序编辑软件上编写程序并保存好程序源文件。

（4）通过编译、连接和运行等步骤来验证程序的正确性。

（5）在查看运行结果后，我们将对其进行详细分析，如果发现结果不符合预期，将及时对程序进行修正，直至达到预期效果。

（6）为了便于他人理解和使用，应撰写相应的程序文档。

对于初学者，程序设计核心目标是培养算法设计能力。下一节，我们将进一步介绍算法的相关知识，为后续的编程实践奠定坚实的基础。程序设计任务如图1-6所示。

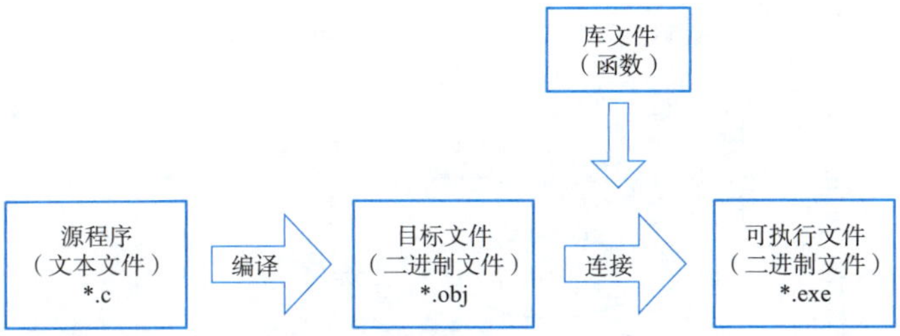

图1-5 C程序运行过程

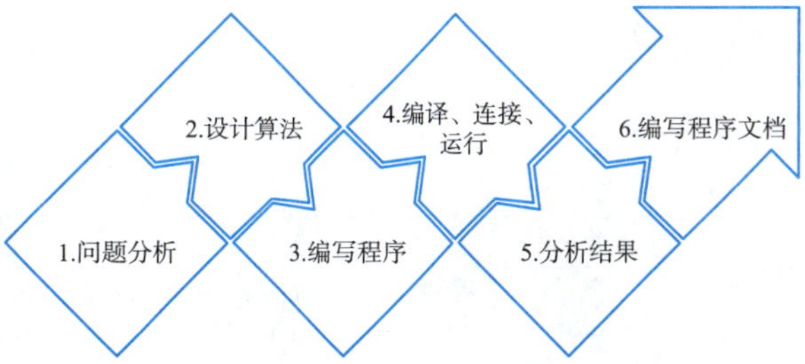

图1-6 程序设计的任务

1.3 C Environment Installation and Running C环境安装与运行

Through the learning of the previous two sections, we have understood what computer languages and the C language are, and have gained a preliminary understanding of the structure of C programming, the operating principles of C programs, and the tasks of programming. However, a compiler is essential software for C programming. So, how is the compiler installed and run? Let's explore this together.

In this section, we will officially enter the world of C programming through learning the download and usage of RedPanda C++.

通过前两节的学习，我们了解到了什么是计算机语言和C语言，也初步了解到了C程序设计中C程序的结构与运行以及程序设计的任务。编译器是C程序设计必不可少的软件，那么编译器是如何安装和运行的呢？接下来让我们一起来探个究竟吧。

本节通过学习小熊猫 C++的下载和使用来正式走进C程序。

1.3.1 Installation and Startup of RedPanda C++（小熊猫 C++的安装与启动）

1. Installation Process（安装过程）

(1) Open RedPanda C++ website. Click “Download” to start downloading, as shown in Figure 1-7.

(2) After downloading RedPanda C++, you'll get an installation package(.exe file). Double-click it to begin installation.

（1）打开小熊猫C++官方网站，单击Download开始下载，如图1-7所示。

（2）小熊猫 C++ 下载完成后会得到一个安装包（.exe 程序），双击该文件即可开始安装。

图1-7 RedPanda C++下载

First, load the installer(which may take a few seconds), as shown in Figure 1-8. Then, the installation process begins. RedPanda C++ supports multiple languages, including English and Simplified Chinese. Here, we choose English as shown in Figure 1-9.

首先加载安装程序（大概需要几十秒），如图1-8所示。随后开始安装，小熊猫C++支持多种语言，包括英文和简体中文，这里我们选择英文（English），如图1-9所示。

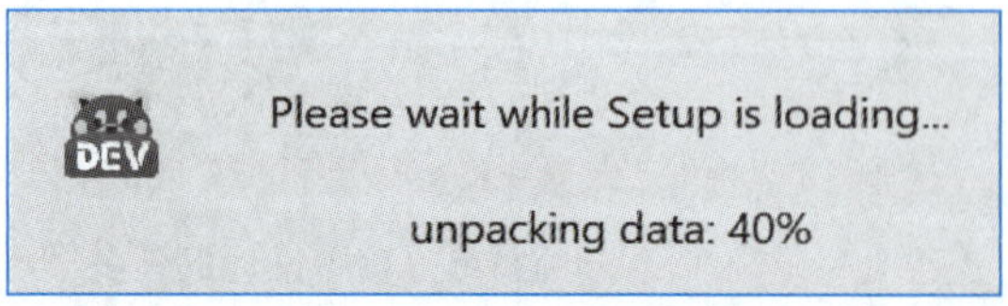

图1-8 RedPanda C++安装

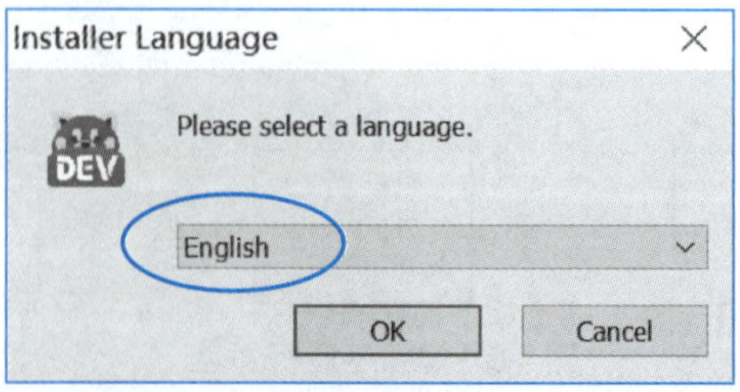

图1-9 RedPanda C++安装过程

(3) Follow the prompts, clicking options like "I agree", "Install for anyone using this computer", and "Next" to complete the installation.

（3）根据指引单击 I agree、Install for anyone using this computer、Next等选项，按照提示的安装过程逐一操作，即可完成安装。

(4) When choosing the installation path, you can install RedPanda C++ anywhere, but it's recommended not to use Chinese characters in the path for debug, as shown in Figure 1-10.

（4）在选择安装路径时，可以将小熊猫C++安装在任意位置，但是路径中不建议包含中文，如图1-10所示。

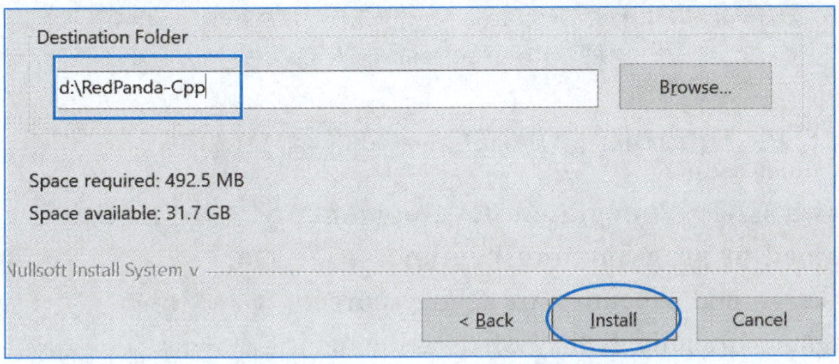

图1-10 Red Panda C++安装过程

Finally, finish the installation, as shown in Figure 1-11.

最后，完成安装，如图1-11所示。

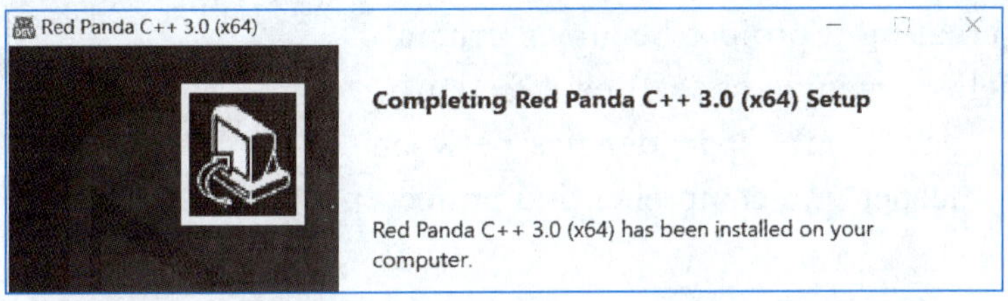

图1-11 RedPanda C++安装完成

2. Starting RedPanda C++（启动小熊猫 C++）

When using RedPanda C++ for the first time, simply select a theme and default programming language. The default interface theme of RedPanda C++ is dark, and the default programming language is C++. For clarity, the images below were captured with a light theme and the default programming language set to C, as shown in Figure 1-12.

首次使用小熊猫C++只需简单地选择主题和默认编程语言。小熊猫C++默认的界面主题是深色的，默认的编程语言为C++。为了清晰起见，本书中的图片都是在浅色界面下截屏而得的。且我们选择的默认编程语言为C，如图1-12所示。

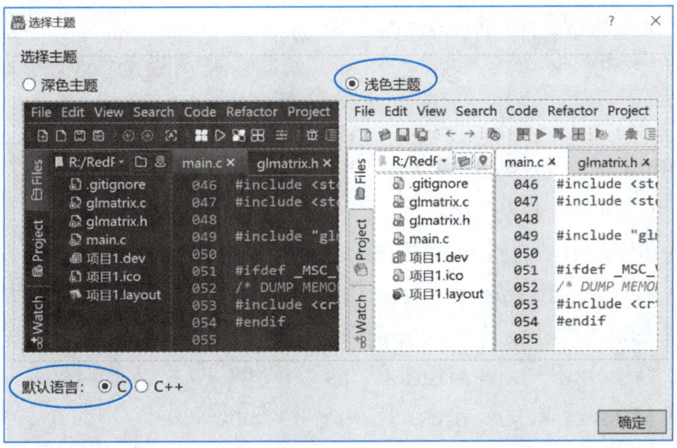

图1-12 RedPanda C++第一次启动

The toolbar is frequently used, and its buttons change based on the current operation state, appearing either "available(bright)" or "unavailable(grayed out)". The functions of the main buttons is shown in Figure 1-13.

工具栏是平时使用最多的，其中各个按钮会随着当前操作状态而改变，处于"可用（明亮状态）"或"不可用（灰色状态）"。几个主要按钮的功能如图1-13所示。

图1-13　RedPanda C++工具栏

1.3.2　RedPanda C++ Tutorial（小熊猫C++的使用）

RedPanda C++ is a C/C++ integrated development environment(IDE) aimed at programming beginners. It supports direct compilation and execution of single source files, which is very beginner-friendly as they often write and debug code in a single source file. In this case, there's no need to create a complex project structure, simply write the code and compile it to run. However, as the code grows and involves multiple source files, creating a project becomes crucial. Project management helps organize code files efficiently, set compilation options, and handle dependencies between source files. We'll delve deeper into comprehensive projects later.

小熊猫C++是一个面向编程初学者的C/C++集成开发环境，支持对单个源文件进行直接编译，这对于初学者来说非常友好，因为他们通常会在单个源文件中进行代码的编写和调试。在这种情况下，无须创建复杂的项目结构，只需编写完代码后直接运行即可。然而，当代码规模扩大，涉及多个源文件时，创建项目变得尤为必要。通过项目管理，可以更高效地组织代码文件，设置编译选项，并处理源文件之间的依赖关系。后续涉及综合项目时再进行介绍。

1. Editing Features（编辑功能）

(1) Click the "New C/C++ file" button on the toolbar(or press 【Ctrl+N】) to create a blank source code file and start programming.

（1）单击工具栏上的"新建C/C++文件"按钮（或者按快捷键【Ctrl+N】），即可创建一个空白的源代码文件，开始编程。

(2) Click the "Open" button on the toolbar(or press 【Ctrl+O】) to edit an existing source code file or project file.

（2）单击工具栏上的"打开"按钮（或按快捷键【Ctrl+O】），就可以打开一个原有的源代码文件或项目文件，进行编辑。

(3) Click the "Save" button on the toolbar(or press 【Ctrl+S】) to save the currently edited file.

（3）单击工具栏上的"保存"按钮（或按快捷键【Ctrl+S】），即可保存当前编辑的文件。

(4) Click the "Close" button on the file tab to close the current file being edited.

（4）单击文件选项卡上的"关闭"按钮，即可关闭当前编辑的文件。

Note（说明）:

Editing files in RedPanda C++ is similar to other text editors, but it offers notable advantages.

在小熊猫C++中编辑文件时，基本操作与其他文本编辑器类似，但有以下突出优点。

(1) RedPanda C++ provides powerful code hinting and autocompletion. When typing, it suggests similar terms based on the initial characters. For example, typing "#in" might prompt "#include" or "#ifndef", as shown in Figure 1-14.If the 【Enter】 key is pressed, the currently highlighted entry will be automatically entered for code completion.

（1）小熊猫C++具有强大的代码提示和补全功能，当用户输入字符时，会自动根据前面已输入的几个字符而提示相似的词条。例如，当用户输入"#in"时，就会自动弹出"#include""#ifndef"等标识符，如图1-14所示。如果按下【Enter】键，则自动输入当前高亮的词条进行代码补全。这可以避免用户因为想要输入方便而选择过于简单的变量名。

(2) RedPanda C++ features real-time syntax error checking. Whenever you press【Enter】, it checks the current line for syntax errors. If found, a red circle appears in the line number area, while a gray circle indicates a warning. Hovering over the line reveals the syntax check information. The diagram in illustrated. in Figure 1-15.

（2）小熊猫C++具有实时语法错误提示功能。每次当用户按下【Enter】键时，它会自动对当前行进行语法检查。如果有语法错误，就会在行号区标记为红色圆圈，行号区的灰色圆圈则表示警告信息，当鼠标悬浮于该行时就会显示语法检查信息。示意图如图1-15所示。

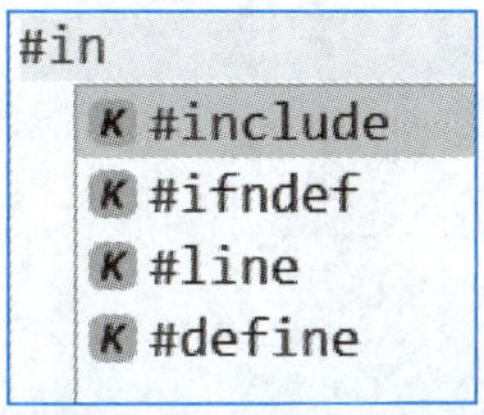

图1-14 RedPanda C++源代码提示和补全

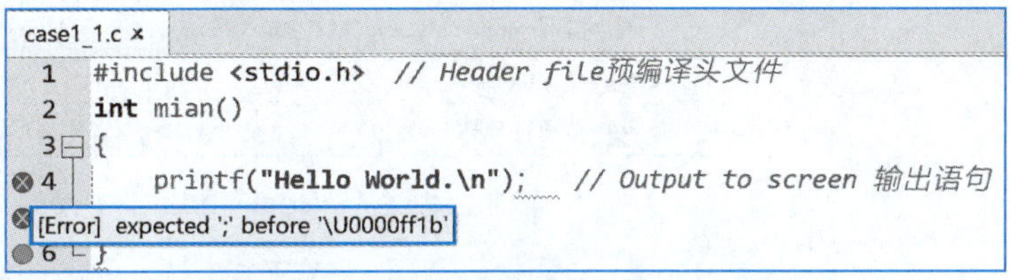

图1-15 RedPanda C++实时语法错误提示

2. Compilation and Running（编译与运行）

After finishing the program, click the "Run" button ▶ on the toolbar to start compilation. It will finish compiling in a moment. Then, the bottom information panel will automatically expand upwards.

程序编写完成之后，单击工具栏上的"编译运行"按钮▶即可开始编译，稍等片刻就会完成编译。这时底部信息面板会自动向上展开。

(1) If there are errors during compilation, the bottom information panel will display the "Issues" page, listing specific compilation error messages, as shown in Figure 1-16.

（1）如果编译出现错误，底部的信息面板会显示"编译器"页，列出具体的编译错误信息，如图1-16所示。

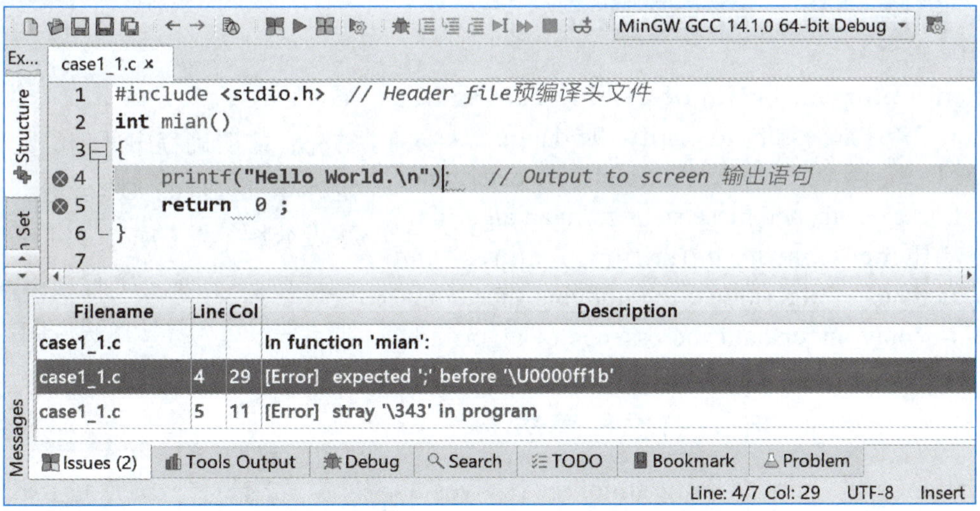

图1-16 RedPanda C++ 显示具体的编译错误信息

(2) If the compilation is successful, the bottom information panel will show the "Tools Output" page, displaying the compilation output information, as shown in Figure 1-17.

（2）如果编译成功完成，底部的信息面板则会显示"工具输出"页，展示编译的输出信息，如图1-17所示。

Afterwards, the system will automatically run the generated program and pop up a console window showing the program's output, as shown in Figure 1-18.

随后，系统将自动运行生成的程序，并弹出一个控制台窗口，显示程序的输出结果，如图1-18所示。

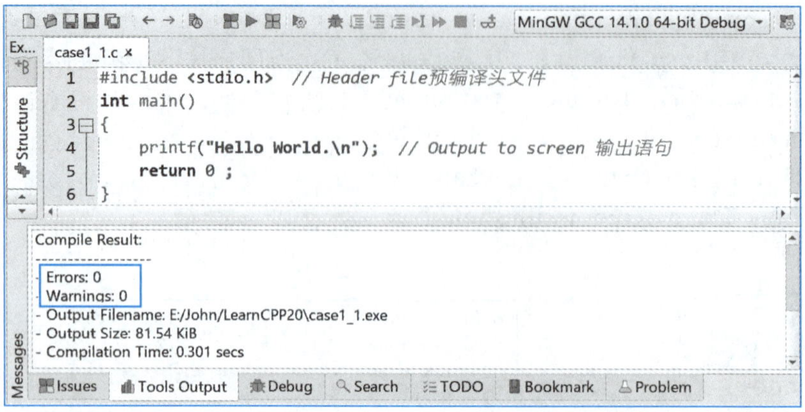

图1-17 RedPanda C++ 编译成功信息

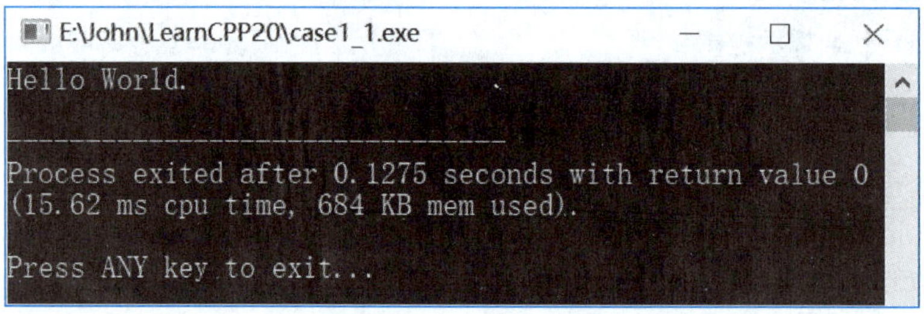

图1-18 RedPanda C++ 运行程序的控制台窗口

1.4 Definition, Characteristics, and Representation of Algorithms 算法定义、特性及表示

1.4节视频

The core of a program lies in describing data and its operations. Firstly, it is necessary to clarify the data processed by the program. Then, a series of operation steps are determined for the computer to execute, and these steps form an algorithm. This section will focus on the definition, features, and representation methods of algorithms. By studying these three aspects, we can deeply understand the essence of algorithms and learn how to use algorithms to solve problems.

程序的核心在于描述数据及其操作。首先，需明确程序处理的数据。接着，确定计算机执行的操作步骤，即算法。本节重点介绍算法的定义、特性及表示方法。通过这三方面，我们能深入理解算法本质，学习运用算法解决问题。

1.4.1 What Is an Algorithm（什么是算法）

The algorithm is the core of programming. It gives an exact and full description of a problem's solution. In other words, algorithms are clear and limited steps to solve a problem. To better grasp the idea of algorithms, let's see a simple example, as shown in Case 1_2.

算法是程序的核心，它是对解决方案进行准确而完整描述的方法，也可以说是解决问题时所采取的明确且有限的操作步骤。为了更好地理解什么是算法，我们可以举一个简单的例子，如案例1~2所示。

Case 1_2（案例 1_2）

To calculate the factorial of 5(5!) which means multiplying from 1 to 5. We need to provide the steps, or the algorithm, to do this.

求 5 的阶乘（5!），即1到5连乘，要求给出算法，即给出具体的操作步骤。

Thinking（解题思路）：

Multiply the numbers in order, using each intermediate result to multiply by the next number until 5.

数据两两相乘，得到的中间结果再乘以下一个数，直到乘到5结束。

Case Solution（案例实现）：

Here's how the algorithm works.

Steps to find 5!:

Step 1: Multiply 1 by 2, getting 2.

Step 2: Multiply the result from Step 1, which is 2, by 3, getting 6.

Step 3: Multiply 6 by 4, getting 24.

Step 4: Multiply 24 by 5, reaching the final answer of 120.

Following these defined and limited steps, we've successfully solved the problem. This is how algorithms work in practice.

下面介绍算法。

求5!的步骤如下：

步骤1：先求1×2，得到结果2。

步骤2：将步骤1得到的乘积2再乘以3，得到结果6。

步骤3：将6再乘以4，得24。

步骤4：将24再乘以5，得120。

在上面算法描述中，经过确定的、有限的操作步骤，问题得以解决，这就是算法。

1.4.2 Features of Algorithms（算法的特性）

As a precise tool for problem-solving, algorithms exhibit the following distinct and essential characteristics.

(1) Finiteness: An algorithm has a limited number of steps. It doesn't get stuck in an endless loop, ensuring that every problem can be definitively solved.

(2) Clarity: Each step is clear and unambiguous, containing no vagueness or ambiguity. This allows the algorithm to execute accurately without error.

(3) Inputs and outputs: An algorithm can accept zero or multiple inputs, which serve as the starting point for solving the problem. Additionally, it produces one or more outputs as a precise response to the input questions.

(4) Effectiveness: The execution of an algorithm is not only valid but also capable of producing a definite result within a finite time. This guarantees the algorithm's efficiency and practicality.

In summary, the essence of an algorithm lies in its set of correct, clear, and finite operational steps designed to efficiently and accurately solve specific problems and produce clear output results.

算法作为解决问题的精确工具，展现出以下鲜明且不可或缺的特征。

（1）有穷性：算法的操作步骤是有限的，它不会陷入无休止的循环，确保每个问题都能得到明确的解决。

（2）明确性：每个步骤都清晰明了，不含任何模糊或歧义之处，这使得算法的执行能够准确无误。

（3）输入与输出：算法可以接收零个或多个输入，这些输入是解决问题的起点；同时，它会产生一个或多个输出，作为对输入问题的精确回应。

（4）有效性：算法的执行不仅是有效的，而且能够在有限的时间内得出明确的结果，这保证了算法的高效性和实用性。

综上所述，算法的本质在于它是一套正确、明确且有限的操作步骤，旨在高效、准确地解决特定问题，并产生清晰的输出结果。

1.4.3 Representation Methods of Algorithms（算法的表示）

After understanding the definition and features of algorithms, we can't help but wonder: How can we efficiently and clearly represent algorithms? Next, we'll introduce several common representation methods of algorithms to help readers better grasp and apply them.

在深入理解了算法的定义及其特性之后，我们不禁会思考一个问题：如何以高效且清晰的方式表示算法呢？接下来，将阐述几种常见的算法表示方法，帮助读者更好地掌握和应用算法。

1. Introduction to Algorithm Representation Methods（算法的表示方法介绍）

(1) Natural language description. Describing algorithms in natural language, as shown in Case1_2, is intuitive and easy to understand. However, when describing algorithms, especially those involving multiple branches or looping operations, it can sometimes be difficult to accurately and unambiguously express complex logical structures. Thus, despite its convenience, natural language has limitations in precisely describing algorithms.

(2) Computer language method. Algorithms can be directly represented using computer languages. Skilled programmers may choose this method, but strict adherence to all grammatical rules is essential.

(3) Pseudocode. Pseudocode combines the features of natural language and computer language. It retains the simplicity of natural language while embodying the logical structure of computer language. For example, when describing an algorithm to output the absolute value of x, we can use pseudocode: "If *x* is positive, output *x*; otherwise, output −*x*." The goal of pseudocode is to facilitate the easy conversion of algorithm descriptions into any programming language(such as Pascal, C, Java, etc.). Therefore, pseudocode must have a clear structure, concise code, strong readability, and strive to maintain the simplicity of natural language. When writing pseudocode, we should focus on clarifying the core logic of the algorithm rather than getting bogged down in specific implementation details.

(4) Graphical representation. Graphical representation is another intuitive and effective method for representing algorithms. It visually demonstrates the execution flow of algorithms through graphics. As the saying goes: "A picture is worth a thousand words." highlighting the significance of graphical representation in algorithm description. This method mainly includes ANSI flowcharts and N-S flowcharts. In subsequent sections, we will focus on introducing these two commonly used graphical representation methods.

（1）自然语言描述。自然语言描述算法如案例1_2所示。虽然直观易懂，但在描述算法时，尤其是在涉及多分支或循环操作时，往往难以准确表达，容易产生歧义。因此，尽管自然语言有其便利性，但在精确描述算法方面却存在局限性。

（2）计算机语言方法。可以直接使用计算机语言来表示算法，对于编程熟练人员可以直接选择该方法，但必须严格遵循所有的语法规则。

（3）伪代码。伪代码是一种介于自然语言和计算机语言之间的表达方式，它既能保留自然语言的通俗易懂，又能体现计算机语言的逻辑结构。例如，描述输出*x*的绝对值的算法时，我们可以使用伪代码："若*x*为正，则输出*x*；否则，输出−*x*"。伪代码的目的在于使算法的描述能够轻松转化为任何编程语言（如Pascal、C、Java等）的实现。因此，伪代码必须结构清晰、代码简洁、可读性强，并尽量接近自然语言。在书写伪代码时，我们无须过分关注具体实现细节，而应着重阐明算法的核心逻辑。

（4）图示法。图示法是另一种有效的算法表示方法，它通过图形直观地展示算法的执行流程。俗话说："一图胜千言。"图示法在算法描述中发挥着举足轻重的作用。图示法主要包括传统流程图和N-S流程图两种方法。后续将重点介绍两种图示方法。

2. ANSI Flowchart（传统流程图）

The American National Standards Institute(ANSI) has defined several commonly used flowchart symbols, such as start/end box, decision box, process box, flow line, input/output box, and connector, as shown in Figure 1-19.

美国国家标准化协会规定了一些常用的流程图符号，如起止框、判断框、处理框、流向线、输入/输出框、连接点，如图1-19所示。

图1-19 常用流程图符号

C language, as a structured programming language, centers around its three basic structures: sequential structure, selection structure, and loop structure. Actually, these three structures exist in all structured programming languages. Mastering these structures is beneficial for building programming logic. To better understand these structures, we can visually demonstrate their operations with the help of ANSI flowcharts.

C语言，作为一种结构化编程语言，其核心在于其三大基本结构：顺序结构、选择结构和循环结构。实际上，这三种结构在所有结构化编程语言中均存在，掌握这些结构有利于构建编程逻辑。为了更好地理解这些结构，我们可以借助传统流程图来直观展示它们的运作方式。

(1) Sequential structure. The sequential structure is the most fundamental structure, indicating that operations in a program are executed in the order they are written or defined. In a flowchart, the sequential structure is usually represented by a series of sequentially arranged boxes, with each box representing an operation or step, and arrows indicating the direction of execution. As shown in Figure 1-20, A and B represent two operations, the boxes are process boxes that contain specific operation statements, and flow lines represent the execution flow.

（1）顺序结构。顺序结构是最基础的结构，它表示程序中的操作按照书写或定义的顺序依次执行。在流程图中，顺序结构通常表现为一系列顺序排列的框图，每个框图代表一个操作或步骤，箭头指示执行的流向。如图1-20所示，其中A、B是两个操作，方框是处理框，将具体操作语句放入框内，用流向线表示执行的流程。

(2) Selection structure. The selection structure allows the program to execute different code blocks based on the results of conditional judgments. In a flowchart, the selection structure is often represented by a diamond or decision box, with the conditional expression written inside. Depending on the truth of the judgment, the flow will follow different paths, which may be connected to different operation boxes. As shown in Figure 1-21, P represents the condition. If Y(condition is true), then statement A is executed; otherwise, statement B is executed. This demonstrates a scenario where both branches perform specific operations. It can also be a single branch executing operational statements, where if the condition is true, A is executed; otherwise, nothing is done.

（2）选择结构。选择结构允许程序根据条件判断的结果来执行不同的代码块。流程图中的选择结构通常以一个菱形或决策框表示，菱形中写入条件判断表达式。根据判断结果的真假，流程会分别流向不同的路径，这些路径上可能连接着不同的操作框图。如图1-21所示，P代表条件，Y代表条件成立，则执行语句A，如果不成立则执行语句B，这是两个分支均执行具体操作的情况，也可以是单分支执行操作语句，如果条件成立则执行A，否则什么都不做。

(3) Loop structure. The loop structure enables the program to repeatedly execute a certain code block until a specific termination condition is met. In a flowchart, the loop structure is usually represented by a loop box that contains the code block to be repeatedly executed, and the loop condition determines when the loop ends. The exit of the loop box typically has a conditional judgment, and the loop terminates only when the condition is not met.

（3）循环结构。循环结构则使得程序能够重复执行某段代码，直到满足特定的终止条件。在流程图中，循环结构通常以循环框表示，循环框内包含要重复执行的代码块，而循环的条件则决定了循环何时结束。循环框的出口通常有一个条件判断，只有当条件不满足时，循环才会终止。

A common way to draw a loop structure flowchart is shown in Figure 1-22. It judges the loop condition p1. If it's true, it executes A; after completion, it returns to continue judging. If it's still true, it repeats executing A until it becomes false, and then the loop terminates. This is called a "while" loop structure. Another type of loop structure is to first execute A, then judge the condition. If it's false, it continues executing A until the condition becomes true, which is called a "do...while" loop structure.

常用的循环结构流程图画法如图1-22所示。判断循环条件P1，若成立则执行A；完成后返回继续判断，若仍成立则重复执行A，直至不成立则循环终止。此即为当型循环结构。另一种循环结构为先执行A，再判断条件，若不成立则继续执行A，直至条件成立方终止，称为直到型循环结构。

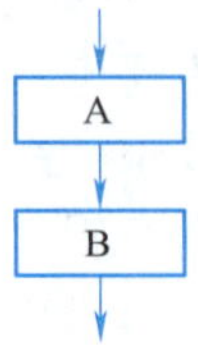

图1-20 顺序结构流程图

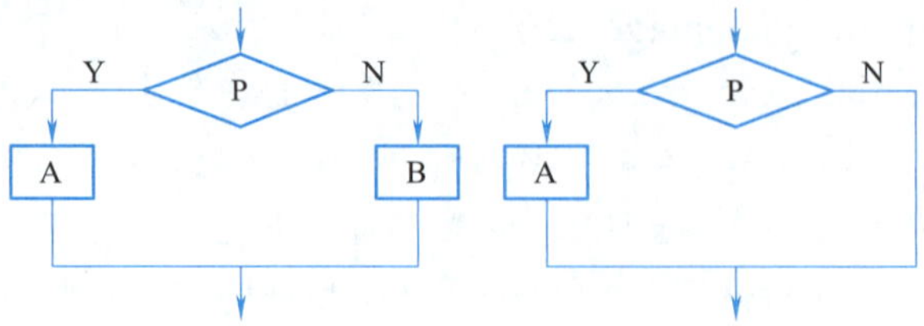

（a）多分支选择结构　（b）单分支选择结构

图1-21 两种选择结构流程图

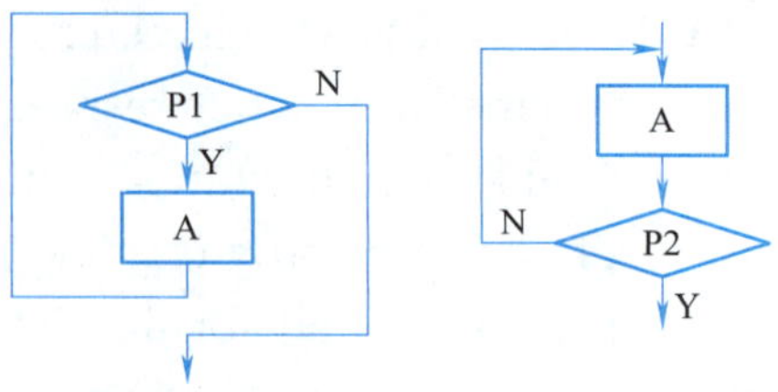

（a）当型循环结构　（b）直到型循环结构

图1-22 两种循环结构流程图

Case 1_3（案例 1_3）

Please use an ANSI flowchart to describe the algorithm for calculating 5!.

使用传统流程图描述求5！的算法。

Thinking（解题思路）：

Use the three introduced structures to draw an ANSI flowchart. This involves a repeated multiplication process, so a loop structure can be used, specifically a “do...while” loop structure. Attention should be paid to standardizing the drawing of different ANSI flowchart symbols.

使用介绍的三种结构绘制传统流程图，这里涉及反复乘数的过程，可以使用循环结构，可以采用直到型循环结构。注意要规范绘制不同传统流程图符号的画法。

Case Solution（案例实现）：

The drawn ANSI flowchart is shown in Figure 1-23. Firstly, let’s introduce the meaning of each variable. t is used to store intermediate results, which is the current accumulation with an initial value of 1. i represents the data to be multiplied, with an initial value of 2. Firstly, calculate 1×2=2, that is, t*i=>t, and then let i increase by 1 to prepare for multiplying the subsequent data 3. This process continues until the value of i is greater than 5, at which point the loop no longer continues. The final result of t is 120. Output the result of t, and the algorithm ends.

绘制的传统流程图如图1-23所示。首先介绍各个变量的含义，t为存放中间结果即当前的累积，初始值设为1，i为将要乘上的数据，初始值设为2。首先计算 1 × 2 = 2，即t*i=>t，随后令i加1，为乘上后续的数据3做准备，直到i的值大于5则不再继续循环，最终t结果为120，输出t的结果，算法结束。

Case 1_4（案例 1_4）

Enter the student IDs and grades of 50 students, and print the IDs and grades of those who failed. Use the ANSI flowchart method to show how to solve this.

输入50名学生的学号和成绩，并输出不及格学生的学号和成绩。使用传统流程图方法描述解决该问题的算法。

Thinking（解题思路）：

This is a complex task. For every student, we do the same thing: enter their ID and grade. So, we can use a loop. The loop will run until we’ve entered info for 50 students. To print the IDs and grades of students who failed, we check each student’s grade. If they fail, we print their info. If they pass, we don’t.

该问题综合性较强，输入学生学号和成绩，每个学生的操作都是一样的，故可以选用循环结构，循环条件设为学生人数小于50即可；输出不及格学生的学号和成绩，也需要循环判断每个学生，且这里涉及选择结构，若不及格才输出，及格则不输出。

Case Solution（案例实现）：

See Figure 1-24. To make it clear, we split it into two parts. First, we enter the data, then we check it. The symbol ① in Figure 1-24 means the same point. If there are other points, we use ②, ③, etc. If the flowchart is long or crosses over, using these symbols makes the ANSI flowchart look nicer and clearer.

如图1-24所示，为了结构清晰，分成了上下两部分。先输入后再依次取出判断。图1-24中使用了连接符号①，代表的是同一个点，若还有其他的连接点，依次画②、③等。若流程图较长或有交叉的情况，建议使用连接符，所呈现的传统流程图会更美观、更清晰。

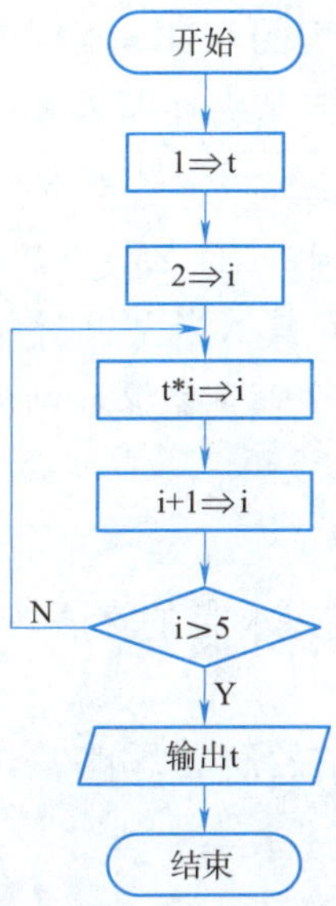

图1-23 求5!的传统流程图

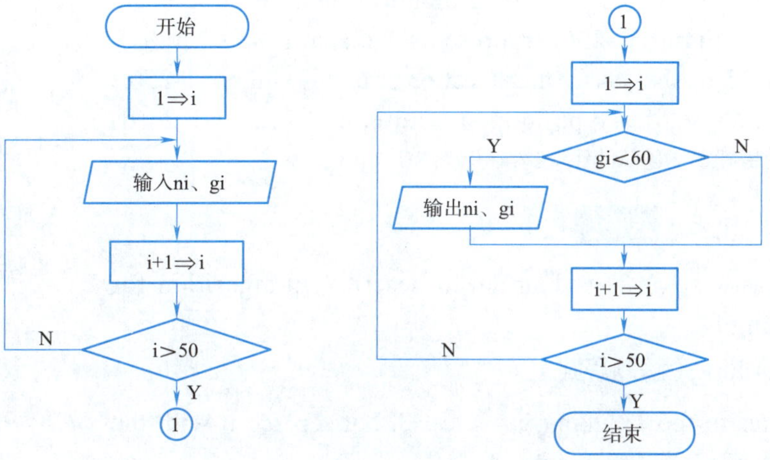

图1-24 传统流程图

3. N-S Flowchart（N-S流程图）

The Nassi-Shneiderman diagram, also known as the box diagram, is a visual modeling method in structured programming. It was introduced by American scholars I. Nassi and B. Shneiderman in 1972. The aim was to eliminate the flow lines in traditional flowcharts, allowing the entire algorithm to fit within a rectangular box that can contain other boxes. The N-S diagram includes three basic structures: sequence, selection, and loop, which correspond to the fundamental structures of a program. The sequence structure represents the sequential execution flow of the program; the selection structure allows different paths to be executed based on conditions; and the loop structure enables code to be repeatedly executed until specific conditions are met.

The N-S diagram has the following advantages.

(1) It clearly demonstrates nested relationships and module hierarchies, aiding in the understanding of complex program structures.

(2) It facilitates the reuse of structures, improving code readability and maintainability.

(3) It is intuitive and visually appealing, with clearly defined functional areas, enhancing the clarity of algorithms and program flows.

In N-S diagram design, boxes typically represent various operations connected through specific relationships. Unlike traditional flowcharts, there are no explicit flow lines in N-S diagrams. This simplifies the complexity of the flowchart, avoids arbitrary control transfers, and ensures the structure of the program.

N-S图也称盒图，是结构化编程中的一种可视化建模方式。它于1972年由美国学者I. Nassi和B. Shneiderman提出，旨在去除传统流程图中的流程线，使全部算法都写在一个矩形阵内，并在框内可以包含其他框。N-S图包括顺序、选择和循环三种基本结构，这与程序的基本结构相对应。顺序结构代表程序的顺序执行流程；选择结构允许根据条件判断执行不同的路径；循环结构则实现了代码的重复执行，直到满足特定的条件。

N-S图具有以下优点。

（1）能够表示嵌套关系以及模块的层级关系，有助于理解程序的复杂结构。

（2）能够实现结构的复用，提高了代码的可读性和可维护性。

（3）形象直观，功能域明确，具有良好的可见度，使得算法和程序的流程更加清晰。

在使用N-S图进行程序设计时，通常会用盒子来表示各种操作或步骤，盒子之间可以通过特定的关系进行连接。但与传统流程图不同的是，N-S图中没有明确的流程线。这种设计方式不仅简化了流程图的复杂度，还避免了随意的控制转移，保证了程序的良好结构。

Although the presentation differs, the underlying logic of ANSI flowcharts and N-S diagrams is consistent. Below, in Figure 1-25,we use N-S diagrams to illustrate three basic structures: Figure 1-25(a) represents the sequence structure, Figure 1-25(b) showcases the selection structure, Figure 1-25(c) depicts the "while" type loop structure, and Figure 1-25(d) represents the "do...while" type loop structure.

尽管传统流程图和N-S流程图的表现形式有所不同，但其内在逻辑相一致。下面使用N-S图描述三种基本结构。图1-25（a）代表顺序结构，图1-25（b）为选择结构，图1-25（c）展示当型循环结构，而图1-25（d）则呈现直到型循环结构。

Case 1_5（案例 1_5）

Please use an N-S diagram to describe an algorithm for calculating 5!.

使用N-S图描述求5！的算法。

Thinking（解题思路）:

Refer to the Thinking in Case1_3, but replace it with the structure of an N-S diagram.

参考案例1_3的解题思路，这里替换成N-S图的结构。

Case Solution（案例实现）:

The drawn N-S diagram is shown in Figure 1-26. Compared to the ANSI flowchart in Case1_3, the execution logic and structures included are consistent.

绘制的N-S图如图1-26所示。与案例1_3的传统流程图进行对比，执行逻辑一致，包含的结构一致。

Case 1_6（案例 1_6）

Enter the student IDs and grades of 50 students, and print the IDs and grades of those who failed. Use the N-S flowchart method to show how to solve this.

输入50名学生的学号和成绩，并输出不及格学生的学号和成绩。使用传统流程图方法描述解决该问题的算法。

Thinking（解题思路）:

This case requires entering information for multiple students, so a loop structure is needed. The case also requires outputting the student IDs and scores of those who fail, so a selection structure is also required. By judging whether the student's score is above or below 60, we can choose whether to output the student's ID and score.

该案例需要输入多名学生的信息，因此需要使用循环结构，案例要求输出不及格学生的学号和成绩，因此还需要使用选择结构。通过判断学生成绩大于60或者小于60来选择是否输出该学生的学号和成绩。

Case Solution（案例实现）:

Compared to the ANSI flowchart in Case1_4, the N-S flowchart is divided into two parts. The upper part involves input and output, including a "do...while" loop structure. The lower part judges whether a student has failed, using a "do...while" loop structure that contains a selection structure within it. The specific details are as shown in Figure 1-27.

与案例1_4的传统流程图对比学习，N-S流程图分为两部分，上部分是输入和输出，包含一个直到型循环结构，下半部分判断是否不及格，使用直到型循环结构，循环结构内包含一个选择结构。具体如图1-27所示。

Summary 小 结

This chapter explores the basic concepts of computer languages and programs. It introduces the development, features,

本章探讨了计算机语言和计算机程序的概念，随后引出C语言的发展、特点及应用，

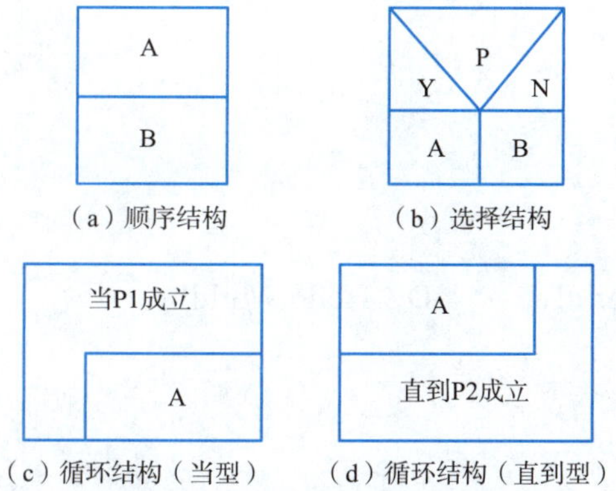

（a）顺序结构 （b）选择结构 （c）循环结构（当型） （d）循环结构（直到型）

图1-25 各种结构的N-S流程图

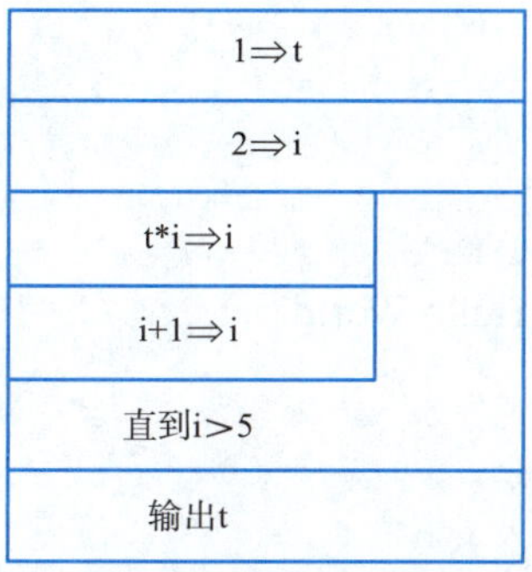

图1-26 N-S流程图

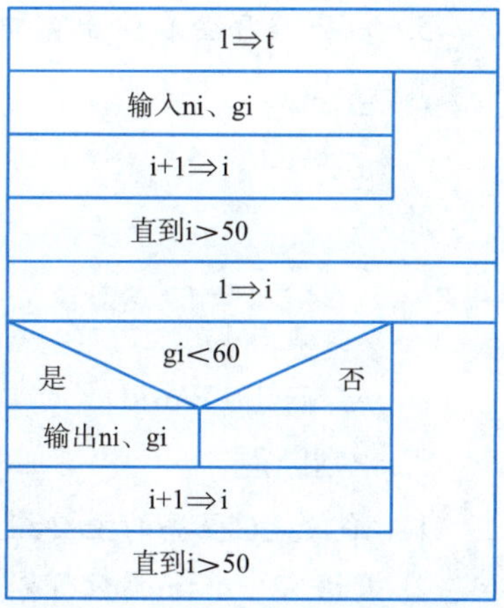

图1-27 案例1-6 N-S流程图实现

and applications of the C language. It explains the basic structure and operation of C programs and gives readers a basic understanding of C language standards, compilers, and compiler environments. Additionally, it covers the core tasks of computer programming. Furthermore, it focuses on explaining the concept of algorithms and their five core characteristics: finiteness, definiteness, input, output, and effectiveness, deeply revealing the essence of algorithms. Finally, it focuses on various representation methods of algorithms, showcasing natural language, pseudocode, ANSI flowcharts, and N-S flowcharts, emphasizing the important role of graphical methods in representing sequential, selection, and loop structures.

Through this series of learning, readers not only master the basics of the C language but also deepen their understanding of algorithms and their representation methods, laying a solid foundation for subsequent programming practices.

介绍了C语言程序结构与运行过程，了解了C语言标准、编译器及编译环境。了解计算机程序的任务。此外，重点讲解了算法的概念及其五个核心特性：有穷性、确定性、输入、输出和有效性，深入揭示了算法的本质。最后，聚焦于算法的表示方法，介绍了自然语言、伪代码、传统流程图和N-S流程图等多种方式，特别强调了图示法在表示顺序、选择和循环结构中的应用。

通过本章学习，读者不仅可以掌握C语言基础，可以更深化对算法及其表示方法的理解，为后续的编程实践打下了坚实基础。

Exercise
习 题

一、单选题

1. C语言属于（ ）。

   A. 机器语言　　B. 汇编语言　　C. 低级语言　　D. 高级语言

2. 下列不属于C语言基本数据类型是（ ）。

   A. int　　B. main　　C. double　　D. long

3. C程序的主要组成单位是（ ）。

   A. 文字　　B. 程序　　C. 函数　　D. 语句

4. 以下注释正确的是（ ）。

   A. //This is a comment　//这是注释
   B. /* This is a comment　/* 这是注释
   C. /* This is a comment/　/* 这是注释 /
   D. /* This is a comment/*　/* 这是注释 /*

5. 以下程序会在输出窗口输出（　　）。

```
#include <stdio.h>
int main()
{
    printf("Hello World!\n");
    return 0;
}
```

A. Hello World　　B. Hello World!　　C. Hello World!\　　D. Hello World!\n

二、判断题

1. 函数是C程序的主要组成部分。（　　）
2. C语言中main函数可以有多个。（　　）
3. int main()可以写成int main(void)。（　　）
4. 要在所有函数的末尾写上“return 0;”。（　　）
5. 算法步骤可以有无限多步。（　　）

三、填空题

1. 在每个数据声明和语句的最后必须要有一个____________。
2. 程序总是从____________函数开始执行。
3. C语言编译后生成的可执行程序的扩展名为____________。
4. 引用自定义头文件Hello.h的语句为____________。
5. 可以打印出“%”的语句为____________。

四、简答题

1. 什么是程序？什么是程序设计？
2. 为什么需要计算机语言？
3. 高级语言有什么特点？
4. 什么是算法？算法有什么特性？
5. C语言的基本结构有哪些？有什么特点？
6. 程序设计的基本任务有哪些？
7. C语言程序为什么需要编译和连接过程？
8. 绘制1+2+3+…+100的传统流程图和N-S流程图，体会两种绘图方法的不同和特点。
9. 绘制求3个数的最大值问题的传统流程图和N-S流程图，体会两种绘图方法的不同和特点。
10. 绘制求10个数的最大值问题的传统流程图和N-S流程图，体会两种绘图方法的不同和特点。

Chapter 2

Sequential Programming

第 2 章

顺序程序设计

Introduction　主要内容

C has three basic structures: sequence, selection, and loop. This chapter mainly introduces sequential structure to help beginners correctly design and implement sequential structure programs. The main content includes examples of sequential programming, data types, constants and variables, operators and expressions, and C statements(such as input and output statements).

C语言有三大基本结构：顺序、选择和循环。本章主要介绍顺序结构，帮助初学者能够正确设计并实现顺序结构程序。主要内容有顺序程序设计举例及数据类型，常量与变量，运算符及表达式和C语句（如输入输出语句）。

Objectives　学习目标

Students should master the steps and methods of sequential structure programming, paying attention to grammatical rules and code readability. Additionally, students must understand the correct definition and usage of data types, constants, and variables, as well as expressions. They should also be proficient in using input and output statements to correctly perform input and output operations.

掌握顺序结构程序设计的步骤和方法，注意语法规则和代码可读性；掌握数据类型、常量与变量、表达式正确定义和使用，熟练运用输入输出语句正确编写输入输出操作等。

2.1 Sequential Programming Examples and Program Components

顺序程序设计举例及程序组成成分

2.1、2.2节视频

This section demonstrates sequential programming through an example, providing an intuitive understanding of its form and giving you a preliminary introduction. Through this example, we illustrate the essential components of a program, such as constants, variables, data types, operators, expressions, and input/output statements. After mastering these key elements, you will

本节通过实例展示顺序程序设计，直观呈现其形态，让读者有一个初步认知。并通过该实例指出程序的重要组成成分，如常量、变量、数据类型、运算符、表达式及输入输

be able to write simple sequential programs, laying a strong foundation for learning more complex structures in the future.

出语句。掌握这些重要的组成成分后，读者可编写出简单的顺序程序，为后续复杂结构学习奠定坚实基础。

2.1.1 Sequential Programming Example（顺序程序设计举例）

Case 2_1（案例 2_1）

Input the radius of a circle to calculate its area.

输入圆的半径，计算圆的面积。

Thinking（解题思路）:

To write a correct code, we need to learn how to analyze the problem first and form a solution strategy. We start by defining the variables r(for radius) and area(for the circle's area). Then, we determine the input and output of the program. "radius" is the input, and "area" is the output. In the program, we calculate the area of the circle based on "radius". In the early stages of learning programming, it's best to describe the algorithm before starting to code. The algorithm description can be illustrated using an N-S diagram, as shown in Figure 2-1.

要编写一个正确的代码，首先要学会分析问题，并形成解决问题的思路。首先定义变量r和area,分别代表半径和面积；随后确定程序的输入和输出，r是输入，area是输出，在程序中根据r计算出圆的面积area。在学习编程的早期阶段，应先描述算法，再动手去编程，算法描述可使用N-S图，如图2-1所示。

The program operates in three steps, known as IPO: Input, where we input the value of the radius r; Process, where we calculate the area of the circle; and Output, where we display the value of the area. These three steps are sequential. Since C programs don't have the mathematical symbol π, we use the constant PI(=3.14) for it. We define and initialize the variable r with a value of 4.0, and similarly, we define and initialize the variable area with a value of 0.0. Then, we calculate the area using the formula area =PI*r*r. Finally, we output the area.

程序有三个操作，分别是输入半径r的值，计算圆的面积area，输出area的值，这三个操作是顺序的。注意程序是没有数学上的π的，我们用符号常量PI来表示3.14，定义变量r和area，并指定r的值为4.0，面积area使用area=PI*r*r计算而来，最后输出面积值。

Program Code（程序代码）:

Case2_1.c

```
1   #define PI 3.14                        //定义符号常量
2   #include <stdio.h>
3   int main()
4   {
5       float r,area;
6       /*定义 r 和 area 为单精度浮点型变量;
7       C 变量要先定义后使用 */
8       r=4.0;                             //指定 r 的值
9       area=PI*r*r;                       //计算 area 的值
10      printf("area=%f\n",area);          //输出 area
11      return 0;
12  }
```

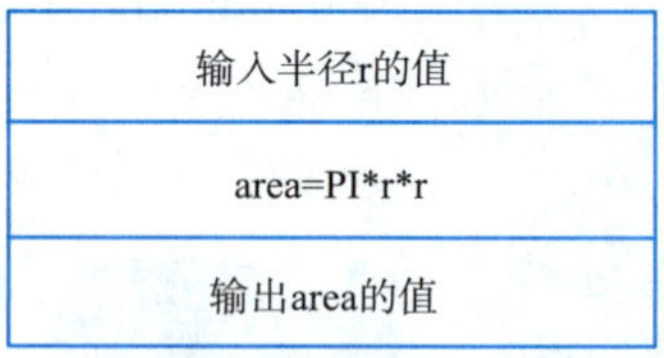

图2-1 N-S流程图

Program Output（运行结果）:

```
area=50.240002
```

Case Analysis（案例解析）:

Line 5: Const declaration. This line declares a constant named PI and initializes it to the mathematical value of π. The

第5行：常量声明。这行代码声明了一个名为PI的常量，并将其初始化为数学值π。

const keyword ensures that PI remains unchanged throughout the program's execution.

Lines 7, 8: Variable definition and initialization. In these lines, we define and initialize two floating-point variables. The variable is set to 4.0. The area variable is initialized to 0.0 to prevent it from containing a random or garbage value before the actual area calculation.

Line 9: Area calculation. Here, we calculate the area of the circle using the formula PI*r*r. The result is assigned to the area variable, overwriting its initial value of 0.0.

Line 10: Output. This line uses the printf function to display the calculated area on the screen. The %f is a placeholder for the floating-point value stored in the area variable.

const关键字确保PI在整个程序执行过程中保持不变。

第7、8行：变量定义和初始化。在这两行中，我们定义并初始化了两个浮点型变量。变量被设置为4.0。area变量被初始化为0.0，以防止在实际计算面积之前它包含随机值或垃圾值。

第9行：面积计算。在这里，我们使用公式PI*r*r来计算圆的面积。结果被赋给area变量，覆盖了其初始值0.0。

第10行：输出。这行代码使用printf函数将计算出的面积显示在屏幕上。%f是存储在area变量中的浮点型值的占位符。

2.1.2 Program Components（程序组成成分）

Through example Case 2_1, we begin to explore the important components of a sequential program.

Firstly, all constants and variables used in a program have data types. In the program, const double PI and 4.0 are constants, where double is a data type, and 4.0 is also a numeric constant processed as a double type. On the other hand, float r; float area; declares variables radius and area as float types. Besides data types, constants, and variables, expressions are also used, such as the initialization expression float r=4.0 and the arithmetic expression PI*r*r.

Additionally, algorithms require input and output, so programs often include input and output statements. In Case 2_1, the data is directly initialized with "r=4.0;" However, the program can also get input from the keyboard, which will be introduced in subsequent chapters. Output is displayed on the screen using the printf function.

Through the above analysis, we can see that a program generally includes data types, constants and variables, expressions, and input and output statements. We will introduce each component of the program in the following chapters.

案例2_1中给出了简单的程序设计举例，下面通过该案例，探讨顺序程序的重要组成成分。

首先程序中使用了常量和变量，而常量和变量是有数据类型的。如float r,area;中的float即为数据类型，此时声明变量r和area为float类型。程序中的4.0和PI为常量，4.0为被处理成double类型的数值常量，而PI为符号常量，在编译阶段会被替换为3.14。除了数据类型、常量和变量之外，还需要表达式，如赋值表达式r=4.0，算术表达式PI*r*r等。

另外算法的特性是需要输入输出，故编写的程序往往需要输入输出语句，在案例2_1中数据是直接在程序中赋值的，如"r=4.0;"。而程序同样也可以从外部输入，将在后续章节介绍。输出使用printf显示在运行界面上。

通过上述程序的分析，程序一般包含数据类型、常量和变量、表达式及输入输出语句。后续我们将陆续介绍程序的各个组成成分的作用和用法。

2.2 Basic Data Types 基本数据类型

2.1、2.2节视频

In this section, we will learn about the first part of a sequential C program, known as basic data types. First, let's understand why computers need to distinguish data types when processing data and what types are included.

本节将介绍C语言顺序程序的第一个组成成分，即基本数据类型。首先理解为什么计算机处理数据需要区分数据类型，包含哪些类型。

2.2.1 Why Data Types Are Necessary（为什么需要数据类型）

There are two main reasons why C needs data types. First, data types help define how much memory a variable will use and how it will be stored. This ensures that different kinds of data, like integers or characters, have the right amount of storage space. This helps the compiler use memory efficiently and makes the program run faster. Second, data types define what operations can be done on a variable. For example, integer types can do arithmetic, while character types can be used for string operations. Using data types not only uses memory better but also makes data processing more accurate and the program more stable.

C语言需要数据类型主要有两个原因。首先，数据类型有助于定义变量将使用多少内存以及将如何存储，这确保了不同种类的数据（如整数或字符）具有正确数量的存储空间。这有助于编译器高效地使用内存并使程序运行得更快。其次，数据类型定义了可以对变量执行哪些操作。例如，整数类型可以进行算术运算，而字符类型可以用于字符串操作。使用数据类型不仅更好地利用了内存，而且使数据处理更加准确，程序更加稳定。

2.2.2 Basic Data Types（基本数据类型）

C language has three basic data types: integer, floating-point, and character. Integer data is identified by the keyword int. Floating-point data can be declared using float , double, or long double. The keyword char represents character data. These basic types are widely used in programming. Next, we will introduce these types in detail.

C语言基本数据类型主要包括三类：整型、浮点型和字符型。整型数据通过系统关键字int来标识；浮点型数据则可使用float、double或long double来声明；而字符型数据则通过系统关键字char来表示。这些基本类型在程序设计中广泛运用，接下来我们将依次对这些类型进行详细介绍。

1. Integer（整型）

Integer data types are used to store whole numbers, both positive and negative. The basic type is marked by the keyword int. Its size and range depend on the specific computer system and compiler. On many systems, int typically occupies 4 bytes(32 bits), with the highest bit(the sign bit) indicating whether the number is positive or negative. So, a signed int often ranges from -2^{31}~$2^{31}-1$. Unsigned integers, denoted as unsigned, range from 0~$2^{32}-1$, as shown in Table 2-1.

C99 also provides other integer types such as short, long, and long long. Type short often uses 2 bytes (16 bits). long is at least as large as int, but could be bigger. On many current systems, it's the same size as int. The long long type uses at least 64 bits, thus its range is at least from -2^{63}~$2^{63}-1$(signed) or 0~$2^{64}-1$(unsigned).

In addition, C99 has a type called size_t. It's an unsigned integer type, often used to show the size of an object or as an index in an array. Its exact size depends on the computer system, but on many systems, it's represented as unsigned long long.

Keep in mind that the C standard does not strictly specify the sizes of these types. Their lengths can vary depending on the compiler and the target computer system.

整型数据类型用于表示整数，其基本类型由int关键字标识，占用4字节（32位）。其中，最高位为符号位，用于表示正负，其余31位用于存储数据。因此，int型有符号整数的范围是-2^{31}~$2^{31}-1$。如需表示无符号整数，可在int前加unsigned，其范围则为0~$2^{32}-1$，见表2-1。

C99还提供了其他整数类型，如short、long和long long。short 类型通常使用2字节（16 位）。long类型至少与int一样大，但可能更大。在许多当前系统上，它与int的大小相同。long long类型至少使用64位，因此其范围至少是-2^{63}~$2^{63}-1$（有符号）或0~$2^{64}-1$（无符号）。

此外，C99有一个名为size_t 的类型。它是一个无符号整数类型，常用于表示对象的大小或作为数组中的索引。它的确切大小取决于计算机系统，但在许多系统上，它被表示为unsigned long long。

请注意，C标准并没有严格指定这些类型的大小。它们的长度可能会因编译器和目标计算机系统而异。

表 2-1 整型数据常见的存储空间和取值范围

分 类	类 型	字 节 数	取值范围
基本整型	int	4	-2^{31}~$2^{31}-1$
	unsigned int	4	0~$2^{32}-1$
双长整型	long long	8	-2^{63}~$2^{63}-1$
	size_t / unsigned long long	8	0~$2^{64}-1$
短整型	short	2	-2^{15}~$2^{15}-1$
	unsigned long	2	0~$2^{16}-1$

2. Floating-point（浮点型）

Floating-point data is used to represent real numbers with decimal points. It is stored in the computer in exponential form, allowing the decimal point to "float". In C99, there are three types of floating-point numbers: float, double, and long double. Type float is a single-precision floating-point number, occupying 4 bytes with only about 6 significant digits. It is suitable for very few memory-limited and low-precision occasions. Type double is a double-precision floating-point number, taking up 8 bytes and having approximately 15 significant digits, making it suitable for general mathematical operations. Type long double, a long double-precision floating-point number, uses 16 bytes and has about 18 significant digits, ideal for high-precision scientific computations as shown Table 2-2.

In C, to ensure the accuracy of calculation results, float data is automatically converted to double for computation, and floating-point constants like 1.0, 2.2, etc., are also recognized as double by default. Therefore, in most cases, it is recommended to use the double type to ensure the reliability of calculation results.

浮点型数据用于表示带小数点的实数，它以指数形式存储在计算机中，因此小数点可以“浮动”。C99中，浮点类型分float、double、long double。float是单精度浮点数，占4字节，有效数字仅约6位，仅适合极少数内存受限且精度要求较低的场合；double是双精度浮点数，占8字节，有效数字约15位，适用于一般数学运算。long double是长双精度浮点数，占16字节，有效数字约18位，适用于高精度科学运算，见表2-2。

在C语言中，为保证运算结果的准确性，float数据会自动转换为double进行计算，且浮点数常量如1.0、2.2等也默认被识别为double类型。因此，在多数情况下，推荐使用double类型以确保运算结果的可靠性。

表 2-2 浮点型数据常见的存储空间及有效位数

类 型	字 节 数	有效位数
float（单精度）	4	6
double(双精度)	8	15
long double(长双精度)	16	18

3. Character（字符型）

Imagine how computers understand and store the letters, numbers, and symbols we use daily. The answer is ASCII, which is short for American Standard Code for Information Interchange. It assigns a specific integer to each character. These integers occupy only 1 byte, or 8 bits, making it very convenient and efficient. It is like how schools give each student a unique ID. Although each student has a unique name, for easy management, schools still give each student a unique ID. Similarly, computers use ASCII to "label" each character, making storage and processing simple and fast.

字符型数据在计算机内部是以其对应的ASCII（美国信息交换标准代码）码值进行存储的。ASCII码实质上是一个整数体系，它赋予每个字符一个唯一的数值标识。这些数值通常占用1字节（即8位）的存储空间。这一过程可以类比学生的学号，每个学生都有一个独特的姓名，但为了便于管理和存储，我们通常会为他们分配一个唯一的学号来标识不同的学生。类似地，计算机为字符分配ASCII码值，不仅方便了数据的存储，也简化了数据的处理过程。

In C, character data can be represented by signed char and unsigned char. By default, signed char is used and abbreviated as char. Signed char can represent values between −128 and 127. But in real use, we focus on 0 to 127 because these numbers match the standard ASCII set, which includes letters, numbers, and punctuation. Character data helps us store and show text symbols. The key difference between signed char and unsigned char is the range of numbers they can handle.

Table 2-3 show you the types and ranges of signed char and unsigned char, Table 2-4 gives the ASCII values for common digits and letters. Interestingly, characters are actually stored as integers in computers. For example, there is a fixed difference of 48(0x30, the ASCII code for `'0'`) between integer numbers 0-9 and their corresponding characters `'0'-'9'`. Similarly, there is a fixed difference of 32(0x20, the ASCII code for a space) between uppercase and lowercase letters. That is why we can use integer operations to process character data. It is important to emphasize that character constants(such as `'0'`) and numeric constants(such as 0) are different. Character constants are enclosed in single quotes, while numeric constants are not. They differ not only in type but also in how computers store them. For example, the character constant `'0'` is actually stored as the ASCII value 48(0x30, the ASCII code for `'0'`), while the numeric constant 0 directly means zero.

在C语言中，字符型数据可以用signed char和unsigned char两种关键字来表示。通常情况下编译器默认使用signed char，并且简写为char。对于signed char类型，它表示的是有符号字符，范围通常是−128~127。然而，在实际应用中，由于字符代码是用来表示具体的字符的，所以其值通常不会为负数。因此，在实际字符编码中，我们主要关注的是0~127这部分的值，它们对应着标准的ASCII字符集。字符型数据主要用于存储和表示文本字符，如字母、数字、标点符号等。

singed char 和unsigned char的数据类型所占字节数和取值范围见表2-3。表2-4给出了常用的数字字符和字母字符的ASCII码值。字符是以一个字节的整数存储和表示的，整型0~9和字符0~9相差48，大写字母和小写字母之间相差32。注意字符用单引号引起来，如`'0'`和0是不同的类型，前者是字符型常量，后者是数值型常量，它们之间相差48。

表 2-3 字符型数据的存储空间及取值范围

类　型	字 节 数	取 值 范 围
signed char	1	−128~127，即 $-2^7 \sim 2^7-1$
unsigned char	1	0~255，即 $0 \sim 2^8-1$

表 2-4 常用字符与 ASCII 代码对照

ASCII	字　符	ASCII	字　符	ASCII	字　符
0x30　48	0	0x41　65	A	0x61　97	a
0x31　49	1	0x42　66	B	0x62　98	b
0x32　50	2	0x43　67	C	0x63　99	c
0x33　51	3	0x44　68	D	0x64　100	d
…	…	…	…	…	…
0x39　57	9	0x5A　90	Z	0x7A　122	z

2.3 Constants and Variables 常量与变量

2.3节视频

Constants and variables are building blocks of programming. Constants are fixed values, providing a stable data reference for programs. Variables, on the other hand, can dynamically change

常量与变量是程序设计的基石。常量是固定不变的值，为程序提供稳定的数据基准；而变量则能随程序运行动态变化，

as the program runs, handling data and interactive logic. This section delves into the concepts and characteristics of both.

处理数据和交互逻辑。本节深入解析这两者的概念与特性。

2.3.1 Constants（常量）

Constants are values in a program that cannot be changed. When writing programs, we often need to use some fixed values, such as π, the size of an array, or a string. These values do not change during the program's operation, so they are called constants. Constants can be divided into numeric constants(including integers and floating-point numbers), character constants, string constants, and boolean constants based on their data types.

常量是程序中值不能被改变的量。在编写程序时，我们常常需要用到一些固定不变的值，如圆周率π、数组的大小、字符串等，这些值在程序运行期间不会发生改变，因此称为常量。常量可以根据其数据类型分为数值型常量（包括整型和浮点型）、字符型常量、字符串常量和布尔型常量。

1. Numeric Constants（数值型常量）

Numeric constants can be further divided into integer constants, like 100, 125, and −4, and real constants, also known as floating-point constants, such as 3.14 and −0.125.

数值型常量又可以分为整型常量和实型常量，整型常量如100、125和−4等；实型常量即浮点型常量，如3.14和−0.125。

2. Character Constants（字符型常量）

Character constants refer to single characters enclosed in single quotes, such as 'a' or '2'.

Escape characters are a special form of character constants that start with a backslash '\'. They represent characters that cannot be directly inputted. For example, '\n' stands for a new line, and '\t' represents a tab. Escape characters are used in programs to denote special or control characters like new lines, carriage returns, or tabs. By using escape characters, we can more easily control character input and output, enhancing the program's readability and maintainability. Table 2-5 lists commonly used escape characters and their corresponding values.

字符型常量是指使用单引号引起来的单个字符，如'a'、'2'等。

转义字符是字符型常量的一种特殊形式，以反斜杠'\'开头，转义字符用于表示不能直接输入的字符。例如，'\n'表示换行符，'\t'表示制表符。转义字符用于在程序中表示一些特殊字符或控制字符，如换行、回车、制表等。通过使用转义字符，我们可以更方便地控制字符的输入和输出，提高程序的可读性和可维护性。常用的转义字符及其对应的字符值见表2-5。

表 2-5 转义字符及其对应字符值

转义字符	字 符 值	转义字符	字 符 值
\'	单引号	\f	换页符（FF）
\"	双引号	\n	换行符（LF）
\?	问号字符	\r	回车符（CR）
\\	反斜杠	\t	水平制表符（HT）
\a	响铃（BEL）	\v	垂直制表符（VT）
\b	退格符（BS）		

3. String Constants（字符串常量）

String constants are sequences of characters enclosed in double quotes, such as "a" and "ab12". At the end of each string, the compiler appends a null character '\0' as a terminator automatically. Here, '\0' is also an escape character and is enclosed by single quotes. It's important to note that

字符串常量是用双引号引起来的字符序列，如"a"和"ab12"，字符串末尾系统会为其自动加上空字符'\0'作为结束标志，'\0'同样是转移字符，是单个字符，用单引号引起来。特别注意：字符串"a"和字

the string "a" and the character 'a' are different constants. The former is a string constant terminated by a null character '\0', while the latter is a single-character constant.

符'a'是不同的常量，前者是字符串常量，末尾系统会自动加上'\0'结束标志，后者是只有单个字符的字符常量。

2.3.2 Variables（变量）

A variable is a value that can change while the program is running. The compiler assigns a certain amount of memory space to each variable. The size of this space depends on the data type of the variable. Every memory unit has a unique address, and the variable name is linked to the first address of this memory space. So, we can access this memory unit through the variable name to read or change the stored data.

变量是在程序运行中值可以变化的量。且编译系统会为每个变量分配内存空间，根据变量所属数据类型开辟不同长度的存储空间。每个内存单元均有地址，变量名关联内存空间首地址，故可以通过变量名找到这个内存单元，从而存取数据。

For example, in Case 2_1, the line float r declares a float type variable named r, and assigns 4 bytes of memory space to it. Also, this line initializes the variable r to 4.0. Because the variable name r is linked to the first address of this memory space, we can access and modify the data in this memory unit through r. The diagram is shown in Figure 2-2.

如案例2_1程序中的float r声明变量r为float型，分配4个字节内存单元，r是变量名，“r=4.0”赋值为4，变量名r关联存储单元地址，则通过r可以找到内存单元。示意图如图2-2所示。

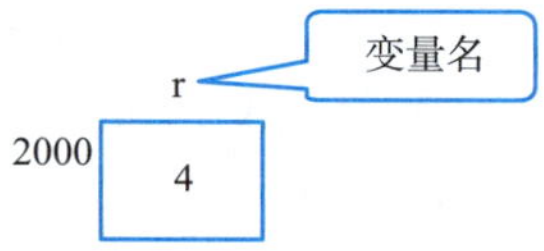

图2-2 变量空间示意图

Note for Variables Use 变量使用注意事项

Variables play a important role in programming, allowing us to store and manipulate various data during program runtime. However, improper use of variables can lead to program errors or difficulties in maintenance. Here are some key points to consider when using variables in C99.

在程序设计中，变量的使用是非常频繁的，它们允许我们存储程序运行时的临时数据。然而，不当的变量使用也可能导致程序出错或难以维护。以下是一些关于C语言变量使用的注意事项。

(1)Definition principles.Variables must be defined before use. This includes specifying the data type, naming the variable, and providing any necessary initialization values. Proper definition helps the compiler accurately understand code intentions and prevents potential errors caused by undefined variables.

（1）定义原则。变量在使用前必须先定义。这意味着需要明确变量的类型、名称以及初始值（如果需要）。这有助于编译器理解所编写的代码，并防止因未定义变量而导致的错误。

(2)Naming rules.Variable names should follow identifier naming conventions, meaning they can only contain letters, numbers, and underscores. The first character must be a letter or underscore, but it is highly unrecommended to start custom variable names with an underscore to avoid naming conflicts with library functions or system macros.

（2）命名规则。变量名应遵循标识符命名规则，即只能包含字母、数字和下划线，且首字符必须是字母或下划线，但强烈不推荐以下划线开头来命名自定义变量，以避免与库函数或系统宏命名冲突。

(3)Other requirements.

（3）其他要求。

①For code readability and maintainability, variable names should be clear, concise, and descriptive, reflecting their purpose or meaning.

①变量名应清晰、简洁，尽可能见名知意。变量名应能够准确反映变量的用途或含义，可提升程序的可阅读性。

②C language distinguishes between uppercase and lowercase. Note that uppercase and lowercase represent different variables, e.g., “int count, Count;” refers to two different variables.

②C语言严格区分大小写。注意大写和小写是不同的变量，如int count, Count;是两个不同的变量。

③Generally, C language variable names use snake_case style, connecting lowercase words with underscores.

③通常，C语言变量名使用snake_case命名方式，即将小写单词用下划线连接。

Case 2_2（案例 2_2）

Valid variable names:
sum, total, student_name, word_counter
Invalid variable names:
M.D.John（contains illegal character '.'）
￥123（contains non-ASCII characters）
#3D（contains illegal character '#'）
3D64、5ss（start with a digit）

合法变量名示例：

非法变量名示例：
包含非法字符点号
包含非ASCII字符￥
包含非法字符井号
以数字开头

Note for Assigning Values to Character Variables 字符型变量赋值注意事项

Character variables(char) are used to store single characters. When assigning values to character variables, special attention should be paid to the following two points.

(1)ASCII code assignment. Character variables can be directly assigned ASCII code values. For example, “char c = 63;” is valid because 63 corresponds to the question mark character '?' in ASCII.

(2)Value range limitations.Typically, the value range for unsigned char variables is from 0 to 255, while for signed char variables, it’s usually from −128 to 127. Assigning values beyond this range can lead to data overflow or truncation, potentially producing unexpected results. For instance, “char c = 263;” is unsafe because 263 exceeds the standard value range for character variables.

字符型变量用于存储单个字符。在赋值时，需要注意以下几点。

（1）ASCII码赋值。字符型变量可以赋值为字符的ASCII码值。例如，char c = 63;是正确的，因为63是问号字符'?'的ASCII码。

（2）取值范围限制。通常，无符号字符变量（unsigned char）的值范围是0~255，而有符号字符变量（signed char）的值范围通常是−128~127。为这些变量分配超出此范围的值可能会导致数据溢出或截断，从而产生意外的结果。例如，“char c = 263;”是不安全的，因为263超出了字符变量的标准值范围。

2.3.3 Symbolic Constants（符号常量）

Symbolic constants are identifiers used to represent specific constant values. To distinguish them from ordinary variables, they are typically represented using all uppercase letters. In the first line of Case 2_1, the #define PI 3.14 preprocessor directive is used to define a symbolic constant. However, in C99 and subsequent versions, it is recommended to use the const keyword to define symbol constants instead of the older #define preprocessor directive. Using const makes the code clearer and provides programmers with type checking, scope control, and better debugging support.

Therefore, in case 2_1, we can replace “#define PI 3.14” with “const double PI = 3.14;”. Here, const tells the compiler that PI is a constant and its value won’t change during the program. Also, we clearly say the constant is of type double, which helps with type checking. Plus, constants defined with const have a clear

符号常量是使用标识符来代替一个具体的常量值，为了与普通变量区分，通常采用全部大写字母来表示。在案例2_1的第1行中，通过#define PI 3.14使用了#define预处理命令来定义符号常量。然而，实际上在C99及以后的版本中，推荐使用const关键字来定义符号常量，而不是使用旧的#define预处理命令。使用const可以使代码更加清晰，同时为程序员提供类型检查、作用域控制以及更好的调试支持。

故在案例2_1中，我们可以将“#define PI 3.14”替换为“const double PI = 3.14;”，这里“const”告诉编译器PI是一个常量，其值在程序运行期间不会改变。同时明确指定了常量的类型为“double”，这有助于类型检查。此外，const定义的常量具有明确的作用域，使代码更加有条理

scope, making the code more organized and easier to fix. During debugging, const constants are easy for the debugger to find and check their values.

且更易于修复。在调试过程中，const常量易于被调试器轻松识别和检查其值。

Case 2_3（案例 2_3）

Knowing the quantity and price per item of goods, we can calculate the total cost.

已知商品的数量和单价，计算商品的总价钱。

Thinking（解题思路）：

We define a symbolic constant to represent the unit price using the const keyword. Here, let's use const to define the symbolic constant PRICE of data type size_t and assume its value is 30. The variable num stands for quantity, and total represents the total cost, both with the data type set to size_t.

使用const关键字来定义一个符号常量表示单价。在这里，使用const来定义数据类型为size_t的符号常量PRICE，并假设其值为30。变量num代表数量，而total代表总价格，这两个变量的数据类型都设定为size_t。

Program Code（程序代码）：

Case2_3.c

```
#include <stdio.h>
int main()
{
    const size_t PRICE=30; // 定义符号常量 PRICE 来表示单价，数据类型为无符号的 size_t

    size_t num=10;                                        // 假设商品的数量是 10
    size_t total=0;                                       // 用于存储总价格
    total=PRICE*num;                                      // 计算总价格
    printf("The total price of the goods is: %zu\n", total);     // 输出总价格
    //%zu 是对应 size_t 的格式说明符
    return 0;
}
```

Program Output（运行结果）：

```
The total price of the goods is: 300
```

Benefits of Using Symbolic Constants 使用符号常量的好处

(1) Clear and descriptive meaning from the name. When reading the code, we can know it's the circumference ratio just from PI. Try to give symbolic constants names that are easy to understand.

（1）"见名知意"，含义清楚。看程序时从PI就可以知道是圆周率，在定义符号常量时尽可能见名知意。

(2) Improves code maintainability. When we need to change the same constant used in many places in the program, we can change it everywhere at once. For example, if PI appears many times in the program and we need to change its value to 3.14159265 for more precision, we only need to change the value when defining it. The new value will replace PI when the program is recompiled.

（2）提高代码的可维护性。在需要改变程序中多处用到同一个常量时，能做到一改全改。如PI如果程序中多处出现，当需要修改其值为3.14159265以增加精度时，不需要在程序中作多处修改，修改其定义时的数值即可，程序再次编译时将用新的值替换PI。

Note（说明）：

Symbolic constants defined by the old #define have no type information and only do simple text replacement. This can lead to unnoticed errors like type mismatches.

旧的#define定义的符号常量没有类型信息，只做简单的文本替换，因此可能引入类型等各种错误而不易被察觉。

In summary, using variables correctly is key to writing high-quality programs. By following the above suggestions, you can write clearer, more reliable, and easier to maintain code.

综上所述，正确使用变量是编写高质量程序的关键之一。遵循上述注意事项，才能写出更加清晰、可靠和易于维护的代码。

2.4 C Operators and Expressions C运算符及表达式

2.4节视频

Key parts of a program include data types, constants and variables, expressions, and input/output statements. In this section, we will focus on the third part: expressions. Expressions are a vital concept in programming. They show how data is calculated and manipulated. We use constants, variables, and operators to build expressions in programs, allowing us to perform various computations.

程序的重要组成成分包括数据类型、常量与变量、表达式和输入输出语句。本小节将重点介绍第三个组成成分——表达式。表达式是程序中的一个关键概念，它代表了数据的计算和操作过程。在程序中使用常量、变量以及运算符来构建表达式，进而可以执行各种运算。

2.4.1 Operator Overview（运算符概述）

C has a lot of operators. Besides input/output and basic controls, most actions are considered operations. For example, "=" is for assignment, and "[]" is used as an index operation. There are a total of 34 operators with 15 levels of precedence. Table 2-6 presents a summary of commonly used C operators, including arithmetic operators, relational operators, assignment operators, and type casting operators. Other operators will be introduced in later chapters.

C运算符十分丰富，除了输入输出和基本控制外均看成运算，如"="是赋值运算，"[ ]"作为下标运算，一共有34种运算符，15个级别。表2-6给出了常用的C运算符一览表，有算术运算符、关系运算符、赋值运算符、强制类型转换符。其他运算符在后续学习过程中陆续介绍。

表 2-6 常用的 C 运算符一览表

运算符类别	运　算　符	运算符类别	运　算　符
算术运算符	+、-、*、/、%、++、--	条件运算符	,
关系运算符	>、<、==、>=、<=、!=	指针运算符	*、&
赋值运算符	＝及其扩展	求字节运算符	Sizeof
强制类型转换	(类型)	成员运算符	.、->
条件运算符	?、:	下标运算符	[ ]
逻辑运算符	!、&&、||		

2.4.2 Common Operators and Expressions（常用运算符及表达式）

Expressions are formed by combining operators and operands according to rules in C. Operands can be constants, variables, function returns, or other expressions. Operators are used to specify what action to perform on the operands.

在C语言中，表达式是由运算符和运算对象按照一定语法规则组合而成的。运算对象可以是常量、变量、函数返回值或其他表达式。运算符用于指定对运算对象执行何种操作。运算对象往往也称操作数。

1. Arithmetic Operators and Expressions（算术运算符及其表达式）

Arithmetic operators in C include addition(+), subtraction(−), multiplication(*), division(/), and modulus(%), as shown in Table 2-7. These operators have different priorities and associativity. Generally,

C语言中的算术运算符包括加法（+）、减法（-）、乘法（*）、除法（/）和取余（%）等，见表2-7。这些运算符具有

multiplication, division, and modulus have higher priority than addition and subtraction, which have the same priority. Among operators of the same priority, calculations follow left-to-right associativity.

不同的优先级和结合性。一般来说，乘、除、取余的优先级高于加法和减法，而加法和减法的优先级相同。在相同优先级的运算符之间，运算按照从左到右的结合性进行。

表 2-7 算术运算符及表达式

算术运算符	含　义	优先级	结合方向
*	乘法运算符	高	从左至右
/	除法运算符		
%	求余运算符		
+	加法运算符	低	
-	减法运算符		

Important Note 重要说明

(1) When writing C language expressions, it's crucial to pay attention to the completeness and accuracy of operators. For example, in mathematical expressions, we can omit the multiplication operator and write 3*x*, but in C language, we must explicitly write the multiplication operator, which is 3*x.

(2) When both the divisor and dividend are integers in division operator(/), C99 performs integer division, truncating the decimal part towards zero instead of rounding. This means that 1/2, 1/(−2), and(−1)/2 will result in 0 in C99, not the mathematical values of 0.5, −0.5, −0.5.

(3) The modulus operator(%), also known as the remainder operator, is used to calculate the remainder after dividing two integers. This operator requires both operands to be integer values, and the result returned is also an integer. For example, 7%3 gives a result of 1. When performing modulus operations on negative numbers, the sign of the remainder is the same as the sign of the dividend. For instance,(−7)%3 and(−7)%(−3) will give a result of −1, while 7%(−3) yields 1. This approach ensures consistency and predictability in the results of modulus operations in C.

(4) C language allows direct arithmetic operations between character data and integers, enabling various flexible character processing functions. This is because characters are stored in the form of ASCII codes in computer. So, character variables actually store their corresponding ASCII integer values. For example, the ASCII value of the character 'a' is 97(0x61 in hexadecimal). If we perform a=a+1, the value stored in the variable 'a' will change to 98(0x62 in hexadecimal), which corresponds to the character 'b'.

（1）在编写C语言表达式时，必须注意运算符的完整性和准确性。例如，在数学表达式中，可以省略乘法运算符并写成3*x*，但在C语言中必须明确写出乘法运算符，即3*x。

（2）关于除法运算符（/），当除数和被除数均为整数时，C语言会执行整数除法，结果会截断掉小数部分，而不是进行四舍五入或其他形式的舍入。这意味着1/2、1/(−2)和(−1)/2的结果在C99中为0，而不是数学上的0.5、−0.5、−0.5。

（3）取余运算符（%）也称模运算符，用于计算两个整数相除后的余数。该运算符要求两个操作数都必须是整数值，并且返回的结果也是一个整数。例如，7%3的结果是1。当对负数进行取余运算时，余数的符号与被除数的符号相同。例如，(−7)%3和(−7)%(−3)结果将是−1，而7%(−3)的结果是1。这种处理方式确保了C语言中的取余运算结果具有一致性和可预测性。

（4）C语言允许字符数据与整数进行直接的算术运算，这可以实现各种灵活的字符处理功能。这是因为字符在计算机内部是以ASCII码的形式存储的。所以，字符变量实际上存储的是它们对应的ASCII整数值。例如，字符'a'的ASCII值是97（十六进制为0x61）。如果执行a=a+1，那么变量'a'中存储的值将变为98（十六进制为0x62），这对应于字符'b'。

In summary, understanding and correctly using operators and expressions in C language is one of the keys to writing high-quality programs. When writing expressions, special attention needs to be paid to issues such as operator precedence, associativity, and data type matching and conversion.

总的来说，理解和正确使用C语言中的运算符和表达式是编写高质量程序的关键之一。在编写表达式时，需要特别注意运算符的优先级、结合性以及数据类型的匹配和转换等问题。

2. Increment and Decrement Operators（自增、自减运算符）

In C language, the increment operator(++) and decrement operator(--) are used to increase or decrease the value of a variable by 1. These operators can only be applied to variables, not constants or expressions. Increment and decrement operators are mainly used for integer types in C, but they can also be used for pointer types to move the pointer. For other types, these operators are generally not applicable. Understanding the behavior of these operators is crucial for writing correct and efficient C code.

在C语言中，自增运算符（++）和自减运算符（--）作用是使变量的值加1或减1。这两个操作符只能用于变量，不能用于常量或表达式。自增和自减运算符在C语言中主要用于整型，但它们也可以用于指针类型。对于其他类型，通常不能使用这些运算符。理解这些操作符的行为对于编写正确和高效的C语言代码至关重要。

Increment and decrement operators have two forms: the prefix and postfix forms.

自增和自减运算符有两种形式：前缀形式和后缀形式。

（1）Prefix form（前缀形式）（前置）。

Represented as（形如）：++i或--i

Meaning: First, increase or decrease variable i by 1, then use the updated value of i.

含义：先将变量i加1或减1，然后使用更新后的i值。

（2）Postfix form（后缀形式）（后置）。

Represented as（形如）：i++或i--

Meaning: First use the current value of variable i, and then increase or decrease i by 1.

含义：先使用变量i的当前值，随后再将i加1或减1。

The main difference between these two forms lies in the order of their combination with variables, which affects their behavior in expressions. For example: Assuming i=3,

这两种形式的主要差异在于它们与变量结合的顺序，进而影响在表达式中的行为。例如：假设i=3。

For j=++i; is equivalent to executing two lines of code: “i=i+1; j=i”;. That is, i is incremented first, and then the new value is assigned to j. The result is i=4, j=4. In this example, the value of j is 4 because ++i will first increment i by 1, and then assign the new value of i to j.

对于 j=++i; 相当于执行两行代码：“i=i+1; j=i”；。即，首先增加i，然后将新值赋给j。结果是i=4, j=4。在这个例子中，j的值为4，因为++i会首先使i增加1，然后将新的i值赋给j。

For k=i++; this is equivalent to two lines of code: “k=i; i=i+1”;. That is, first assign the current value of i to k, and then increment i by 1. The final result is i=4, k=3. Because i++ will first assign the current value of i(which is 3 at this time) to k, and then increment i.

对于k=i++; 这相当于两行代码：“k=i; i=i+1”；。即，首先将i的当前值赋给k，然后i再加1，最终结果是i=4, k=3。因为i++会首先将i的当前值（此时为3）赋给k，然后i再自增。

3. Assignment Operators and Expressions（赋值运算符及其表达式）

The equal sign “=” is used as the assignment operator for programming. By mastering the assignment operator and its expressions, we can write more efficient and concise code. The

在编程中，我们使用等号“=”作为赋值运算符，通过理解和运用赋值运算符及其表达式的特性，我们可以编写出更加

unique feature of the assignment operator is that its left side must be a variable, which stores the result of the expression on the right side. For example, x=4 means assigning the value 4 to the variable x.

高效、简洁的代码。赋值运算符的特别之处在于，它的左侧必须是一个变量，用于存储右侧表达式的计算结果。例如，x=4表示将数值4赋给变量x。

The assignment operator does more than just simple assignments. When combined with binary operators like arithmetic operators, it can form compound assignment operators, simplifying the code.

赋值运算符的功能并不仅限于简单的赋值。它还可以与二目运算符结合，形成复合赋值运算符。这种组合使得代码更加简洁高效。例如，算术运算符就可以与赋值运算符结合。

Here are some examples of compound assignment expressions.

以下是一些复合赋值表达式的例子。

(1) a+=3 is equivalent to a=a+3, meaning increase the value of variable a by 3.

（1）a+=3等价于a=a+3，它表示将变量a的值增加3。

(2) x*=y+5 is like x=x*(y+5). Note the operation priority: first calculate y+5 in parentheses, then multiply it with x, and finally assign the result to x.

（2）x*=y+5等价于x=x*(y+5)，这里要注意运算的优先级，括号内的y+5先计算，然后乘以x的值，再将结果赋给x。

(3) x%=2 is the same as x=x%2, indicating taking the modulus of variable x with 2 and storing the result back in x.

（3）x%=2等价于x=x%2，它表示将变量x的值与2取模，并将结果存回x。

A crucial aspect of assignment expressions is that they always calculate a numerical result and assign it to the variable on the left side of the equal sign. Thus, the result of an assignment expression matches the value of the left-side variable.

赋值表达式的一个重要特性是，它总是能够计算出一个结果，这个结果是一个数值，并且这个数值会被赋给等号左侧的变量。因此，赋值表达式的结果与最左侧变量的值是相同的。

For instance, in the expression a=3*5, 3*5 evaluates to 15, which is then assigned to variable a. Hence, both the value of variable a and the expression's result are 15.

举个例子，a=3*5这个表达式中，3*5的计算结果是15，这个值会被赋给变量a，所以变量a的值和表达式的结果都是15。

Additionally, it's important to note that the associativity of the assignment operator is from right to left. This means in complex assignment expressions; the rightmost assignment is evaluated first. For example, in a=(b=5), due to parentheses having the highest priority, b=5 is calculated first, assigning 5 to b, and then b's value(i.e., 5) is assigned to a. Therefore, variables a, b, and the entire expression all have a value of 5.

此外，赋值运算符的结合性是从右至左的。这意味着在复杂的赋值表达式中，会先计算最右侧的赋值操作。例如，在a=(b=5)这个表达式中，由于括号的优先级最高，所以先计算b=5，将5赋给b，然后将b的值（也就是5）赋给a。因此，变量a、b以及整个表达式的值都是5。

2.4.3 Type Conversions and Casts（类型转换）

In calculations, data types can be changed if needed. These changes can happen automatically or can be forced. Automatic changes often occur when we mix different types of data in calculations. Let's learn more about it.

在运算过程中，数据类型可以根据需要进行转换，可以分为自动类型转换和强制类型转换。自动类型转换一般发生在不同类型数据间的混合运算中，下面进行介绍。

1. Automatic Type Conversions(Implicit Conversions)（自动类型转换）

When different types of data are used in calculations, the compiler automatically performs type conversions to ensure consistency and smooth operation. These conversions follow certain rules to maintain accuracy and efficiency.

在编程中，当涉及不同类型的数据进行混合运算时，系统会自动进行类型转换，以确保操作数具有相同的类型，从而能够顺利进行运算。这种转换遵循一定的规律，使得运算结果既准确又高效。

(1)Character to integer conversions: When a character (char) type is used in an operation with an integer(int) type, the character is treated as its ASCII value. For instance, the ASCII value of 'A' is 65(0x41 in hexadecimal), so when used in calculations with integers, the value 65 represents the character 'A'.

(2)Float to double promotions: During addition, subtraction, multiplication, or division, if any operand is of type float, the compiler automatically promotes it to double to ensure high precision in the calculations. Even if only one operand is of type float, the result of the operation will be of type double.

(3)Conversions involving floating-point and integral Types: When integral types(such as int or char) are used in operations with floating-point types(float or double), the integral types are implicitly converted to double. As a result, regardless of the original data types involved, the outcome of the operation will be of type double.

These conversions are handled automatically by the compiler without any explicit instructions from the programmer. However, there are situations where the programmer might want to perform a specific type of conversion, which requires type casting.

（1）当char型数据与int型数据一起运算时，系统会将char型数据视为其对应的ASCII码值，然后与整型数据进行运算。例如，如果char变量存储的是字符'A'，其ASCII码值为65(0x41)，那么在与整型数据运算时，会使用这个数值进行计算。

（2）在加、减、乘、除运算中，如果操作数中包含float型数据，系统会自动将这些float型数据转换为double型数据，然后进行运算。这样做的目的是确保运算结果具有尽可能高的精度。因此，即使运算中只涉及一个float型数据，最终的结果也会是double型。

（3）当非浮点型数据（如int或char）与浮点型数据（float或double）混合运算时，系统同样会自动进行类型转换。在这种情况下，非浮点型数据会被转换成double型，然后再参与运算。这样做的结果是，无论原始数据的类型如何，运算结果都将是double型。

这些转换都是系统自动进行的，用户无须手动干预。然而，在某些情况下，用户可能需要进行特定的类型转换以满足特定的需求。这时，可以使用强制类型转换运算符。

2. Type Casts(Explicit Conversions)（强制类型转换）

Type casting refers to the explicit conversion of one data type to another by the programmer. It is done by specifying the target type in parentheses before the expression to be converted, such as(target_type)(expression).

For example,(double)(a) casts the expression a to type double, while(int)(x+y) casts the result of x+y to type int. It's essential to exercise caution when performing type casts, as they can potentially lead to loss of precision or data truncation. Furthermore, type casts have higher precedence than other operations. For instance,(int)(x+y) performs the addition first and then casts the result to an integer, whereas(int)x+y casts x to an integer before performing the addition with y. Therefore, when writing code involving type conversions, it's crucial to be mindful of the order of operations and the expected outcome.

强制类型转换的基本形式是在需要转换的表达式前加上括号，并在括号内指定目标类型形如：(数据类型)(表达式)。

举个例子，(double)(a) 将表达式a的值转换为double型，而(int)(x+y)则是将x+y的结果转换为整型。需要注意的是，强制类型转换可能会导致数据精度损失或截断，因此在使用时应谨慎考虑。强制类型转换的优先级高于其他运算。例如，(int)(x+y)与(int)x+y的运算顺序和结果是不同的。前者是先进行x+y的运算，然后再将结果转换为整型；而后者则是先将x转换为整型，再与y进行运算。因此，在编写涉及类型转换的代码时，务必注意运算的顺序和结果是否符合预期。

2.5 C Statements C语句

2.5节视频

After understanding data types, constants and variables,

在深入了解了C语言的数据类型、常量与

operators, and expressions in C language, we will move on to learn about C statements. By learning the structure and proper usage of C statements, we will be able to create powerful and well-structured programs.

变量、运算符与表达式之后，我们即将进入C语句的学习。通过掌握C语句的结构和用法，我们将能够编写出功能丰富、逻辑清晰的程序。

2.5.1 What Are C Statements（什么是C语句）

In C language, statements are the basic units that make up a program, and each statement ends with a semicolon ";". There are many types of C statements, including control statements, function call statements, expression statements, and compound statements. These statements together form the logical framework of the program, allowing the program to execute according to a predetermined process.

在C语言中，语句是构成程序的基本单元，每个语句都以分号“;”作为结尾。C语句的种类繁多，包括控制语句、函数调用语句、表达式语句以及复合语句等。这些语句共同构成了程序的逻辑框架，使得程序能够按照预定的流程执行。

Control statements like if, switch, for, and while are key to controlling the program flow. They can execute corresponding code blocks based on conditional judgments or loop conditions. Function call statements are used to call defined functions to achieve specific functions. Expression statements perform operations on the evaluated results of expressions, such as assignment operations. Compound statements are a group of statements enclosed in curly braces and executed as a whole. An expression followed by a semicolon constitutes an expression statement, while an assignment statement is an assignment expression followed by a semicolon, such as: "c1='A'; a=b=c=3";.

控制语句如if、switch、for、while等，是程序流程控制的关键所在，它们能够根据条件判断或循环条件来执行相应的代码块。而函数调用语句则用于调用已定义的函数，实现特定的功能。表达式语句则是对表达式求值后的结果进行操作，如赋值操作等。复合语句则是用大括号括起来的一组语句，它们作为一个整体被执行。表达式加上分号后就是表达式语句，赋值语句，就是在赋值表达式加上分号，如：“c1='A';a=b=c=3”;。

Here is a simple example of a C statement.

下面给出一个简单的C语句示例。

Case 2_4（案例 2_4）

Write a program to change the uppercase letter 'A' to lowercase 'a'.

编写程序将大写字母A输出为小写字母a。

Thinking（解题思路）:

In this case, we use an assignment expression connected by the assignment operator. We add 32(the ASCII code difference between lowercase and uppercase letters) to the ASCII code value of the uppercase letter 'A' to get the ASCII code value of the lowercase letter 'a'. Then, we assign this value to the variable lower_a. After that, we use the printf function to display the value of lower_a on the screen. Finally, we use the "return 0;" statement to end the program.

在这个示例中，我们使用了赋值运算符连接起来的赋值表达式语句，将大写字母A的ASCII码值加上32（小写字母与大写字母之间的ASCII码差值），得到小写字母a的ASCII码值，并将其赋值给变量c。然后，我们调用输出函数printf将变量c的值输出到屏幕上。最后，使用“return 0;”控制语句结束程序的执行。

Program Code（程序代码）:

Case2_4.c

```
#include <stdio.h>
int main()
{
```

```
    char upper_a='A';                  //初始化一个字符变量 upper_a 为大写字母 'A'
    char lower_a=' ';                  //声明变量保存小写字母

    lower_a=upper_a+32;        //通过增加 32 将大写字母 'a' 转换为小写字母 'a'
    printf("The lowercase of %c is %c\n", upper_a, lower_a);    //小写字母 'a'
    printf("The ASCII value is %d\n", lower_a);                 //'a' 的 ASCII 码值
    return 0;
}
```

Program Output（运行结果）:

```
The lowercase of A is a
The ASCII value is 97
```

Through this case, we can see the important role of C statements in programming. They not only process and compute data but also control the program's flow. Mastering the grammar and usage of C statements can help us write more efficient and stable programs.

通过这个示例，我们可以看到C语句在程序中的重要作用。它们不仅实现了数据的处理和计算，还控制了程序的执行流程。因此，熟练掌握C语句的语法和用法，有利于提高编写程序的高效性和稳定性。

2.5.2 Output Statements（输出语句）

In C programming, input and output statements are crucial and need to be mastered. This section focuses on output statements, mainly achieved through the printf function. The printf function is part of the standard library and is declared in the stdio.h header file. So, when using this function, we need to include the header file #include <stdio.h>.

The basic format of the printf function is: printf(format control string, output list);

Format control string: This string controls the specific output format of the output list and needs to be enclosed in double quotes. It mainly determines the data type and specific format of each output item in the output list. The format string consists of ordinary byte characters(except %), which are copied unchanged into the output stream, and conversion specifications. Each conversion specification has introductory % character, and Table 2-8 lists commonly used conversion specifier.

C语句中，比较重要的且需要数量掌握的是输入输出语句。本小节着重介绍输出语句，输出语句使用printf函数实现，printf函数是标准库函数，在头文件stdio.h中声明，故使用该函数，应加上#include <stdio.h>。

printf函数的使用格式为：printf(格式控制字符串,输出表列);

格式控制字符串：是控制输出表列输出的具体格式，用双引号引起来。主要控制输出表列中输出项的数据类型、具体输出格式等。格式字符串由普通字符（除了%）和转换指示构成，普通字符直接输出，而每个转换指示以“%”开头，常用的格式字符见表2-8。

表 2-8 printf 格式字符

格式字符	作　用	举　例
d	以十进制形式输出带符号整数（正数不输出符号）	int a=3; printf("%d",a); // 输出 3
zu(C99)	以十进制形式输出无符号整数类型 size_t	size_t z=3; printf("%zu", z); // 输出 3

f	以小数形式输出单、双精度实数，默认输出六位小数	float f=1.2;printf("%f",f); // 输出 1.200000
		double d=2.3;printf("%f", d); // 输出 2.300000

c	以字符形式输出，只输出一个字符	printf("%c",'?'); // 输出?
s	输出字符串	char *s="hello";printf（"%s",s）; // 输出 hello

e 或 E	以标准指数形式输出单、双精度实数（e 或 E 代表以 10 为底）	printf("%e",123); // 输出 1.230000e+002
		printf("%E",123); // 输出 1.230000E+002
		// 等价于数学上的科学记数法表示，1.23 × 102

Output list: This part can include constants, variables, expressions, or function calls, separated by commas. Importantly, the format control characters must strictly correspond to the items in the output list in terms of quantity and type.

输出表列：可以是常量、变量、表达式或者函数，用逗号隔开。要求格式字符与输出表列的输出项数量和类型一一对应。

Format Control Characters Can Be Added with Additional Characters, Such as Field Minimum Width, Precision, etc. 格式控制附加字符格式

The general form is: %additional_characters format_character

格式控制字符可以加上附加字符，如最小宽度、保留小数点位数等。一般形式为：%附加字符 格式字符

For example（举例）:

```
printf("%d\n%5d\n%5zu\n",1,-345,12);
```

Program Output（运行结果）:

```
1
 -345
   12
```

In this example, three types of format specifiers are used. ①%d: This format specifier is for printing a signed integer. In this example, it will print the first argument, which is 1. ②%5d: This is also a format specifier for a signed integer. The output number is aligned to the right by default, and 5 sets the width of the output. If the integer(including the negative sign, if any) has less than 5 characters, it will be padded with spaces on the left to make it 5 characters wide. Here, it will print the second argument, −345, with one leading space because −345 already takes up 4 characters(including the negative sign), resulting in "−345". ③%5zu: This format specifier is for printing an unsigned integer of type size_t. The z indicates the size_t type, u denotes unsigned, and 5 sets the output width. In this case, it will print the third argument, 12, padded with three leading spaces to achieve a 5-character width, like this: "12". Note that if the specified field width is less than the actual data, the actual output will take precedence and the set field width will be ignored.

在这个例子中，使用了三种类型的格式说明符。①%d：这个格式说明符用于打印有符号整数。在这个例子中，它将打印第一个参数，即1。②%5d：这也是一个有符号整数的格式说明符。输出的数字默认向右对齐，5设置了输出的宽度。如果整数（包括负号，如果有的话）少于5个字符，它将在左边填充空格，使其宽度为5个字符。在这里，它将打印第二个参数，−345，前面有一个空格，因为−345已经占据了4个字符（包括负号），结果是"−345"。③%5zu：这个格式说明符用于打印类型为size_t的无符号整数。z表示size_t类型，u表示无符号，5设置输出宽度。在这种情况下，它将打印第三个参数，12，前面填充三个空格以达到5个字符的宽度，像这样："12"。请注意，如果指定的字段宽度小于实际数据，则实际输出将优先，并且设置的字段宽度将被忽略。

%m.nf is used to control the output format for float or double data types. Its function is to retain n decimal places, and the data will occupy a total of m columns. Please note that the decimal point also counts as one column. For example, %7.2f retains two decimal places and the data occupies a total of 7 columns.

%m.nf是控制float或者double型数据的输出格式，作用是：保留小数点后n位，该数据一共占m列，注意，小数点也占一列。如%7.2f，保留小数点后两位，数据一共占7列。

For example（举例）:

```
printf("%f\n%7.2f\n%.2f\n",1.2345,12.673,-2.6);
```

Program Output（运行结果）:

```
1.234500
```

```
  12.67
-2.60
```

In this example, we use three different output format specifiers to control how floating-point numbers are displayed. ①%f: This specifier is used to print a floating-point number with six decimal places by default. So, 1.2345 will be printed as 1.234500. ②%7.2f: This specifier means the output floating-point number should have a width of at least 7 characters and retain two decimal places. If the number is less than 7 characters wide, it will be padded with spaces on the left. Therefore, 12.673 will be formatted as 12.67(note the two leading spaces to meet the 7-character width). ③%.2f: This specifier indicates to print only two decimal places without specifying the overall width. Thus, -2.6 will be printed as -2.60.

在这个例子中，我们使用了三种不同的输出格式说明符来控制浮点数的显示方式。①%f：这个说明符用于打印浮点数，默认情况下保留六位小数。因此，1.2345将被打印为1.234500。②%7.2f：这个说明符意味着输出的浮点数至少应具有7个字符的宽度，并保留两位小数。如果数字的宽度小于7个字符，则将在其左侧填充空格。因此，12.673将被格式化为12.67（注意前面有两个空格以满足7个字符的宽度）。③%.2f：这个说明符表示只打印两位小数，而不指定整体宽度。因此，-2.6将被打印为-2.60。

Attention for Using printf　printf 使用注意事项

(1) C language lacks a strict data type checking mechanism, requiring programmers to check themselves. Otherwise, the syntax check may pass, but it may output unwanted results.

(2) In format string, besides format specifiers(like %d, %f) being replaced by values from the output list, and escape characters(e.g., \n for newline) converted to their meanings, other characters in the string are outputted as they are. For example, “int a = 3; printf("a=%d\n", a);” This code will print "a=3" followed by a new line.

(3) If the output item is an expression, printf calculates its value first and then outputs it. For variables, it directly outputs their values. For instance, “printf("%d", 1+2);” prints 3, while “printf("%d", a);” prints the value stored in variable a.

(4) The number of format specifiers in the format string must exactly match the quantity and types of output items in the output list. For example, “int a = 3; double d = 4.0; printf("a=%d, d=%f", a, d);”. Here, %d corresponds to the integer a, and %f corresponds to the double d.

(5) To output the character %, use two consecutive %% in the format string. For example, printf("c=%f%%", 1.0/3); This will output c=0.333333%(assuming this is the floating-point precision).

（1）C语言对于数据类型的检查并不严格，这就要求编程者必须自行确保数据类型匹配。否则，即使程序可以通过编译，运行时可能会输出意料之外的结果或产生运行时错误。

（2）printf的格式控制字符串中的数据，除了格式控制符被输出表列的输出项替换之外，以及转移字符转换成对应含义外，字符串中的其他字符原样输出。如“int a=3;printf("a=%d\n", a);”输出a=3和换行。

（3）如果输出项是表达式，printf会先计算该表达式的值然后再输出。如果是变量，则直接输出该变量的值。例如，“printf("%d", 1+2);”会输出3，而“printf("%d", a);”则会输出变量a中的数值。

（4）格式控制字符串的格式控制符个数与输出表列的输出项类型和数量要保持一一对应的关系。如“int a=3,double d=4.0; printf("a=%d, d=%f", a, d);”，其中%d对应a，%f对应d。

（5）若想输出字符%，用连续两个%%控制。举例：printf("c=%f%%", 1.0/3); 结果为：0.333333%。

2.5.3 Input Statements（输入语句）

In Case2_1, the variable r was initialized directly with a constant. However, in practical applications, we often need to read data from external sources, such as the keyboard. In C, we use the standard library function scanf to read data for variables from the keyboard.

在案例2_1中，变量r是直接用常量进行初始化的。然而，在实际应用中，我们经常需要从外部源（如键盘）读取数据。在C语言中，我们使用标准库函数scanf来完成从键盘为变量读取数据的操作。

The general format of the scanf function is: scanf(format control string, address list);

Here the format control string has a similar meaning to that in the printf function. But note that when inputting data of type double, we should use %lf as the format specifier(where 'l' is the lowercase form of the letter L). For long double type data, both input and output format specifier are %Lf.

In the scanf function, we must provide the addresses of variables so that the function can store the inputted data in correct memory location. The address representation format is &variable_name, where ‘&’ is the address-of operator. For example, if we want to read data for the “radius” variable from the keyboard, we can write: scanf("%f", &radius);. When the user enters 3 from the keyboard and presses Enter, the r variable will be assigned a value of 3.0.

scanf函数的一般格式为：scanf(格式控制字符串, 地址表列);

这里的格式控制字符串与printf函数中的格式控制字符串含义基本相同。但需要注意的是，当输入double类型的数据时，应使用%lf作为格式控制符（其中l是字母L的小写形式）。而对于long double类型的数据，输入输出格式字符均为%Lf。

在scanf函数中，我们必须提供变量的地址以便函数能够将输入的数据存储到正确的内存位置。地址的表示格式为&变量名，其中&是取地址运算符。例如，如果我们想从键盘为radius变量读入数据，可以这样写：scanf("%f", &radius);。当用户从键盘输入3并按Enter键后，r变量将会被赋值为3.0。

For example（举例）:

```
scanf("%d%zu%lf", &a, &z, &d);
```

This line reads three values from keyboard and assigns them to variables a of type int, z of type size_t, and d of type double. When entering input, these values should be separated by spaces, tabs, or newlines. For example, users can enter:

这行代码可以从键盘读取三个值，并分别将它们赋给int类型的变量a、size_t类型的变量z和double类型的变量d。在输入时，这些值应由空格、制表符或换行符分隔。例如，用户可以输入：

```
1 2 3
```

or

或者

```
1
2
3
```

Assign values to these three variables, a gets 1, z gets 2, and d gets 3.0.

It’s important that the format string of scanf should only have format specifiers like %d, %zu, %lf, etc. It’s not a good idea to include other characters, like spaces or commas, unless they are part of what you’re typing. This is because scanf reads what you type based on the format string. If the format string has extra characters, then what you type must also have these characters to match correctly.

Additionally, when using scanf to input numeric data, if the user enters non-numeric characters(such as letters or special symbols), scanf will think the number input is done. This might make the next input

均可以为这三个变量赋值。变量a赋值为1，z赋值2，d赋值3.0。

需要注意的是，scanf的格式控制字符串应仅包含格式控制符（如%d、%zu、%lf等），而不建议包含其他字符，如空格或逗号，除非这些字符是输入数据的一部分。这是因为scanf会根据格式控制字符串来解析输入，如果格式控制字符串中包含了额外的字符，那么输入数据中也必须包含这些字符才能正确匹配。

另外，当使用scanf输入数值数据时，如果用户输入了非数值的字符（如字母或特殊符号），scanf会认为该数据的输入已经结束，并可能导致后续的输入操作失败或数据读取不完整。因此，在使用scanf

fail or not read all the data. So, when using scanf, make sure you type in the expected format.

进行输入时，应确保用户按照预期的格式输入数据。

Example（例如）：scanf("%d%c%zu", &a, &c, &z);

Enter（键入）1234a16e12

Then（则）a=1234, c='a', z=16。

Example（例如）：char c1, c2, c3; scanf("%c%c%c", &c1, &c2, &c3);

Enter（输入）：abc (no spaces 无空格分隔)

Then（则）c1='a'， c2='b'， c3='c'

Enter（输入）：a b c (with spaces有空格分隔)

Then（则）c1='a'， c2=' '， c3='b'

In this example, spaces are valid characters and can be used as input for subsequent char variables. When using %c format to input characters, spaces and escape characters are valid inputs for scanf.

在这个例子中空格为有效字符，可以作为数据输入给后面的char型变量。使用%c格式输入字符时，空格和转义字符都是有效的输入字符。

Attention for Using scanf scanf 使用注意事项

(1) Matching the format control string with the address list.

When using scanf, the format control string should match the following address list(the addresses of variables) precisely in both type and quantity. For example, for int a, char c, size_t z; you should use "scanf("%d%c%zu", &a, &c, &z);". Here, %d corresponds to the int type, %c corresponds to the char type, and %zu corresponds to the size_t type.

（1）格式控制字符串与地址表列的匹配。

在使用scanf时，格式控制字符串应与后续的地址表列（即变量的地址）在类型和数量上精确对应。例如，对于int a, char c, size_t z; 应使用"scanf("%d%c%zu", &a, &c, &z);"。其中，%d对应int类型，%c对应char类型，%zu对应size_t类型。

(2) Format specifier for double data type.

For data type of the double, %lf should be used as the format conversion specifier(where l is the lowercase letter L). This is different from the usage in the printf function, where %f can be used for both float and double types.

（2）double类型的格式控制符。

对于double类型的数据，应使用%lf作为格式控制符（其中l是小写字母L）。这与printf函数中的用法不同，printf中%f既可以用于float也可以用于double类型。

(3) Requirements for the address list.

The address list for the scanf function must only contain the addresses of variables, and cannot be constants or expressions. This is because the function of scanf is to read data from the keyboard and store it in the specified memory address. Therefore, it is necessary to obtain the address of the variable through the & operator.

（3）地址表列的要求。

scanf函数的地址表列必须只包含变量的地址，不可以是常量或表达式。这是因为scanf的功能是从键盘读取数据并将其存储在指定的内存地址中。因此，必须通过&运算符获取变量的地址。

2.5.4 Single-character Input and Output（单个字符的输入和输出）

Besides the scanf and printf functions introduced in the previous two sections, which use %c for character input and output, C language also provides specialized functions for handling the input and output of single characters. Let's learn how to use them.

除了上小节介绍的scanf和printf函数，结合%c输入输出字符外，还有专门的函数用来处理单个字符的输入和输出。下面介绍其使用方法。

1. Single-character Input Function（单个字符的输入函数）——getchar

The function getchar is in the C language standard library, declared in the <stdio.h> header file. This function is used to read a character from standard input(usually the keyboard). If

getchar是C语言中的一个标准库函数，定义在<stdio.h>头文件中，它用于从标准输入（通常是键盘）读取一个字符，若读

the read is successful, it returns the ASCII code value of the character. Note that the getchar function only reads one character at a time.

取成功，则返回该字符的ASCII值。注意，getchar 函数一次只读取一个字符。

Here's a simple example demonstrating how to use the getchar function to read a single character from the keyboard and print out the character and its corresponding ASCII value.

以下是一个简单的示例，展示了如何使用getchar函数从键盘读取字符，并打印出该字符及其ASCII值。

```
#include <stdio.h>
int main()
{
    char ch=' ';
    printf("Please enter a character: ");                    //请输入一个字符:
    ch=getchar();                //Read a character and store it in ch 读取字符并存入 ch 中
    printf("The character you entered is: %c\n", ch);       //你输入的字符是:%c\n
    printf("The ASCII value of the character is: %d\n", ch);    //该字符的 ASCI 值是:%d\n
    return 0;
}
```

First, the program includes the stdio.h header file, which is necessary for standard input and output in C. In the main function, a character variable ch is defined and initialized as a space character. The program then uses the printf function to display a prompt, asking the user to enter a character. Next, the getchar function is called to read a character from standard input(usually the keyboard) and store it in the variable ch. After that, the program prints the entered character and its ASCII value using the printf function. This is done by passing the character variable ch as a %c(character) and %d(integer) type parameter to the printf function.

在这个例子中，我们首先包含了<stdio.h>头文件，然后定义了一个main函数。在main 函数中，我们声明了一个char类型的变量ch，用于存储从getchar读取的字符。然后，我们使用printf 函数提示用户输入一个字符，并使用getchar函数读取该字符。随后打印出该字符及其ASCII值。

2. Single-character Output Function（单个字符的输出函数）——putchar

putchar is a standard library function in C language, declared in the <stdio.h> header file. This function is used to output a single character to the standard output device, usually the screen. It's important that putchar can only output one character at a time. When using it, put the character you want to output as a parameter inside the parentheses, for example: "putchar('A');" this code will output the character 'A' to the screen.

putchar是C语言中的一个标准库函数，定义在<stdio.h>头文件中，用于将单个字符输出到标准输出设备（通常是屏幕）。注意，putchar函数一次只输出一个字符。要输出的字符写在括号内，如 "putchar('A');"。

The following is a simple example of how to use the putchar function to output a single characters.

以下是一个简单的示例，展示了如何使用 putchar函数输出字符。

```
#include <stdio.h>
int main()
{
    char ch='A';               //定义并初始化字符变量 ch 为 'A'
    putchar(ch);               //输出字符 ch，即输出 'A'
    putchar('\n');             //输出一个换行符
    return 0;
}
```

In the program we define a character variable ch and initialize it to 'A'. Then we use the putchar function to display the character stored in ch(which is 'A') on the screen. After that, we call putchar again to print a newline character, so that

在这个例子中，我们定义了一个char类型的变量ch，并将其初始化为字符'A'。然后，我们调用putchar函数，将ch存储的'A'字符输出到屏幕上。接着，我们再次

any following output starts on a new line. Finally, the main function returns 0, showing that the program has successfully finished.

Using the putchar function can easily print a single character to the screen. However, if you need to output an entire string or perform formatted output, it is recommended to use the printf function, which provides more complex and flexible output options.

调用putchar函数输出换行符。

使用putchar函数可以方便地输出单个字符到屏幕上，但如果你需要输出整个字符串或格式化输出，推荐使用 printf 函数。

2.6 Comprehensive Cases 综合案例

综合案例 2_6_1 数据类型的表达能力。

案例描述：编写程序计算以下问题。

（1）表达式 11111*11111 的值是多少？把5个1改成6个1呢？改成9个1呢？

（2）把上面的数都换成浮点数，结果如何？

（3）如果这5个1是小数，即0.11111*0.11111呢？把5个1改成6个1呢？改成9个1呢？

程序代码：

Case2_6_1.c

```c
#include <stdio.h>
int main()
{
    size_t z51 = 11111, z61 = 111111, z91 = 111111111;
    printf("    var51 = %zu\t  var61 = %zu\t   var91 = %zu\n", z51, z61, z91);
    printf(" z51*z51 = %zu\t z61*z61 = %zu\t z91*z91 = %zu\n", z51*z51, z61*z61, z91*z91);

    int i51 = 11111, i61 = 111111, i91 = 111111111;
    printf(" i51*i51 = %d\t i61*i61 = %d\t i91*i91 = %d\n\n", i51*i51, i61*i61, i91*i91);

    float f51 = 11111, f61 = 111111, f91 = 111111111.0f;
    printf(" f51*f51 = %.0f\t f61*f61 = %.0f\t f91*f91 = %.0f\n", f51*f51, f61*f61, f91*f91);

    double d51 = 11111, d61 = 111111, d91 = 111111111;
    printf(" d51*d51 = %.0f\t d61*d61 = %.0f\t d91*d91 = %.0f\n\n", d51*d51, d61*d61, d91*d91);

    f51 = d51 = 0.11111;
    f61 = d61 = 0.111111;
    f91 = d91 = 0.111111111;

    printf("    var51 = %.18f\t    var61 = %.18f\t    var91 = %.18f\n", d51, d61, d91);
    printf(" f51*f51 = %.18f\t f61*f61 = %.18f\t f91*f91 = %.18f\n", f51*f51, f61*f61, f91*f91);
    printf(" d51*d51 = %.18f\t d61*d61 = %.18f\t d91*d91 = %.18f\n", d51*d51, d61*d61, d91*d91);

    return 0;
}
```

运行结果：

```
   var51 = 11111                 var61 = 111111                 var91 = 111111111
z51*z51 = 123454321          z61*z61 = 12345654321          z91*z91 = 12345678987654321
i51*i51 = 123454321          i61*i61 = -539247567           i91*i91 = 1653732529
```

```
f51*f51 = 123454320           f61*f61 = 12345654272         f91*f91 = 12345679481405440
d51*d51 = 123454321           d61*d61 = 12345654321         d91*d91 = 12345678987654321

  var51 = 0.11111000000000000    var61 = 0.11111100000000001    var91 = 0.11111111099999999
f51*f51 = 0.01234543230384588 2  f61*f61 = 0.01234565395861864 1  f91*f91 = 0.01234567910432815 6
d51*d51 = 0.011234543210000000 d61*d61 = 0.01234565432100000 0  d91*d91 = 0.01234567898765432 0
```

案例解析：

（1）各数据类型的结果准确性分析：

①对于5个、6个和9个1相乘的结果都是正确的。

②5个1相乘的结果是正确的，而6个和9个1相乘的结果出现了溢出，是错误的。

③对于整数和小数的5个1相乘的结果最后一位数字差1，而整数和小数的6个和9个1相乘，误差很大，结果不正确。

④对整数和小数的5个和6个1相乘，结果正确，对整数和小数的9个1相乘的结果最后一位数字差1，结果基本正确。

（2）综合结论：在当前普遍采用64位CPU系统情况下，优先考虑使用size_t（对于非负整数）或double（对于需要小数点的计算）。

综合案例 2_6_2 求圆柱体的体积和表面积。

案例描述：求一个圆柱体的体积，要求：（1）从键盘读入数据。（2）结果保留两位小数。通过该案例掌握基本的顺序程序编程方法。

解题思路：首先定义变量，这里可以处理成double型数据，至少包括底圆半径、圆柱体的高、体积等，从键盘读入半径和高的数据，通过公式计算体积和表面积。

程序代码：

Case2_6_2.c

```
#include <stdio.h>
int main()
{   const double PI=3.1415926;
    double radius=0.0, height=0.0;           //初始化为 0，变量声明即初始化是好习惯
    double volume=0.0, surface_area=0.0;     //初始化为 0，表示尚未计算
    printf("Enter the radius and height of the cylinder(separated by space):\n");
    scanf("%lf %lf", &radius, &height);
    volume=PI*radius*radius*height;                      //圆柱体的体积
    surface_area=2*PI*radius*(radius+height);            //表面积
    printf("Volume is: %.2f\n", volume);                 //输出体积
    printf("Surface area is: %.2f\n", surface_area);     //输出表面积
    return 0;
}
```

运行结果：

```
Enter the radius and height of the cylinder(separated by space):
1.1 2
Volume is: 7.60
Surface area is: 21.43
```

案例解析：

（1）若是double类型，scanf要求使用格式控制符%lf，而printf的格式控制符仍为%f。

（2）使用scanf等输入函数时，多个输入数据之间应该通过空格、制表符或回车键进行分隔。

（3）通过printf函数输出浮点数时，使用%.2f可以使结果保留两位小数。

（4）编写顺序程序设计的程序框架，可以总结如下：

```
#include <stdio.h>
int main()
{    //顺序程序设计的 IPO 框架：
    //（1）定义并初始化变量或键盘输入。
    //（2）对数据进行计算或处理。
    //（3）输出结果。
}
```

综合案例 2_6_3 求一个数的逆序数。

案例描述： 从键盘读入一个3位数，输出该3位数的逆序数。比如读入123，输出321。

解题思路： 首先定义一个整型变量，使用scanf读入该整型变量，使用“%”和“/”运算符与数据10进行运算，从而完成数据的各个数位的分离操作，最后再计算出逆序数。

程序代码：

Case2_6_3.c

```
#include <stdio.h>
int main()
{
    int number=0;                                          //define and initialize
    printf("Please enter a three-digit number: ");
    scanf("%d", &number);
    int units=number%10;                                   //个位 123%10=3
    int tens=(number/10) % 10;                             //十位 123/10=12
    int hundreds=number/100;                               //百位 123/100=1
    int reversed=units*100+tens*10+hundreds;               //计算逆序数
    printf("The reversed number is: %d\n", reversed);
    return 0;
}
```

运行结果：

```
Please enter a three-digit number: 123
The reversed number is: 321
```

案例解析：

（1）分离一个整数的各个数位，需要使用求余和除法运算符，求余符号每次分离出个位，而除法可以删除个位，如123%10的结果为3，123/10的结果为12，这是因为计算机识别出两个操作数均为整数，故结果也为整数，则实现了去掉个位的运算。

（2）可以将分开的数字以相反的顺序组合起来：个位数字乘以100，十位数字乘以10，百位数字保持不变。最后，将它们相加得到并输出反转后的数字。

Summary
小 结

This chapter mainly explained the key components of sequential programming, including data types, constants and variables, operators, expressions, and C statements. It's important to master basic data types such as int, size_t, double, and char. In terms of constants, one should clearly distinguish between characters(enclosed in single quotes) and strings(enclosed in double quotes), and understand that escape

本章主要讲解了顺序程序设计的重要组成部分，包括数据类型、常量与变量、运算符、表达式以及C语句。应重点掌握基本数据类型，如整型（int）、大小类型（size_t）、双精度浮点型（double）和字符型（char）。在常量方面，需要明确区分字符（使用单引号括起来）和字符串（使用

characters represent special constants. Master commonly used operators and expressions, such as arithmetic, increment, decrement, and assignment. Additionally, the chapter introduced automatic and explicit data type conversions, as well as input/output statements like printf, scanf, and the usage of single-character processing functions getchar and putchar.

双引号括起来），并理解转义字符代表的特殊常量。同时，还需要掌握常用的运算符及表达式，如算术运算、自增自减运算、赋值运算等。此外，本章还介绍了自动类型转换与强制类型转换，以及输入输出语句如printf、scanf，还有单个字符处理函数getchar和putchar的用法。

Exercise 习　　题

一、单选题

1. 设有int k=0;则以下四个表达式中与其他三个值不同的是（　　）。

A. k++　　B. k+=1　　C. ++k　　D. k+1

2. 设a=6，b=4，则表达式b*=a+3的值为（　　）。

A. 3　　B. 18　　C. 27　　D. 36

3. 下列程序输出的结果为（　　）。

```
#include <stdio.h>
int main()
{
    int m=6,n=8;
    printf("%d,%d\n",m++,n--);
    return 0;
}
```

A. 7,8　　B. 8,7　　C. 5,9　　D. 7,7

4. 在C99中下列属于整数类型的是（　　）。

A. int、short、float、double　　B. int、double、long、char

C. char、float、long、short　　D. short、int、long、char

5. 在以下的变量定义中，合法的是（　　）。

A. float 3_f=3.5　　B. int abc=9

C. double a=1+5e3.9　　D. short go=20

6. 在C语言中，以下哪个是合法的字符常量？（　　）

A. '123'　　B. "A"　　C. \101'　　D. 'ab'

7. 在C语言中，以下哪个是合法的标识符？（　　）

A. 123abc　　B. _variable　　C. int#　　D. float$value

8. 在C语言中，关于赋值运算符（=），以下说法正确的是（　　）。

A. 赋值运算符（=）是左结合的，即 a = b = c 是先将 c 赋值给 b，再将 b 的值赋值给 a

B. 赋值运算符（=）的左侧可以是一个常量

C. 赋值运算符（=）只能用于给变量赋值，不能用于表达式

D. 在 C 语言中，赋值运算符（=）和等于运算符（==）是相同的

9. 已知字母A的ASCLL码为十进制的65，则下面程序输出为（　　）。

```
#include <stdio.h>
int main()
{
```

```
    char cha1,cha2;
    cha1='A'+5-2;
    cha2='A'+8+5;
    printf("%d , %c\n",cha1,cha2);
    return 0;
}
```

A. D, N　　B. D,78　　C. 68,78　　D. 68,N

10. 已知int x,y;double z;，则以下语句中正确的函数调用是（　　）。

A. scanf ("%d,%x,%e", &x,&y,&z);

B. scanf ("%1d,%d,&1f", &x,&y,&z);

C. scanf ("%x,%d,%o", &x,&y);

D. scanf ("%d,%d,%lf", &x,&y,&z);

二、填空题

1. 下面程序的运行结果为____________。

```
#include <stdio.h>
int main()
{
    int i,j,m,n;
    i=8;
    j=8;
    m=++i;
    n=j++;
    printf("%d,%d,%d,%d",i,j,m,n);
    return 0;
}
```

2. 算术表达式a=b+3中，常量为____________，变量为____________，运算符为____________。

3. 若int x = 10;，则表达式x /= 2;执行后，x的值是____________。

4. 若float f = 3.14;，则要将f转换为整数类型，语句为____________。

5. 以下程序的输出结果为____________。

```
#include <stdio.h>
int main()
{
    float a=5.034;
    double b=3.62169;
    printf("%.2f  %0.3lf\n",a,b);
    return 0;
}
```

三、编程题

1. 编写一个程序，输入两个整数，计算并输出它们的和、差、积和商。

2. 编写一个程序，输入一个华氏温度值，将其转换为摄氏温度值并输出。

3. 输入长方体的长、宽、高，计算并输出长方体的体积。

4. 编写一个程序，求圆锥的底面半径为2、高为3的体积。

5. 编写一个程序，将输入的整型数据各个数位分离，并依次输出，如输入12345整型数据，输出如下（每个数据占5个位宽输出）：

```
    1    2    3    4    5
```

Chapter 3

Selective Structure Programming

第3章

选择结构程序设计

Introduction 主要内容

C language includes three basic structures: sequential, selective, and looping. After deeply exploring the sequential structure, we will learn about the selective structure. The selective structure is a crucial part of C language programming. The if statement and the switch statement are two commonly used ways to implement the selective structure. They allow the program to execute corresponding code blocks under different conditions, thus enhancing the flexibility of the program.

C语言包含顺序、选择和循环三大基本结构。在深入探讨了顺序结构后，我们接下来将学习选择结构。选择结构是C语言编程中的关键部分，if语句和switch语句是实现选择结构的两种常用方式，它们允许程序在不同条件下执行相应的代码块，从而增强程序的灵活性。

Objectives 学习目标

Students are required to master statements such as if, if...else, if...else if...else, and switch to implement the selective structure. They should avoid common programming pitfalls. Through practical training, students should be able to flexibly use the selective structure to build efficient and logically rigorous code to solve practical programming problems. Meanwhile, we aim to cultivate students' logical thinking ability and problem-solving ability, laying a solid foundation for subsequent programming learning.

要求掌握if、if...else、if...else if...else、switch等实现选择结构的语句，避免常见编程错误。通过实践训练，学生能够灵活运用选择结构，构建高效且逻辑严谨的代码，解决实际编程问题。同时，培养学生的逻辑思维能力和问题解决能力，为后续的编程学习奠定坚实基础。

3.1 Relational, Logical Operators and Expressions 关系、逻辑运算符及其表达式

3.1节视频

First, we need to understand why programs need selection structures. Understanding this will help us build

首先要明确为什么程序需要选择结构，理解该问题有助于构建编

programming logic. In selection structures, we need to learn how to write conditional statements, which often involve the use of relational and logical expressions. This section will introduce these concepts in turn.

程逻辑。选择结构需要学会如何编写判断条件，一般使用关系表达式和逻辑表达式，本节将依次介绍。

3.1.1 Why Do We Need Selection Structures（为什么需要选择结构）

Selection structures play a crucial role in programming. They allow programs to execute different code blocks based on different conditions, enabling them to handle complex and varying requirements. Whether it's user menu selection, grade level determination, file existence checks, or network status judgments, selection structures play a key role. They give programs greater flexibility, making them easier to maintain and understand. Just as we make decisions based on conditions in real life, computers also rely on selection structures for logical judgments and choices.

For example, if it's sunny today, we might choose to go to the park; if it's not sunny, we might choose to stay home and read a book. In this example, "Is it sunny today?" is a conditional statement. If the condition is true, we execute the plan to go to the park; if the condition is false, we stay home and read a book. Such logical judgments are common in life. The ultimate goal of programming is to solve practical problems, and selection structures are necessary to achieve this goal. For everyone, choice is equally important. When faced with difficulties and problems, do we face them head-on or back down? Different choices will lead to different outcomes. Therefore, let's strive to improve our cognition and increase our knowledge reserves to make the right choices. Let's be like the Chinese Women's Volleyball Team, never backing down and giving it our all, even when faced with strong opponents.

选择结构在编程中至关重要，它使程序能依据不同条件执行不同代码块，应对复杂多变的需求。无论是用户菜单选择、成绩等级判定、文件存在性检查还是网络状态判断，选择结构都发挥着关键作用。它使程序更灵活、易维护且易于理解。就像我们在现实生活中根据条件做出决策一样，计算机也需要通过选择结构实现逻辑判断和选择。

举例：若今天晴天，则去公园玩，否则留在家里看书。其中“今天是晴天吗？”是判断条件。若条件成立，则执行去公园玩的计划，若条件不成立则留在家里看书。这样的例子生活中比比皆是。程序最终是为了解决实际问题，选择结构是必要的。对于每个人来说，选择同样重要，在遇到困难和问题时，我们是迎难而上还是退缩不前？不同的选择将收获不同的结果。让我们努力提高认知、知识储备，尽可能做出正确的抉择。让我们像中国女排一样，遇到再强大的对手，也不退缩、全力以赴。

3.1.2 Relational Operators and Expressions（关系运算符及其表达式）

In selection structures, condition judgments are crucial. Therefore, we need to first master how to correctly represent these conditions. In programming, conditions are usually reflected through relational expressions. A relational expression is a formula composed of two operands connected by a relational operator. C provides us with six relational operators, as shown in Table 3-1. These relational operators are used for the comparison of values between two operands.

选择结构需要进行条件判断，故先要学会如何正确地表示条件。程序中的条件可以使用关系表达式。关系表达式是由关系运算符连接两个操作数的式子。C语言提供了六种关系运算符，见表3-1。这些关系运算符用于比较两个操作数的大小关系。

表 3-1 关系运算符

关系运算符	含 义	数学中的表示	优 先 级
>	大于	>	高
<	小于	<	

续表

关系运算符	含　义	数学中的表示	优 先 级
>=	大于或等于		高
<=	小于或等于	≤	
==	等于	=	低
!=	不等于	≠	

Please note that the relational operators used in programming may differ from the symbols in mathematics. Table 3-1 compares the two, so please make sure to write them correctly. It's worth mentioning that the operator for judging whether two operands are equal is the double equal sign "==", while the single equal sign "=" is the assignment operator. Don't confuse the two. In terms of priority, the first four relational operators(>, <, >=, <=) in the table have the same priority, and their priority is relatively high; while equal "==" and not equal "!=" have another same priority, which is relatively low.

注意程序中表示的关系运算符与数学上不同，表中给出了对比，请正确书写。其中判断是否相等的运算符是两个等号"=="，一个等号"="是赋值运算符，不要混淆。表中前四个关系运算符（>、<、>=、<=）优先级相同，优先级相对高，等于"=="和不等于"!="两个优先级相同，优先级相对低一些。

A relational expression is an expression that connects operands using the relational operators. Its result is a logical value, namely "true" or "false". When the condition is met, the expression evaluates to "true"; otherwise, it evaluates to "false". Inside the computer, "true" is usually represented by the numerical value 1, while "false" is represented by 0. For example, the expression 5>3 evaluates to "true", and its value is represented as 1 inside the computer; while the expression 5<3 evaluates to "false", and its value is represented as 0.

关系表达式就是用上述关系运算符连接起来的式子，关系表达式的值为逻辑值"真"或"假"，若条件成立则为"真"，若条件不成立则为"假"。计算机用值1代表"真"，用值0代表"假"。举例说明：5>3表达式为真，值为1。5<3表达式为假，值为0。

3.1.3　Logical Operators and Expressions（逻辑运算符及其表达式）

Conditions can also be expressed using logical expressions. A logical expression is an expression that connects relational expressions or other logical quantities using logical operators, describing more complex judgment conditions. C provides three logical operators: logical AND(&&), logical OR(||), and logical NOT(!). Among them, "&&" and "||" are binary operators, which means they require two operands; while "!" is a unary operator, needing only one operand. Generally, unary operators have higher priority than binary operators. Here, "!" has the highest priority, followed by "&&", and finally "||". The value of a logical expression, like that of a relational expression, is either "true" or "false", represented by the numerical values 1 or 0 inside the computer.

判断条件也可用逻辑表达式来表示。逻辑表达式是用逻辑运算符连接起来的式子，其可以连接关系表达式或其他逻辑量描述较为复杂的判断条件。C语言有三种逻辑运算符，逻辑与（&&）、逻辑或（||），逻辑非（!）。其中"&&""||"是双目运算符，即连接两个操作数，"!"是单目运算符，即只有一个操作数。一般单目运算符优先级高于双目运算符，这里"!"优先级最高，"&&"优先级高于"||"。逻辑表达式的值与关系表达式的值一样，也为"真"或"假"，用数值1或0存储。

The truth table for logical operators is shown in Table 3-2. In this table, we assume a and b are two operands, which can be relational expressions or other legal logical quantities.

逻辑运算符的运算真值表见表3-2，假设a和b是两个操作数，代表关系表达式或者是其他合法的逻辑量。

表 3-2 逻辑运算的运算真值表

a	b	a && b	a || b	! a
真	真	真	真	假
真	假	假	真	假
假	真	假	真	真
假	假	假	假	真

For the expression a&&b, the logical expression is true only when both a and b are true; for the expression a||b, the logical expression is false only when both a and b are false. "!" is the NOT operation. If a is true, then "!a" is false; if a is false, then "!a" is true. It's important to note that in C, any non-zero value is considered "true", following the principle of "non-zero is true".

对于a&&b表达式，a、b都为真时逻辑表达式才为真；对于a||b表达式，只有a、b都为假时逻辑表达式才为假。"!"是取反运算，a真"!a"为假，a假"!a"为真。特别注意，在逻辑表达式中C语言将一个非零数值认作为"真"，即"非零即为真"。

Case 3_1（案例 3_1）

Given a=4 and b=5, calculate the results of the following logical expressions.

!a、a&&b、a||b、!a||b、4&&0||2。

!a：Following the principle of "non-zero is true", the value of a is 4(non-zero), so its logical value is 1(true). After logical NOT operation, the value of the expression is 0(false).

a&&b：Since "non-zero is true", the values of a and b are 4 and 5(both non-zero), so the logical values of a and b are both 1(true). According to the truth table of logical AND, the value of this expression is 1(true).

a||b：Since the logical values of both a and b are 1(true), according to the truth table of logical OR, as long as one operand is true, the expression is true. So the value of this expression is 1(true).

!a||b：In this expression, the logical NOT "!" has the highest priority. First, perform the !a operation, and the result is 0(false). Then perform the logical OR operation with b(true). According to the truth table of logical OR, the value of this expression is 1(true).

4&&0||2：Since the priority of logical AND "&&" is higher than logical OR "||", we first calculate 4&&0, and the result is 0(false). Then perform the logical OR operation with 2(true). According to the truth table of logical OR, the result is 1(true).

若a=4，b=5，请计算下列逻辑表达式的结果。

遵循"非零即真"的原则，变量a的值为4（非零），所以它的逻辑值为1（真）。经过逻辑非操作后，表达式的值为0（假）。

既然"非零即真"，变量a和b的值分别是4和5（都是非零），所以变量a和b的逻辑值都是1（真）。根据逻辑与的真值表，这个表达式的值也是1（真）。

由于变量a和b的逻辑值都是1（真），根据逻辑或的真值表，只要有一个操作数为真，表达式就为真。所以，这个表达式的值是1（真）。

在这个表达式中，逻辑非"!"具有最高优先级。首先执行!a操作，结果是 0（假）。然后用这个结果与 b（真）进行逻辑或操作。根据逻辑或的真值表，这个表达式的值是 1（真）。

由于逻辑与"&&"的优先级高于逻辑或"||"，我们首先计算"4&&0"，结果是0（假）。然后与2（真）进行逻辑或操作。根据逻辑或的真值表，结果是1（真）。

Case 3_2（案例 3_2）

How to express the condition to judge whether a person's age is between 10 and 18?

判断一个人的年龄是否在10~18岁之间，如何表示该判断条件？

Thinking（解题思路）:

First, define a variable to represent a person's age, assuming int age; The question requires satisfying two conditions simultaneously:"age>=10" and "age<=18". Therefore, it is necessary to use the logical AND operator "&&" to connect these two conditions.

首先定义一个变量，表示某个人的年龄，假设为int age; 题目涉及两个条件："age>=10"以及"age<=18"，这两个条件需要同时成立才能满足要求，故使用逻辑运算符"&&"进行连接。

Case Solution（案例实现）:

The final expression is "age>=10 && age<=18". When age is 9, this condition is not met; if age is 10, this condition is met.

最终的表达式为"age>=10&&age<=18"，当age为9时条件不成立，若age为10则条件成立。

Case Analysis（案例解析）:

Beginners may mistakenly write the mathematical expression "10<=age<=18", which is incorrect. In C language, conditional expressions cannot be written in this way. Although such an expression is syntactically correct, its meaning is different from what is expected. According to the associativity of relational operators from left to right, the expression "10<=age" will be calculated first. If age is 9, the expression is false(0). Then, this value(0) will be compared with 18, that is, to judge "0<=18". This expression is true(1), so the result of the entire expression "10<=age<=18" is true(1), which is obviously inconsistent with our expectations. The schematic diagram of the operation process is shown in Figure 3-1.

对于初学者往往会写成数学上的表达"10<=age<=18"，这是错误的，特别注意C语言的条件不能连写。虽然该表达式符合语法，但含义却截然不同。按照关系运算符自左至右的结合性，会先计算"10<=age"这个表达式，若age为9，则该表达式为假，值为0。接着参与运算的是这个值0与18进行比较，即判断"0<=18"，该表达式为真，则"10<=age<=18"表达式为真，这显然与实际结果不相符合。运算过程示意图如图3-1所示。

10<=age<=18 ——真
0

图3-1 表达式运算过程示意图

Conclusion（结论）:

When multiple conditions are involved, each condition should be written independently and connected using logical operators such as logical AND(&&), logical OR(||), and logical NOT(!). Do not write them together.

当涉及多个条件时，每个条件应该独立书写，并使用逻辑运算符，如逻辑与（&&）、逻辑或（||）和逻辑非（!）进行连接，不要连写。

Case 3_3（案例 3_3）

Decide if a year is a leap year using a logical expression.

判断某一年是否是闰年，用逻辑表达式表示该条件。

Thinking（解题思路）:

we first need to know what makes a leap year. A leap year meets one of these two conditions:

(1) It can be divided by 4 but not by 100, like the year 2008.

(2) It can be divided by 400, like the year 2000.

We define a variable called "year" to represent the year we want to check. Now, for condition(1), based on Case3_2, we need to check each small condition separately. To see if "year" can be divided by 4, we use the modulo operator "%" and check

首先明确闰年的条件，闰年需要满足如下两个条件之一。

（1）能被4整除，但不能被100整除，如2008。

（2）能被400整除，如2000。

定义一个变量year来代表要判断的年份。首先表示条件（1），通过案例3_2的分析与结论，首先我们要独立表示出各个小条件，判断year能否被4整除，可以使用"%"

if the remainder of year % 4 is 0, which means "year % 4 == 0". If this is true, "year" can be divided by 4. Also, "year" should not be divisible by 100, shown as "year % 100 != 0". Both these conditions must be met, so we use the logical AND operator "&&" to connect them: "year % 4 == 0 && year % 100 != 0".

For condition(2), it's simple: year % 400 == 0.

In the end, as long as one of the two conditions is met, the year is a leap year. So, we use the logical OR operator "||" to join them. The final logical expression is:

(year % 4 == 0 && year % 100 != 0) ||(year % 400 == 0)

If this expression is true, the year is a leap year. If it's false, the year is not a leap year. This method lets us accurately decide if any given year is a leap year.

符号，判断"year%4"的余数是否为0来实现，即"year%4==0"，若成立则能整除，否则不能整除。year不能被100整除，表示为"year%100!=0"。这两个条件需要同时满足，因此使用逻辑与（&&）运算符将它们连接起来，即"year%4==0&&year%100!=0"。

条件（2）可表示为year%400==0。

最后，只要满足两个条件中的任何一个，这一年就是闰年。因此，我们使用逻辑或"||"运算符将它们连接起来。最终的逻辑表达式是：

如果表达式的值为真，则表示该年份是闰年；如果表达式的值为假，则表示该年份不是闰年。通过这种方式，我们可以准确地判断任意给定年份是否为闰年。

3.2 Selection Structures Using if and switch Statements if和switch语句实现选择结构

3.2节视频

C language primarily uses if and switch statements to implement selection structures. The if statement is crucial for implementing a two-branch selection structure, while the switch statement is particularly suitable for handling multiple specific selection scenarios.

C语言使用if和switch语句实现选择结构。if语句是实现两分支选择结构的关键，而switch语句则适用于处理多分支选择场景。

3.2.1 Selection Structures Using if Statements（if 语句实现选择结构）

The basic syntax of the if statement is as follows:

```
if(conditional expression)
{
    //Code to be executed when condition is true
}
else
{
    //Optional code to be executed when condition is false
}
```

if语句基本语法如下：

```
if（条件）
{
    //如果条件为真，执行这里的代码
}
else
{
    //如果条件为假，执行这里的代码
}
```

Note（说明）：

if and else are keywords in C language. The condition is typically relational expressions, logical expressions, or numeric expressions. The else part is optional and is executed when the condition is false. The code block usually consists of multiple statements enclosed by curly braces {}. To ensure code clarity and prevent errors, it is recommended to always use {} even if the code block contains only one statement.

if和else是关键字，条件一般是关系、逻辑和数值表达式，else部分有时候可省略。语句块可以由一个或多个语句组成，并且通常用花括号{}括起来。当只有一条语句时，{}是可选的，但为了代码清晰和避免错误，建议总是使用{}括起来。用{}括起来的代码块也称复合语句，代表是一个整体，代码更为清晰。

In practice, there are three commonly used forms of if statements.

在实践中，人们总结出了if语句常用的三种形式。

1. if Statement（if 基本形式）

```
if(condition) { statements }
```

```
if（条件）｛执行语句｝
```

When the condition is true, that is, non-zero or non-empty, the statements within {} will be executed. If the condition is false, namely zero or empty, the execution statement within {} will be skipped. The flowchart of the if basic form is shown in Figure 3-2.

当条件为真时，即非零或非空时，花括号{}内的语句将被执行。如果条件为假，即零或空，那么花括号{}内的执行语句将被跳过。if基本形式的流程图如图3-2所示。

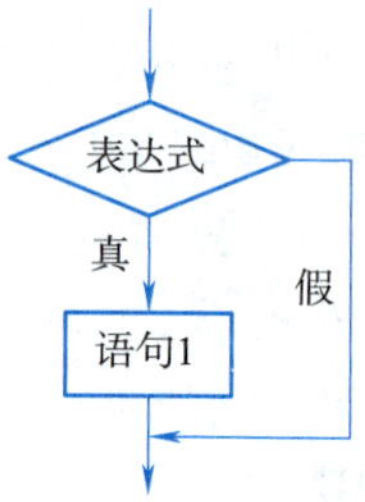

图3-2　if基本形式流程图

For example（举例）:

```
int x;
scanf("%d", &x);
if(x>0)
{
    printf("x is a positive number.\n");
}
```

In this case, if x is read as 5, the condition x > 0 is true, so the output statement within {} is executed. If x is read as −5, the condition x > 0 is false, so the printf statement will not be executed.

在该例子中，若为x读入5，则x>0条件为真，则执行{}内输出语句。若为x读入−5，则x>0条件为假，则不会执行printf语句。

Case 3_4（案例 3_4）

Using the basic form of if statements to find the maximum of two integers.

使用if语句的基本形式，求两个整数的最大值。

Thinking（解题思路）:

Define two integer variables, a and b, to store the two input integers. Then, define an integer variable max to store the maximum value and set its initial value to a. Next, use the if statement to compare the values of max and b. If b is greater than the current maximum value, then update the maximum value to b. Finally, output the value of max.

定义两个整数变量a和b，用于存储两个输入的整数。然后，定义一个整数变量max来存储最大值，并将其初始值设置为a。接下来，使用if语句比较max和b的值。如果b大于当前的最大值，则将最大值更新为b。最后，输出max的值。

Program Code（程序代码）:

Case3_4.c

```
#include <stdio.h>
int main()
{
    int a=0, b=0;                                   //声明及初始化变量
    printf("Please enter two integers: ");          //输入提示信息
    scanf("%d %d", &a, &b);                         //从键盘读入
    int max=a;                                      //声明及初始化 max 为 a
```

```
        if(max<b)                                   //如果b比当前最大值max大
            max=b;                                  //更新最大值max为b
        printf("max=%d\n", max);                    //输出最大值
        return 0;
    }
```

Case Analysis（案例解析）:

The program first prompts the user to enter two integers and then reads them. It assigns one of the integers to the variable max as the initial maximum value. Then, using an if statement, the program compares the two integers and updates max to the value of the second integer if it is greater than the current max. Finally, the program outputs the maximum value found.

程序首先提示用户输入两个整数，并读取这两个整数。它将其中一个整数赋值给变量max作为初始的最大值。然后，使用if语句，程序比较这两个整数，如果第二个整数大于当前的最大值max，则将max更新为第二个整数的值。最后，程序输出找到的最大值。

Program output（运行结果）:

```
Please enter two integers: 5 9
max=9
```

When 5 and 9 are entered from the keyboard and Enter is pressed, it will output “max=9” .

5和9是从键盘读入的数据，按【Enter】键后，则输出max=9。

2. if...else Statement（if...else形式）

```
if(condition) { statements1 }
else { statements2 }
```

```
if(条件) {执行语句1}
else {执行语句2}
```

When the condition is true, the statements1 will be executed. If the condition is false, the statements2 will be executed. The flowchart of the if...else form is shown in Figure 3-3.

In this case, if the condition “x>0” is false, the statement after “else” will be executed, outputting “x is not a positive number”.

当条件为真时，执行语句1会被执行；当条件为假时，执行语句2会被执行。执行流程如图3-3所示。

例如：在下列代码中，x>0条件为假，故执行else后面的语句，输出x不是正数。

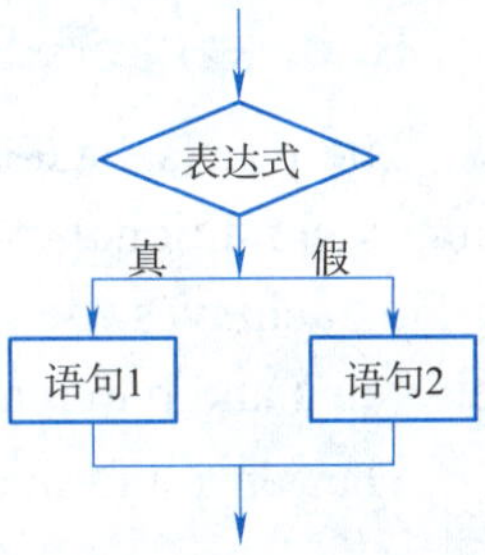

图3-3 if...else结构流程图

```
int x=-3;
if(x>0)
{
    printf("x is a positive number.\n");
}
else
{
    printf("x is not a positive number.\n");
}
```

Case 3_5（案例 3_5）

Using if...else statements to find the maximum of two integers.

使用if...else形式，求两个整数最大值。

Thinking（解题思路）:

Like Case3_4, define variables, input their values, and use

与案例3_4相同，定义变量、输入变量

conditions to judge the value of a and b. Assign the larger value to max to find the maximum of the two.

值，条件可以判断a和b的大小，将较大的值赋值给max，即可求得两者中的最大值。

Program Code（程序代码）:

Case3_5.c

```
1    #include <stdio.h>
2    int main()
3    {
4        int a=0, b=0, max=0;                      //声明及初始化变量
5        printf("Please enter two integers: ");    //输入提示信息
6        scanf("%d %d", &a, &b);                   //从键盘读入
7        if(a>b) max=a;                            //a 如果 a 大于 b，设置 max 为 a
8        else max=b;                               //否则，设置 max 为 b
9        printf("max=%d\n", max);                  //输出最大值
10       return 0;
11   }
```

Program output（运行结果）:

```
Please enter two integers: 5 9
max=9
```

Case Analysis（案例解析）:

In this case, an if...else statement is used, as shown in lines 7, 8 of the code. The condition “a>b” is checked. If true, it executes “max=a”, otherwise, it executes “max=b”. This ensures that max holds the maximum of the two.

在该案例中使用了if...else语句，如第7、8行代码。判断条件为a>b，若该条件成立则执行max=a，否则执行max=b，可确保max存放的是两者中的最大值。

Knowledge Extension（知识拓展）:

The if...else statement in the above case can be simplified using a conditional expression, rewritten as max =(a > b) ? a : b; This achieves the same effect of assigning the larger value between a and b to max.

The general form of a conditional expression is:

```
<expression1> ? <expression2> : <expression3>
```

上述案例中的if...else语句可以使用条件表达式进行简化，可以改写成max=(a>b)?a:b; 同样可以达到取两者的大数赋值给max的效果。

条件表达式通用形式为：

```
<表达式 1> ? <表达式 2> : <表达式 3>
```

Its functionality is equivalent to a complete if statement. In this expression, “?” serves the role of if, while “:” is equivalent to else. Specifically, the conditional expression works as follows: First, it evaluates the value of <expression1>. If the value is true(or non-zero), it calculates and returns the value of <expression2>. If the value of <expression1> is false(or zero), it calculates and returns the value of <expression3>.

Taking max =(a > b) ? a : b as an example, it first determines the result of the expression(a > b). If the condition is true(a > b is valid), it assigns the value of a to max; otherwise, it assigns the value of b to max.

Note that when multiple conditional expressions are used consecutively, such as x = <expression1> ? <expression2> :

其功能和一条完整的if语句等价。在这个表达式中，“?”相当于if的作用，而“:”则相当于else。具体地，条件表达式的工作方式：首先计算<表达式1>的值，如果该值为真（或非零），则计算并返回<表达式2>的值；若<表达式1>的值为假（或为零），则计算并返回<表达式3>的值。

以max =(a > b) ? a : b为例，这里首先判断a是否大于b。如果条件成立（a > b为真），则将a的值赋给max；否则，将b的值赋给max。

注意：当多个条件表达式连续使用时，如 x = <表达式1> ? <表达式2> : <表达式3> ?

<expression3> ? <expression4> : <expression5>, it's important to remember that the conditional operator has right associativity. This means the expression is structurally interpreted as x = <expression1> ? <expression2> :(<expression3> ? <expression4> : <expression5>). The evaluation process goes like this: First, <expression1> is calculated. If <expression1> is true, then the value of <expression2> is computed and returned, without evaluating the subsequent expressions. If <expression1> is false, only then will <expression3> ? <expression4> : <expression5> be evaluated.

For example, given a=1, b=2, c=3, d=4, in the nested conditional expression x = a < b ? a : c < d ? c : d, since the condition a < b is true, the expression directly returns the value of a without the need to evaluate the c < d ? c : d part. Thus, the value of x is 1.

If we have a=4, b=3, c=2, d=1, because a=4 and b=3, the condition a < b evaluates to false. Therefore, we continue to evaluate the subsequent expression c < d ? c : d. Here, with c=2 and d=1, the condition c < d also evaluates to false, so we take the value of d. That is, x=1.

Hence, under the given conditions of a=4, b=3, c=2, d=1, the value of x is still 1. This is because in this conditional expression, only when all preceding conditions evaluate to false, will the rightmost value, namely d, be taken.

<表达式4> : <表达式5>，需要注意条件运算符具有右结合性，即此表达式在语法结构上被解析为 x = <表达式1> ? <表达式2> : (<表达式3> ? <表达式4> : <表达式5>)。因此，表达式的计算过程是：首先计算 <表达式1>。如果 <表达式1> 为真，则计算并返回 <表达式2> 的值，而不会去计算后面的表达式。如果 <表达式1> 为假，那么才会去计算 <表达式3> ? <表达式4> : <表达式5>。

例如，当a=1, b=2, c=3, d=4时，在嵌套的条件表达式x = a < b ? a : c < d ? c : d中，由于a < b的判断为真，因此表达式直接返回a的值，而不需要计算c < d ? c : d部分。所以，x的值为1。

如果给定 a=4, b=3, c=2, d=1时，由于a=4 和 b=3，所以 a < b 的判断为假。故继续计算后面的表达式 c < d ? c : d。在这里，c=2 和 d=1，所以 c < d 的判断也为假，故取 d 的值。即 x=1。

因此，在给定a=4, b=3, c=2, d=1条件下，x的值仍然为1，这是因为在此条件表达式中，只有当所有前面的条件判断都为假时，最终才会取到最右边的值，即d的值。

3. if...else if...else Statement（if...else if...else 形式）

```
if(condition1) { statements1 }
else if(condition2) { statements2 }
...
else { statementsn }
```

```
if(条件 1) {执行语句 1}
else if(条件 2) {执行语句 2}
...
else {执行语句 n}
```

This format lets you run different blocks of code based on multiple conditions. Its core idea is to nest a complete if...else statement after each else, enabling multi-condition judgments. If statements can be nested, meaning you can put a full if statement inside an if or else block, and you can even nest them multiple times. But be careful, the structure must stay intact. For instance, every else must have a matching if, and it pairs with the closest unmatched if.

The execution flow of the if...else if...else structure is shown in Figure 3-4. When Condition 1 is true, Statement 1 runs, and then the if statement ends. If Condition 1 is false, it checks Condition 2. If Condition 2 is true, Statement 2 runs, and the if statement concludes. This pattern continues. If all conditions are false, the final else block's Statement n is executed.

这种形式允许根据多个条件执行不同的语句块。本质上是在每条else后面又嵌套一个完整的if...else语句，进而实现了多条件判断。if语句允许嵌套，可以在if或else后面嵌套完整的if语句，且可多层嵌套，但要注意结构的完整性，如else前需要有配对的if，其与最近的尚未匹配的if进行配对。

if...else if...else结构的执行流程如图3-4所示。当条件1为真时，执行语句1会被执行，该if语句结束；当条件1为假时继续判断条件2，若条件2为真，执行语句2会被执行，该if语句结束；依此类推。如果所有条件都为假，则执行最后的else块中的执行语句n。

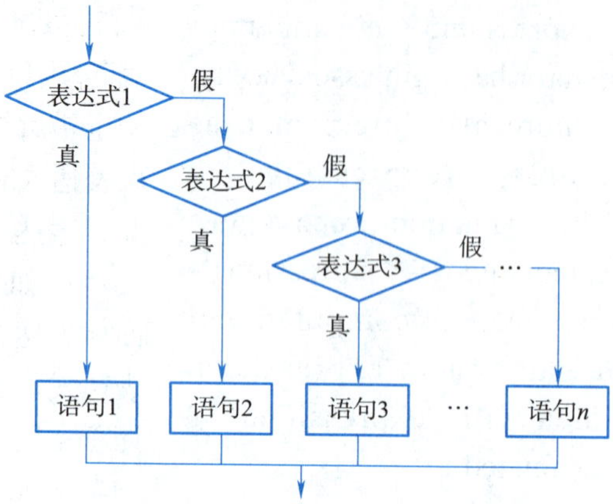

图3-4 if...else if...else结构执行流程图

For example, in the following code, we check the conditions in order. Since x is 2, the first condition x<0 is not true. Then it checks if x==0, which is also not true. So, it will run the code after the "else" part. The output is "x is a positive number".

例如，在下列代码中，顺次判断各个条件，因x为2，故第一个条件x<0不成立，则判断x==0，此时仍然不成立，则执行else后面的语句。结果输出x为正数。

```
int x=2;
if(x<0)
{
    printf("x is a positive number. \n");            //x是负数
}
else if(x==0)
{
    printf("x is zero.\n");                          //x是0
}
else
{
    printf("x is a positive number.\n");             //x是正数
}
```

Case 3_6（案例 3_6）

Students' grades are rated by scores. 85 and above is grade 'A', 70-84 is grade 'B', 60-69 is grade 'C', and below 60 is grade 'D'.

将学生的成绩按分数评定等级。85分及以上为等级'A'，70-84分为等级'B'，60-69分为等级'C'，60分以下为等级'D'。

Thinking（解题思路）:

This involves multiple conditions, which can be achieved using if...else if...else statements.

这里涉及多个条件的判断，可以使用if...else if...else语句来实现。

Case Solution（案例实现）:

Define a variable named "grade" to represent the score, then read this data from the keyboard. Output the corresponding grade based on the score range.

定义一个变量grade表示成绩，然后从键盘读入该数据，根据成绩的范围，输出对应的等级。

Program Code（程序代码）:

Case3_6.c

```
#include <stdio.h>
int main()
{
```

```
    int grade;                                      //定义变量，表示成绩
    scanf("%d",&grade);                             //从键盘读入成绩
    if(grade>=85&&grade<=100) printf("A")           //依次进行条件判断
    else if(grade>=70&&grade<85) printf("B")
    else if(grade>=60&&grade<70) printf("C")
    else printf("D");
    return 0;
}
```

Program Output（运行结果）:

```
80
B
```

Case Analysis（案例解析）:

(1) In the if...else if...else structure, once a condition is true and the corresponding block is executed, the subsequent else if and else parts will not be executed.

（1）在if...else if...else结构中，一旦某个条件为真并执行了相应的语句块，后续的else if和else部分将不会被执行。

(2) The conditions in the if...else if...else structure are judged in sequence. When “grade>=85 && grade<=100” is not true, it will evaluate the condition “grade>=70 && grade<85”. Since the first condition’s falsity implies “grade<85”, the second condition can be simplified to “grade>=70”. Similar simplifications can be made for other conditions. The code can be rewritten as follows.

（2）if...else if...else结构中的各个条件是依次判断的，当“grade>=85&&grade<=100”不成立时，会执行“grade>=70&&grade<85”语句的判断，而第一个条件不成立则隐含了“grade<85”，故第二个条件可以简写成“grade>=70”，其他条件依此类推。故代码可以改写如下：

```
#include <stdio.h>
int main()
{
    int grade;                                      //定义变量，表示成绩
    scanf("%d",&grade);                             //从键盘读入成绩
    if(grade>=85 && grade<=100) printf("A")
    else if(grade>=70) printf("B")
    else if(grade>=60) printf("C")
    else printf("D");
    return 0;
}
```

(3) This case can include an error tolerance mechanism. For example, if the input integer is less than 0 or greater than 100, an error message should be displayed. The code can be rewritten as follows.

（3）该案例可以包含容错机制，比如输入的整数若小于0或者大于100，应该给出提示错误信息。代码可以改写如下：

```
#include <stdio.h>
int main()
{
    int grade;                                      //定义变量，表示成绩
    scanf("%d",&grade);                             //从键盘读入成绩
    if(grade>=85 && grade<=100) printf("A")
    else if(grade>=70) printf("B")
    else if(grade>=60) printf("C")
    else if(grade>=0) printf("D");
    else printf("输入的数据不是0~100之间的成绩数据");
    return 0;
}
```

Program output（运行结果）:

```
-100
输入的数据不是 0~100 之间的成绩数据
```

3.2.2 Using the switch Statement for Multi-branch Selection Structures（switch语句实现多分支选择结构）

In the previous section, we introduced the if statement as a two-branch selection structure. Although nesting if statements can achieve a multi-branch selection structure, it is not very neat. The switch statement handles multi-branch selection structures for integers or characters more cleanly and clearly.

General structure of the switch statement:

```
switch(expression)
{
    case constant1: statement1; break;
    case constant2: statement2; break;
    ...
    case constant n: statement n; break;
    Default: statement n+1;
}
```

在上节中介绍if是双分支选择结构，通过适当的嵌套可以实现多分支选择结构，但不够简洁，switch语句可以处理整型或者字符型的多分支选择结构，更清晰简洁。

switch语句的一般结构：

```
switch(表达式)
{
    //case 常量 1: 语句 1; break;
    //case 常量 2: 语句 2; break;
    ...
    //case 常量 n: 语句 n; break;
    //Default: 语句 n+1;
}
```

Notes（说明）:

①switch, case, and default are keywords. ②The result of the expression after switch can only be an integer or a character. ③case should only be followed by an integer or character constant, and the values after case cannot be the same. Note that there should be at least one space between case and the constant value; they cannot be written together. ④When no case matches, the statements after default are executed, similar to the role of else in an if statement. ⑤Each case statement is often followed by a break statement to ensure that after executing the current case's statements, the program exits the switch structure to avoid entering the next case.

①switch、case和default是关键字；②switch后的表达式的结果只能是整型或字符型；③case后面只能跟一个整型或者字符型的常量值，且case后面的值不能相同，注意case和常量值之间至少有一个空格，不能连写；④当case均未匹配时，执行default后面的语句，相当于if语句中else的作用。⑤每条case语句后往往会跟上break语句。

Case 3_7（案例 3_7）

Using the switch statement, we can output student grades based on their grades. If the grade is A, the score range is 85-100. If the grade is B, the score range is 70-84. If the grade is C, the score range is 60-69. If the grade is D, the score is less than 60.

使用switch语句实现根据等级输出对应学生成绩，若等级为A，则成绩范围为85~100，若成绩为B，成绩范围为70~84，若等级为C，成绩范围为60~69，若等级为D，成绩<60。

Thinking（解题思路）:

Grades can be treated as char variables. For multiple choice situations, the switch statement is preferred. Different grades are followed by different cases, outputting the corresponding score ranges. Remember to use the break statement.

成绩等级可以处理成char型变量，涉及多分支选择情况，优选switch语句，case后面跟上不同等级，输出对应的成绩范围，注意应结合break语句实现。

Program Code（程序代码）:

Case3_7.c

```
#include <stdio.h>
int main()
{
    char grade;
    scanf("%c",&grade);
    switch(grade)
    {
        case 'A': printf("85~100");break;
        case 'B': printf("70~84"); break;
        case 'C': printf("60~69"); break;
        case 'D': printf("<60"); break;
        default: printf("Enter data error!");
    }
    return 0;
 }
```

Program Output（运行结果）:

```
A
85 ~ 100

D
<60

H
Enter data error!
```

Case Analysis（案例解析）:

(1) Grade is a char type data. Note that characters after case should use single quotes to indicate character constants. The character constants after case can use corresponding ASCII values, such as case 'A' is equivalent to case 65.

(2) In this case, the break after case cannot be omitted. The reason is that once a case in the switch matches, it will execute from the matched case until the end of the switch unless a break statement forces it to end. In this case, we need to end after selecting a case, so we need a break to forcibly end the switch statement.

(3) This case can be implemented using if...else if...else statements. In fact, the if statement is more general, and the switch statement can only handle integer or character discrete data. Let's compare their differences and connections.

（1）grade是char型数据，注意case后面的字符应使用单引号，表示字符常量。case后的字符常量可以使用对应的ASCII值，比如case 'A'等价于case 65。

（2）本案例中，case后面的break不能省略，原因是switch的case一旦匹配后，从匹配到的case开始一直执行到switch结束，除非遇到break语句强制结束。本案例选择一个case后则应结束，故需要break强制结束switch语句。

（3）该案例可以使用if...else if...else语句来实现。实际上if语句更具有一般性，switch语句仅能处理整数或者字符型离散的数据。我们对比两者的区别和联系。

```
#include <stdio.h>
int main()
{
    char grade;
    scanf("%c",&grade);
    if(grade=='A')printf("85~100");
    else if(grade=='B') printf("70~84");
    else if(grade=='C') printf("60~69");
    else if(grade=='D') printf("<60");
```

```
        else printf("Enter data error!");
        return 0;
}
```

Let's briefly compare the if and switch statements.

① The if statement is more cumbersome and less clear and readable through nested 3-layer if implementation, and the conditions need to be compared; the switch statement has a simple and clear structure. The switch statement only requires one check.

② However, the switch statement has limitations as it can only handle integer types(including character types). The if statement has generality and universality, so these two statements should be flexibly used based on actual problems.

下面我们对if和switch两种语句进行简单的比较。

①if语句是通过嵌套三层if实现较为烦琐、不清晰易读，条件要经过比较；switch语句结构简洁、清晰明了。而switch语句经过一次检查即可。

②但switch语句具有局限性，它只能处理整数类型（含字符型）的情况，而if具有一般性和普适性，两种语句应根据实际问题灵活运用。

3.3 Comprehensive Cases 综合案例

综合案例 3_3_1　求分段函数值。

案例描述： 根据下列分段函数编写程序，判断x不同的区间范围，选择不同的运算公式。

$$y=\begin{cases} x & (x-1) \\ 3x+1 & (1\leqslant x<6) \\ 3x-1 & (x\geqslant 6) \end{cases}$$

解题思路： 该案例属于多分支选择结果，可采用if...else if...else语句实现。判断x的区间范围，使用算术表达式计算y和x的关系。首先定义变量x和y，这里可定义成float型，允许输入实数。x使用scanf读入，y的值通过x的算术运算赋予其值。最后输出查看结果。

程序代码：

```
#include <stdio.h>
int main()
{
    float x,y;                          //定义变量
    printf("输入x:\n");                  //提示信息
    scanf("%f",&x);                     //输入x的值
    if(x<1) y=x;
    else if(1<=x && x<6)
    {
        y=3*x+1;
    }
    else
        y=3*x-1;
    printf("y=%f",y);
    return 0;
}
```

运行结果：

```
输入x:10
y=29.000000
```

案例解析：

（1）在代码第6行，使用scanf从键盘读入x的数据，注意scanf的格式控制符要与类型一致，float型使用%f，在x前面不要忘记&符号。

（2）第7~13行是一个if...else if...else结构，每个if后面跟一个条件判断，从上至下依次判断条件，若else后面的语句被执行则前面的条件均不成立，隐含x>=6。

（3）若if...else语句只有一条语句时{}可加可不加，若不加花括号，默认只管一条。故12行的else只管语句“y=3*x-1;”，而后面的语句“printf("y=%f",y);”不属于if结构，是正常的顺序操作，if语句结束后，按顺序执行该输出语句。

（4）特别注意，数学表示乘法时可以忽略运算符，但计算机不可以省略任何运算符，需要写完整，如y=3*x-1;不可以写成y=3x-1。

（5）第8行表达x在[1,6)区间上的数据，则应写成1<=x&&x<6，不可以写成1<=x<6。

（6）本案例处理的是float型数据，不能选用switch结构。

综合案例 3_3_2 求三个数的最大值。

案例描述：从键盘输入3个整数数据，求出3个数据的最大值，并输出查看。

解题思路：使用if语句实现3个整数的大小比较关系。首先定义3个int型的数据，然后两两比较，依次判断大小。

案例实现：本案例使用多种解题方法，帮助学生进一步灵活掌握和理解if语句。

方法一：使用嵌套if语句。

C语言允许if语句嵌套，只要结构完整即可，可以嵌套多层，其中特别注意的是，else与最近的未被匹配的if进行配对，若else没有匹配的if语句，则结构不完整，编译将出错。

```
1    #include <stdio.h>
2    int main()
3    {
4        int a,b,c;
5        printf("请输入3个整数:");
6        scanf("%d%d%d",&a,&b,&c);
7        if(a<b)
8        {
9            if(b<c) printf("max=%d\n",c);
10           else printf("max=%d\n",b);
11       }
12       else if(a<c)
13       {
14           printf("max=%d\n",c);
15       }
16       else
17       {
18           printf("max=%d\n",a);
19       }
20       return 0;
21   }
```

运行结果：

```
请输入3个整数:6 9 8
max=9
```

案例解析：

（1）为了结构分明，if结构均使用了{}。第7行的if内嵌套了一个if...else语句。若a<b成立则进入该

if...else语句的执行，若b<c成立，则c大，若不成立，说明b大。

（2）整体上使用了if...else if...else结构。

方法二： 单条if语句。

可以使用单条if语句，分别判断a是最大值、b是最大值和c是最大值的三种情形。

```
#include <stdio.h>
int main()
{
    int a,b,c,max;
    printf("请输入3个整数:");
    scanf("%d%d%d",&a,&b,&c);
    if(b<=c && c<=a || c<=b && b<=a)
        max=a;
    if(a<=c && c<=b || c<=a && a<=b)
        max=b;
    if(a<=b && a<=c || b<=a && a<=c)
        max=c;
    printf("max=%d\n",max);
    return 0;
}
```

案例解析：

（1）三条if是并列的，不属于嵌套关系，分别实现a是最大值、b是最大值和c是最大值的条件判断。注意若a是最大值，则存在两种情况，b<=c&&c<=a或者c<=b&&b<=a。判断b和c是最大值也是两种情况。

（2）求得的最大值最后赋值给max，由最后一条printf语句输出。该解决方案可以训练条件的写法。

方法三： 条件表达式。

可以使用条件表达式进行优化。

案例实现：

```
#include <stdio.h>
int main()
{
    int a,b,c,temp,max;
    printf("请输入3个整数:");
    scanf("%d%d%d",&a,&b,&c);
    temp=a>b?a:b;
    max=temp>c?temp:c;
    printf("3个数中的最大数是%d\n",max);
    return 0;
}
```

案例解析：

（1）定义了一个辅助变量temp存放中间结果。

（2）temp=a>b?a:b; 这条语句中，a>b?a:b为条件表达式，条件运算符的优先级高于赋值运算符“=”，故先进行运算，求出计算出a和b的大数，再赋值给temp。

（3）max=temp>c?temp:c; 这条语句中，将存放a和b中大数的temp与c进行对比，将两者的大数赋值给max，则max存放的即为三个数据中的最大值。

（4）通过此案例，可以观察到条件运算符可以极大地简化代码。可以优先选择该方法。

综合案例 3_3_3 算术计算器。

案例描述： 使用switch语句实现两个实型数据的加减乘除运算，模拟简单的算术计算器功能。

解题思路： 首先定义两个float和一个char型变量，分别存储两个数据和运算符，通过判定运算符的类别选择不同的算术运算表达式，这里涉及判断运算符的多种情况，处理的是char型数据，故可以优选switch

语句实现，令每一个case对应一种情况，结合break实现，选择一种情况后跳出switch语句。

案例实现： op代表运算符,a、b代表要运算的数据，result存放运算结果。

```
1    #include <stdio.h>
2    int main()
3    {
4        float a,b,result;                          //定义变量
5        char op;                                   //存放运算符
6        printf("请输入形如 1.1+2.2 的表达式:\n");
7        scanf("%f%c%f",&a,&op,&b);                 //个数和类型要匹配，& 符号不要忘记
8        switch(op)
9        {
10           case '+': result=a+b; break;
11           case '-': result=a-b; break;
12           case '*': result=a*b; break;
13           case '/': result=a/b; break;
14           default: printf("超出运算范围! \n");
15       }
16       printf("%f%c%f=%f",a,op,b,result);
17       return 0;
18   }
```

运行结果1：

```
请输入形如 1.1+2.2 的表达式:
1.1+2.2
1.100000+2.200000=3.300000
```

运行结果2：

```
请输入形如 1.1+2.2 的表达式:
1.1*2.2
1.100000+2.200000=2.420000
```

案例解析：

（1）第4、5行定义程序所需要的变量，第6、7行输入程序所需要的数据。

（2）第8~15行为switch语句，注意switch语句只能处理整型或者字符型数据，本案例处理的字符型数据，case后面跟上字符常量，由于选择一种情况即结束，故需要跟上break语句。若将所有的break去掉，即代码如下：

```
switch(op)
{
    case '+': result=a+b;
    case '-': result=a-b;
    case '*': result=a*b;
    case '/': result=a/b;
    default: printf("超出运算范围! \n");
}
```

当输入1.1+1.2时，result的结果为0.5，而不是所期望的3.3，原因是若没有break语句，则从匹配的case开始一直执行到switch花括号结尾，后面的default语句也会执行。

完整的运行结果为：

```
请输入形如 1.1+2.2 的表达式:
1.1+2.2
超出运算范围!
1.100000+2.200000=0.500000
```

故本案例中break不可省略。

（3）若将程序中的输入数据稍微改动一下，即先输入两个数据再输入运算符号，其他程序代码不变，是否会有问题呢？即第6、7行改为：

```
printf(" 请输入两个实数:\n");
scanf("%f%f",&a,&b);
printf(" 请输入算术运算符:\n");
scanf("%c",&op);
```

看下运行结果：

```
请输入两个实数:
1.1 2.2
请输入算术运算符:
超出运算范围!
1.100000
2.200000=0.000000
```

请思考为什么结果不是所预期的那样。

原因是我们所输入的数据1.1 2.2后按了【Enter】键，而回车符是有效的字符，被送入到了op中，在执行switch语句时，由于没有匹配到对应的case故执行了default后面的语句，而在输出时运算符处理成了回车符。我们习惯性地在输入数据后输入回车符或者空格符作为分隔符，然而这些隔开符号若后面紧跟着%c即char型数据的读入操作，则会被有效读入。简单的处理方式是：在输入时若后面跟的是char型数据的读入，则不要按【Enter】键或空格键，直接写对应的符号，如1.1 2.2+。

除了上述方法，较为常用的处理方式是：在%c前加一行getchar()语句，吞掉多余的回车符或空格符。

```
printf(" 请输入两个实数:\n");
scanf("%f%f",&a,&b);
printf(" 请输入算术运算符:\n");
getchar();                              // 读一个字符
scanf("%c",&op);
```

运行结果：

```
请输入两个实数:
1.1 2.2
请输入算术运算符:
+
1.100000+2.200000=3.300000
```

拓展：实际上，从键盘读入的数据首先会被存放到输入缓冲区的，当遇到回车或者缓冲区满时才会依次送给程序。缓冲区是内存中开辟的一个区域，为了提高输入输出数据的效率。当1.1 2.2被读走时，缓冲区还有一个回车符，等待被程序读取，使用getchar()从缓冲区读取回车符，则后续再输入“+”运算符时，可以正确输送给op变量。

综合案例 3_3_4　卡车运费问题（灵活运用 switch 示例）。

案例描述：假设运费的计算基于货物质量，并且每个质量区间对应一个固定运费。0~99 kg的运费是100元；100~299 kg的运费是200元；300~599 kg的运费是300元；600 kg以上的运费是400元。

解题思路：switch语句只能处理整型或者字符型数据，但可以根据实际问题进行相应转换。在本案例中，尽管质量是一个连续值，但我们可以通过将质量除以一个固定值（如100 kg）来将其转换为整数，然后利用这个整数作为switch语句的条件。

案例实现：

```
#include <stdio.h>
```

```
2   int main()
3   {
4       int weight;                          //变量定义，尽可能见名知意
5       printf("请输入货物质量(kg):");
6       scanf("%d", &weight);
7       int category=weight/100              //将质量转换为类别
8       int freight=0;                       //运费，初始化为 0
9       switch(category)
10      {
11      case 0:                              //质量为 0~99 kg
12          freight=100;
13          break;
14      case 1:
15      case 2:                              //质量为 100~299 kg
16          freight=200;
17          break;
18      case 3:
19      case 4:
20      case 5:                              //质量为 300~599 kg
21          freight=300;
22          break;
23      default:                             //质量为 600 kg 以上
24          freight=400;
25          break;
26      }
27      printf("运费为:%d元\n", freight);
28      return 0;
29  }
```

案例解析：

（1）定义见名知意的变量，weight代表是质量，category代表类别，对weight整除100后赋值给category。通过题目分析，不同的区间范围运费是不同的，100~199之间的数据整除上100后，结果为1，200~299区间的整数，若整除100则结果为2，通过这样的转换，就可以使用switch语句处理。

（2）case可以连写，如case 1和case 2的情况对应的是100~299 kg的区间，运费是200元。写法可参考第14行和第15行，无论匹配case 1或者case 2，均可以执行语句freight = 200; 同样，"300~599 kg的运费是300元"的表达方式如第18~22行代码所示。

通过综合案例3_3_4的分析可知，当面对连续的区间或范围时，直接使用switch 语句会变得不太直接。如果可以通过某种转换（如通过整数映射）将连续区间映射为离散状态，仍可以使用switch语句。

Summary
小　结

Up to this point, we have introduced the commonly used operators including arithmetic operators, logical operators, relational operators, conditional operators, and assignment operators. The order of precedence from high to low is: logical "!", arithmetic operators, relational operators, logical "&&" and "||", conditional operators, and assignment operators(including compound assignment operators). We also introduced the selection structures, including if statements, if...else statements, if...else...if statements, and multi-branch switch statements. We also discussed the transformation and comparison of multi-branch selection

至此我们已经介绍了常用运算符，包括算术运算符、逻辑运算符、关系运算符、条件运算符和赋值运算符，优先级顺序从高到低依次是，逻辑"！"，算术运算符、关系运算符、逻辑"&&"和"||"、条件运算符、赋值运算符，其中赋值运算符含复合赋值运算符。我们介绍选择结构：if语句、if...else语句、if...else...if语句和多分支switch

structures implemented by switch statements and nested if statements. We know that although if statements are more cumbersome, they have general universality, while switch statements, though concise, have limitations. It is necessary to flexibly apply them according to practical problems.

语句、switch语句和if嵌套实现多分支选择结构的转化及比较。知道if语句虽较烦琐却具有一般普遍性，switch语句虽简洁却有局限性，还需根据实际问题灵活运用。

Exercise 习　题

一、单选题

1. 算术运算符、赋值运算符和关系运算符的运算优先级从高到低依次为（　　）。

   A. 算术运算符、赋值运算符、关系运算符　　B. 算术运算符、关系运算符、赋值运算符

   C. 关系运算符、赋值运算符、算术运算符　　D. 关系运算符、算术运算符、赋值运算符

2. 在以下一组运算符中，优先级最高的运算符是（　　）。

   A. <=　　B. =　　C. %　　D. &&

3. 表示关系X<=Y<=Z的C语言表达式为（　　）。

   A. (X<=Y) AND (Y<=Z)　　B. (X<=Y<=Z)

   C. (X<=Y) & (Y<=Z)　　D. (X<=Y) && (Y<=Z)

4. 若a为int型变量，则下列C语言表达中，不能表达数学关系式10<a<15的是（　　）。

   A. a==11 || a==12 || a==13 || a==14　　B. 10<a<15

   C. a>10 && a<15　　D. !(a<=10) && !(a>=15)

5. 假设a、b、c、d、m、n均为int型变量，且a=5、b=6、c=7、d=8、m=2、n=2，那么逻辑表达式(m=a>b) && (n=c>d)运算后，n的值为（　　）。

   A. 0　　B. 1　　C. 2　　D. 3

6. 若有定义int a = 1,b = 2, c = 3, d = 4，则表达式a<b?a:c<d?a:d的结果为（　　）。

   A. 4　　B. 3　　C. 2　　D. 1

7. 下列说法中正确的是（　　）。

   A. 在 switch 语句中一定要使用 break 语句

   B. 在 switch 语句中不一定要使用 break 语句

   C. break 语句是 switch 语句的一部分

   D. break 只能用于 switch 语句中

8. 下列说法中正确的是（　　）。

   A. break 用在 switch 语句中，而 continue 用在循环语句中

   B. break 用在循环语句中，而 continue 用在 switch 语句中

   C. break 能结束循环，而 continue 只能结束本次循环

   D. continue 能结束循环，而 break 只能结束本次循环

9. 下面程序执行后的输出结果是（　　）。

```
#include <stdio.h>
int main()
{
    int x=1,y=0,a=0,b=0;
    switch(x)
```

```
    {
        case 1: switch(y)
        {
            case 0: a++; break;
            case 1: b++; break;
        }
        case 2: a++; b++; break;
        case 3: a++; b++;
    }
    printf("a=%d,b=%d",a,b);
    return 0;
}
```

A. a=1, b=0　　B. a=2, b=2　　C. a=1, b=1　　D. a=2, b=1

10. 执行下面语句序列时，不可能出现的情况是（　　）。

```
#include <stdio.h>
int main()
{
    int x;
    scanf("%d",&x);
    if(x>200) printf("*");
    if(x<200) printf("##");
    else printf("*");
    return 0;
}
```

A. 显示 **　　B. 显示 *#　　C. 显示 *　　D. 显示 ##

二、程序分析题

1. 阅读下面的程序，分析程序是否能编译通过并正确运行。如果不能，说明原因；如果能，写出运行结果。

```
#include <stdio.h>
int main()
{
    char ch='b';
    switch(ch)
    {
        case 'A': printf("it is A\n");
        case 'B': printf("it is B\n")
        case 'C': printf("it is C\n");
        default: printf("other\n");
    }
    return 0;
}
```

2. 将程序改写成条件运算符形式。

```
#include <stdio.h>
int main()
{
    int a=3;
    int b=6;
    if(a>b) printf("a is max");
    else printf("b is max");
    return 0;
}
```

3. 设程序运行时输入4215，写出输出结果。

```
#include <stdio.h>
int main()
{
    int n1=0,n2=0,n3=0;
    char ch;
    ch=getchar();
    switch(ch)
    {
       case '1':case '3': n1++; break;
       case '2':case '4': n2++; break;
       default: n3++; break;
    }
    printf("%d, %d, %d\n",n1,n2,n3);
    return 0;
}
```

4. 两次运行下面的程序，从键盘上输入6和4，分别写出运行结果。

```
#include <stdio.h>
int main()
{
    int x;
    scanf("%d",&x);
    if(x++ > 5) printf("%d",x);
    else printf("%d\n",x--);
    return 0;
}
```

5. 阅读下列代码，写出运行结果。

```
#include <stdio.h>
int main()
{
    int a=1,b=0;
    if(--a) b++;
    else if(a==0) b+=2;
    else b+=3;
    printf("%d\n",b);
    return 0;
}
```

三、编程题

1. 请编写一个判断偶数的函数，从键盘输入一个数，实现对奇数和偶数的判断。

2. 输入一周中的某一天，判断这一天是工作日还是周末休息日，用1~7表示周一到周日。输入1~7中的某一个数，判断数字对应的是工作日还是休息日。要求使用switch语句实现。

3. 编写程序,实现一个投票表决器的功能。

若输入Y或y，打印agree

若输入N或n，打印disagree

输入其他，打印lose

4. 编写程序，实现如下功能：根据输入的成绩，输出成绩对应等级，等级评定规则为：90~100为A，80~89为B，70~79为C，60~69为D，60以下为E。要求灵活使用switch语句实现。

5. 从键盘输入三个无序的整数，按照从小到大的顺序依次输出三个整数数据。

Chapter 4
Loop Structure Programming
第 4 章
循环结构程序设计

Introduction 主要内容

The C language has three basic structures: sequential, selective and loop, The previous two have introduced the sequential structure and selective structure. This chapter introduces the loop structure. The main contents include the usage of while, do...while and for loops, the break and continue statements that change the execution state of the loop, and the nesting of the loop structure.

C语言有三大基本结构：顺序、选择和循环，前两章已经介绍完了顺序结构和选择结构，接下来本章介绍循环结构。主要内容有while、do...while和for三种循环语句的用法、改变循环执行状态的break和continue语句，以及循环结构的嵌套等。

Objectives 学习目标

Cultivate students' thinking of looping structure programming, flexibly master the three programming methods of while, do...while and for, understand the difference and connection between the three, train students to think rigorously and logically, and be able to recognize common repetitive operations and solve problems efficiently by using looping structure.

培养学生循环结构编程思维，灵活掌握while、do...while和for三种语句编程方法，理解三者之间的区别和联系，训练学生严谨的思维逻辑，能够识别常见的重复操作并使用循环结构高效解决问题。

4.1 Why Do We Need a Loop Structure 为什么需要循环结构

The reasons why C language needs loop structures can be summarized as follows.

C语言需要循环结构的原因可以总结以下几点。

(1) Improve efficiency: Loop structures allow the program to repeatedly execute a certain code segment, which is very useful when dealing with repetitive tasks. Without loop structures, for

（1）提高效率：循环结构允许程序重复执行某段代码，这在处理重复性的任务时非常有用。

operations that need to be repeated, programmers may need to manually write a large amount of repetitive code, which not only increases the workload of programming but may also lead to code redundancy and maintenance difficulties. Through loop structures, programmers can write more concise and efficient code.

如果没有循环结构，那么对于需要重复执行的操作，程序员可能需要手动编写大量的重复代码，这不仅增加了编程的工作量，还可能导致代码冗余和难以维护。通过循环结构，程序员可以编写更简洁、更高效的代码。

(2) Handling data collections: In programming, it is often necessary to handle large amounts of data collections, such as arrays or linked lists(which will be learned later). Loop structures allow programmers to traverse these data structures and perform the same operation on each element.

（2）处理数据集合：在编程中，经常需要处理大量的数据集合，如数组或链表（后续会学习到）。循环结构允许程序员遍历这些数据结构，并对每个元素执行相同的操作。

(3) Implementing algorithms: Many algorithms involve repeatedly executing certain steps until a certain condition is met. Loop structures are key to implementing these algorithms.

（3）实现算法：许多算法都涉及重复执行某些步骤，直到满足某个条件为止。循环结构是实现这些算法的关键。

(4) Simplifying logic: In some cases, using loop structures can simplify the logic structure of the program. By abstracting repetitive tasks into loops, the code can be made easier to understand and maintain.

（4）简化逻辑：在某些情况下，使用循环结构可以简化程序的逻辑结构。通过将重复的任务抽象为循环，可以使代码更易于理解和维护。

In summary, loop structures are an indispensable part of the C language, allowing programmers to handle repetitive tasks more efficiently, implement complex algorithms, and simplify the logic structure of programs.

总的来说，循环结构是C语言中不可或缺的一部分，它使得程序员能够更高效地处理重复性的任务，实现复杂的算法，并简化程序的逻辑结构。

The following is a simple example so that the reader can more intuitively understand the need for loops.

下面通过一个简单的例子，让读者可以更直观地理解循环的必要性。

Case 4_1（案例 4_1）

Assuming that there are 50 students in the class and you are asked to calculate the grade point average of each student, draw a traditional flowchart.

假设全班有50名学生，要求统计每个学生的平均成绩，请绘制传统流程图。

Case Solution（案例实现）:

The flowchart drawn is shown in Figure 4-1.

绘制的流程图如图4-1所示。

Case Analysis（案例解析）:

The i represents the first student, given an initial value of 1. The diagram contains a loop structure containing 4 loop operations: input a student's grade, find a student's average grade, output a student's average grade, and increment i by 1. Each student is required to perform each of these 4 operations. The loop judgment condition is i<=50, if it is greater than 50, then end the loop.

i代表第几个学生，给予初值1，图中包含一个循环结构，包含了4个循环操作：输入一个学生的成绩、求一个学生平均成绩、输出一个学生平均成绩以及i增1。每个学生均需执行这4个操作。循环判断条件为i<=50，若大于50则结束循环。

Through this case we can also intuitively see that the real life need to loop structure. the C language loop structure can be used while, do...while and for statements to achieve, the subsequent will be introduced in turn.

通过这个案例我们也可以直观看出，现实生活中需要循环结构。C语言循环结构可以使用while、do...while和for语句来实现，后续将依次介绍。

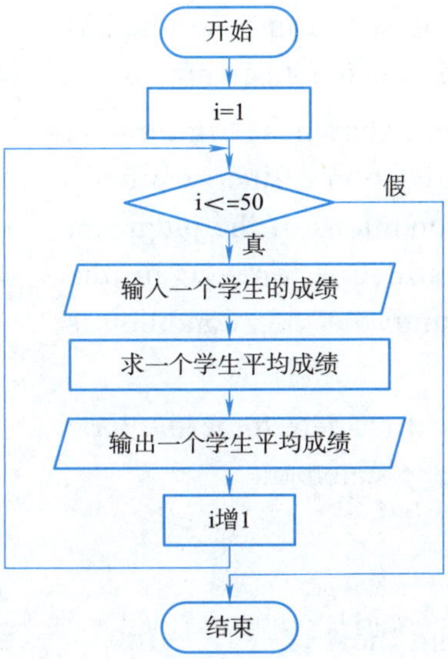

图4-1 案例4_1执行过程的传统流程图

4.2 The while and do...while Statements while和do...while语句

4.2节视频

Let's begin with an introduction to while and do...while statements. How do while and do...while statements control loops? And what are the significant differences between them? In this section, we'll take a deeper look at while and do...while looping statements to get an initial grasp of the programming methodology and logic of looping structures.

首先介绍while和do...while语句。while和do...while语句是如何控制循环的呢？它们之间又有哪些显著的差异？在本节中，我们将深入探讨while和do...while循环语句，初步掌握循环结构的编程方法和逻辑。

4.2.1 The while Statements（while语句）

The general form is:

```
while(cyclic conditional expression)
{
    loop body statement
}
```

一般形式为：

```
while(循环条件表达式)
{
    循环体语句
}
```

Syntax Explanation 语法说明

Parentheses is the loop condition expression, first determine the logical value of the expression, follow the principle of "non-zero is true", when the value of the expression is "true" when the implementation of {} within the loop body statement, when the expression value is "false" then do not execute the loop body statement, that is, the condition is falsewhen the termination of the loop. When the value of the expression is "false", the loop body statement is not executed, that is, the loop is terminated when the condition is false. The loop body statement can be a simple statement or a compound

括号里的是循环条件表达式，首先判断该表达式的逻辑值，遵循的原则仍为"非零即为真"，当表达式值为"真"时执行{}内的循环体语句，当表达式值为"假"时则不执行循环体语句，即条件为假时终while语句执行流程如图4-2所示，当循环条件为真时执行循环体语句，随后回到循环条件的判断操止循环。循环体语句可以是一个简单的语句，也可以是复合语句，若

statement, if there is only one statement then {} can be omitted, if it contains more than one statement then {} can not be omitted.

只有一条语句则{}可以省略，若包含多条语句则{}不可省略。

While statement execution process shown in Figure 4-2, when the loop condition is true, the execution of the loop body statement, and then back to the loop conditions of the judgment operation, if the loop conditions are still true, once again the execution of the loop body, the cycle until the loop condition is false, the end of the loop.

while语句执行流程如图4-2所示。当循环条件为真时执行循环体语句，随后回到循环条件的判断操作。若循环条件仍然为真，再次执行循环体，循环往复直到循环条件为假时，结束循环。

Characteristics of while statement: first determine the loop condition expression, then execute the loop statement.

while语句的特点：先判断循环条件表达式，后执行循环语句。

Case 4_2（案例 4_2）

Use a while loop statement to find the 1+2+3+⋯+100 results.

使用while循环语句，求1+2+3+⋯+100的值。

Thinking（解题思路）:

This case requires the use of while loop to find 1 to 100 cumulative sum, repeated plus a certain data is possible to use the loop structure. Successively add 100 numbers, you need to repeat the addition operation 100 times.

本案例要求使用while循环求1~100累加和，重复加上某个数据是可以使用循环结构的。先后将100个数相加，需要重复100次加法运算。

The algorithm is first described using a traditional flowchart, as shown in Figure 4-3. Define variables i and sum initialized to 0 at the same time, when i<=100, repeat the execution of “sum=sum+i;” and “i++;” two statements, until the loop condition i<=100 does not hold, the end of the loop.

首先使用传统流程图描述算法，如图4-3所示。定义变量i和sum同时初始化为0，当i<=100时，重复执行“sum=sum+i;”和“i++;”两条语句，直到循环条件i<=100不成立，结束循环。

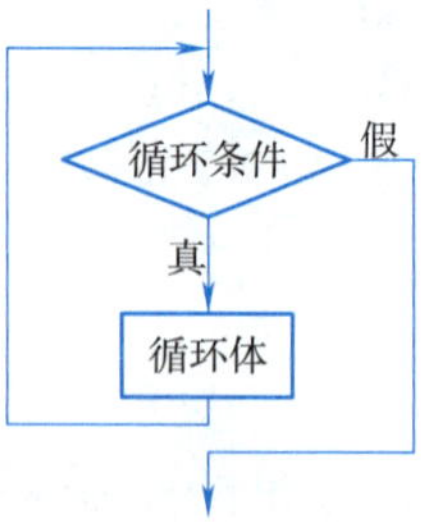

图4-2　while语句执行流程图

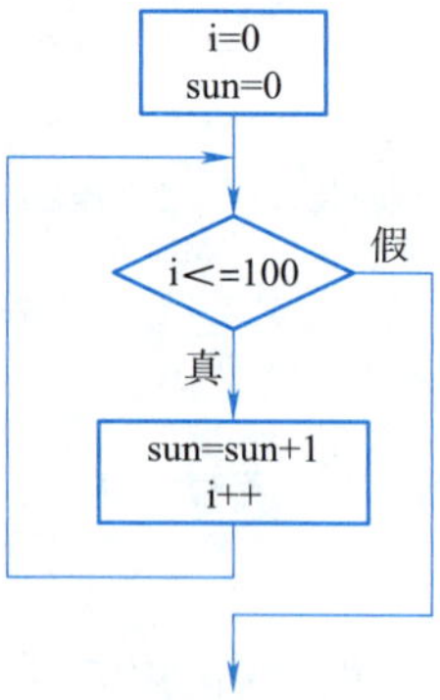

图4-3　求1~100累加和的传统流程图

Program Code（程序代码）:

Case4_2.c

```
#include <stdio.h>
int main()
{
    int i=0,sum=0;                              //初始化
    while(i<=100)                               //while 循环结构
    {
```

```
7           sum=sum+i;                      //求累和
8           i++;                            //调整 i 的值
9       }
10      printf("sum=%d\n",sum);             //输出和
11      return 0;
12  }
```

Program Output（运行结果）:

```
sum=5050
```

Case Analysis（案例解析）:

(1) Line 4 defines the variables i and sum, which are also initialized to 0. Where i is used to denote the data to be summed and sum is used to hold the summed data. Note in particular that i and sum are random values if they are not initialized, then the results are not accurate.

（1）第4行定义了变量i和sum，同时初始化为0。其中i是用来表示即将累加的数据，而sum是用来存放累加和数据。特别注意，i和sum若不初始化，则为随机值，则结果不准确。

(2) The cyclic conditional expression in line 5 is i<=100, and the variable that controls the change in the condition is called a cyclic variable, so i is a cyclic variable. The loop body statement is executed continuously when i<=100 is established. When i is 101, i.e., when the condition does not hold for the first time, the loop ends.

（2）第5行循环条件表达式为i<=100，控制条件变化的变量称为循环变量，故i为循环变量。当i<=100成立时不断执行循环体语句。当i为101时，即条件第一次不成立时，结束循环。

(3) Lines 6 through 9 {} are compound statements enclosed as the body of the while loop. In this case, sum=sum+i indicates the cumulative sum, and i++ makes i increase by 1 to the next number. Special note: the i++ statement must be there, not only to add the next data, but also because it is a loop variable, should be able to make the loop conditions tend to the end, to avoid dead loop.

（3）第6行到第9行{}括起来的是复合语句，作为while循环的循环体。其中，sum=sum+i表示求累和，i++令i增加1，变成下一个数。特别注意：i++语句一定要有，不仅仅是为了加上下一个数据，还因为其是循环变量，应能让循环条件趋于结束，避免死循环。

(4) the 10th statement, is immediately after the loop statement after the output statement, the end of the loop in turn after the implementation of the statement, after the while loop structure, sum value for the cumulative sum of 1 to 100, note that the output statement do not write in the while, does not belong to the loop body part, is a separate statement, after the addition of all the data, and then output.

（4）第10条语句，是紧跟着循环语句后的输出语句，循环结束后顺次执行该语句，经过while循环结构后，sum值为1~100的累加和，注意输出语句不要写在while内，不属于循环体部分，是单独的语句，在所有数据相加后，再输出。

Case Summary（案例总结）:

while loop conditions often require loop variables, which should be initialized before while. And the loop variable should be adjusted within the loop, so that it is adjusted towards the direction that the loop condition is false, i.e. the end of the loop.

while循环条件往往需要循环变量，循环变量应在while前进行初始化。且循环变量在循环体内，应进行调整，令其向着循环条件为假，即结束循环的方向进行调整。

4.2.2 The do...while Statement（do...while语句）

After learning about while statements, we'll move on to do...while statements, which are similar to while statements.

学习了while语句之后，接下来我们来学习与while语句相似的do...while语句。

The general form of the do...while statement:

do...while语句的一般形式为：

```
do
{
    loop body statement(computing)
}while(cyclic conditional expression);
```

```
do
{
    循环体语句
}while(循环条件表达式);
```

Syntax Explanation 语法说明

(1) As with the while statement, the statement enclosed by the {} parentheses is a loop statement. Note that the {} of do...while cannot be omitted even if there is only one statement, and it is a fixed syntax format.

(2) If the expression of the loop condition after while is true, the loop continues, otherwise the loop ends.

(3) Note that the final semicolon “;” cannot be omitted and is part of the grammar.

The do...while statement is characterized by unconditionally executing the loop body statement first, and then determining whether the loop condition is valid.

（1）与while语句一样，花括号{}括起来的是循环体语句，注意do...while的{}即使只有一条语句也不可以省略，属于固定语法格式。

（2）while后的循环条件表达式为真，则继续循环，否则结束循环。

（3）注意最后的分号“;”不能省略，是语法的一部分。

do...while语句的特点：先无条件地执行循环体语句，后判断循环条件是否成立。

Case 4_3（案例 4_3）

Use do...while statements to find the result of 1+2+3+…+100.

使用do...while循环语句，求1+2+3+…+100的值。

Thinking（解题思路）:

Use a do...while loop to find the cumulative sum from 1 to 100, as shown in the traditional flowchart in Figure 4-4. It is still necessary to define i and sum and complete the initialization before the loop, followed by the result of the loop, the loop body is the same as in Case 4-2. Observe from the diagram that the loop condition is placed after the loop body.

使用do...while循环求1~100的累加和，传统流程图如图4-4所示。仍然要定义i和sum并在循环之前完成初始化，随后跟着循环结果，循环体与案例4_2中的一致。从图上观察，循环条件放在了循环体之后。

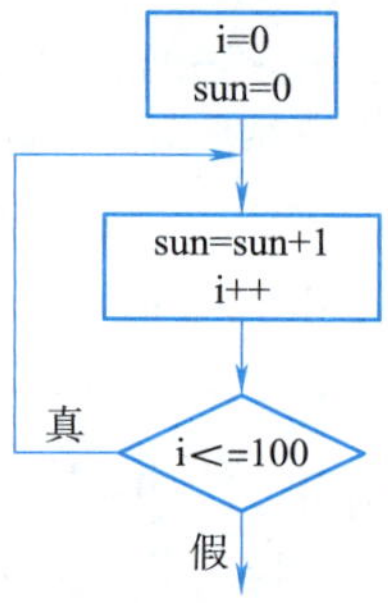

图4-4　求1~100的累加和的流程图

Program Code（程序代码）:

Case4_3.c

```
#include <stdio.h>
int main()
{
    int i=1,sum=0;              //定义循环变量i，存放累加和的sum，同时初始化
    do
    {
        sum=sum+i;
        i++;
    }while(i<=100);             //注意分号不可以省略
    printf("sum=%d\n",sum);
    return 0;
}
```

Program Output（运行结果）:

```
sum=5050
```

From Case 4_2 and Case 4_3 can be learned, the problem of seeking 1 to 100 cumulative sum we can use both the while statement, you can also use the do...while statement to achieve the results of these two practices are the same. So what is the difference between the two? The next subsection explores this question.

从案例4_2和案例4_3可以得知，对求1~100累加和的问题既可以用while语句，也可以用do...while语句来实现，这两种做法得到的结果是相同的。那么，二者有什么区别呢？下面探索该问题。

4.2.3 Comparison of while and do...while（while和do...while的比较）

First we intuitively understand from the flowchart. Figure 4-5 is a do...while flowchart, you can execute the logic can be transformed into the structure of while, as shown in Figure 4-6. The execution process and execution logic of the two diagrams are consistent. In Figure 4-6, the dotted line boxed up for the do...while structure, which shows that the structure of the do...while statement is by a “loop body” plus a while structure. Therefore, the do...while statement executes one more loop statement than the while statement in any case.

首先我们从流程图上直观理解。图4-5是do...while流程图，可以执行逻辑可以转化为while的结构，如图4-6所示。两个图的执行流程和执行逻辑一致。在图4-6中，虚线框起来的为do...while结构，由此看出do...while语句的结构是由一个“循环体”加上一个while结构得到的。所以do...while语句无论如何都比while语句多执行一次循环语句。

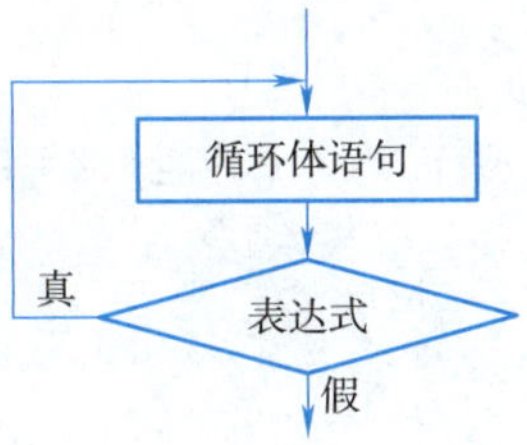

图4-5 do...while流程图

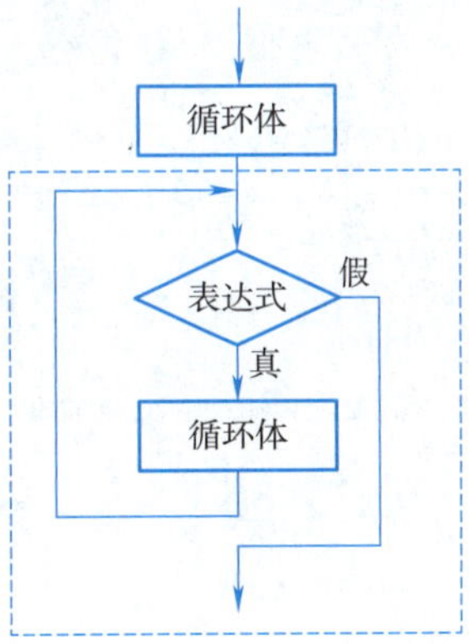

图4-6 使用while实现的do...while流程图

Case 4_4（案例 4_4）

A case comparing while and do...while statements.

The code for the while statement is given first:

while和do...while语句的比较案例。

首先给出while语句的代码：

```
#include <stdio.h>
int main()
{
    int i,sum=0;
    printf("i=");                    //提示语句
    scanf("%d",&i);                  //输入i数据
    while(i<=5)
```

```
    {
        sum=sum+i;
        i++;
    }
    printf("sum=%d",sum);
    return 0;
}
```

The result of the run when we enter i=1 is:

当我们输入i=1时运行结果为：

```
i=1
sum=15
```

When we enter i=6(which neither satisfies the loop condition) the result of the run is:

当输入i=6（既不满足循环条件）时运行结果为：

```
i=6
sum=0
```

The program implemented by the do...while statement is given below:

下面给出do...while语句实现的程序：

```
#include <stdio.h>
int main()
{
    int i,sum=0;
    printf("i=");
    scanf("%d",&i) ;
    do
    {
        sum=sum+i;
        i++;
    }while(i<=5);
    printf("sum=%d",sum);
    return 0;
}
```

The result of the run when we enter i=1 is:

当输入i=1时运行结果为：

```
i=1
sum=15
```

When we enter i=6(which neither satisfies the loop condition) the result of the run is:

当输入i=6时（既不满足循环条件）运行结果为：

```
i=6
sum=6
```

Case Analysis（案例解析）：

When the condition is true at the beginning, using both while and do...while implementations gives the same runtime result of 15. When the condition is false at the beginning, the while code segment sums to 0, while the do...while result is 6. Why do while and do...while statements output different results when the loop condition does not hold from the beginning? This is because do...while executes at least 1 time, while while

当一开始条件为真时，使用while和do...while实现，所得到的运行结果相同，均为15。当一开始条件为假时，while代码段sum为0，而do...while结果为6。为什么循环条件从一开始就不成立时，while和do...while语句所输出的结果不同呢？这是

does not execute once.

因为do...while至少会执行1次，而while一次也不执行。

Case Summary（案例总结）:

(1) When the conditional expression after while is true for the first time, while and do...while have the same result and can be interchanged.

(2) When the conditional expression following while is false for the first time when it is false for the first time, then while and do...while results are different because do...while executes the loop body statement at least once.

(3) When writing a program, if it is required to be executed at least once or to be executed first and then judged, the do...while structure can be preferred, and both can be used in other cases.

（1）while后面的条件表达式第一次为真时，while和do...while结果相同，此时可以互换。

（2）while后面的条件表达式第一次为假时，则while和do...while结果不同，因为do...while至少会执行一次循环体语句。

（3）当在编写程序时，若要求至少执行一次或者先执行后判断的场合，可以首选do...while结构，其他情况两者均可。

4.3 The for Statements for语句

4.3节视频

In the previous section, we have stepped into the journey of learning about looping structures and explored while and do...while statements. We've gotten a better understanding of how while and do...while statements work. Next, we will continue to focus on the for statement and explore in detail its differences and logical connections with the first two, so as to more fully grasp the essence of the looping structure, while helping scholars use scientific thinking to build an accurate logical system.

上节内容中，我们已经步入了循环结构的学习旅程，探讨了while、do...while语句。已经较为深入了解了while和do...while语句的工作原理。接下来，我们将继续聚焦于for语句的学习，并细致探讨它与前两者的差异和逻辑相通之处，从而更全面地掌握循环结构的精髓，同时帮助学习者使用科学的思维构建准确的逻辑体系。

4.3.1 Review and Introduction（回顾与引入）

First, let's review the case of finding the cumulative sum of 1 to 100 from the previous lesson. The code for this is as follows:

首先，我们回顾上节课的求1~100的累加和的案例。其中代码如下：

```
#include <stdio.h>
int main()
{
    int i=0,sum=0;                    //表达式1：循环变量初始化
    while(i<=100)                     //表达式2：循环条件
    {
        sum=sum+i;
        i++;                          //表达式3：循环变量调整器
    }
    printf("sum=%d\n",sum);
    return 0;
}
```

The int i=0 is the initialization of the loop variable, i<=100 is the loop conditional expression, and i++ is the adjustment operation that tends to terminate the loop. In fact, loops generally contain three elements as shown above, summarized as follows:

其中int i=0为循环变量初始化，i<=100为循环条件表达式，i++令循环趋于终止的调整操作。实际上循环一般包含如上述所展示的三个要素，总结如下：

Element 1: Initialization of loop variables. If it is not initialized, the conditional judgment is inaccurate.

Element 2: Loop judgment conditions. An expression that determines whether the loop continues or not, typically represented by a loop variable.

Element 3: Loop variable adjustment. A statement that adjusts a loop variable so that the loop tends to end, otherwise it is a dead loop. Algorithms are finite operation steps, and the loop program written should have an end, so avoid writing a program with a dead loop.

Each loop statement needs to contain the above three elements, in fact, the execution logic or execution process of the loop is the same, so while and for statements can be converted to each other. As shown in Figure 4-7. No matter what kind of loop structure is chosen, the logic of the loop is unchanged.

要素1：循环变量初始化。若不初始化则条件判断不准确。

要素2：循环判断条件。判断循环是否继续，一般是由循环变量表示的表达式。

要素3：循环变量调整。调整循环变量的语句，使得循环趋于结束，否则为死循环。算法是有限的操作步骤，所编写的循环程序应有结束，故避免编写死循环的程序。

每个循环语句均需要包含上述三要素，实际上循环的执行逻辑或者执行流程是一样的，故while和for语句可以相互转化，如图4-7所示。无论选用何种循环结构，循环逻辑是不变的。

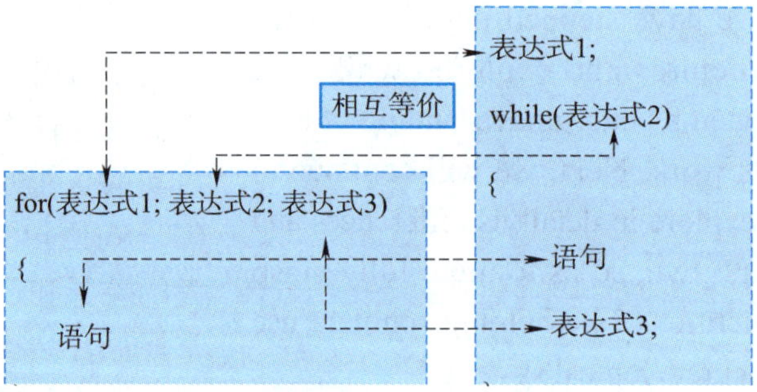

图4-7　for和while相互转换图示

4.3.2 The for Statements to Implement Loops（for语句实现循环）

The general form of the for statement:

```
for(expression 1;expression 2;expression 3)
{
    loop body statement
}
```

for语句的一般形式：

```
for(表达式1;表达式2;表达式3)
{
    循环体语句
}
```

Syntax Explanation 语法说明

(1) The meaning of each expression is as follows:

Expression 1: initialize loop variables. Equivalent to the initialization before the while loop, which is executed only 1 time in the whole for loop.

Expression 2: Loop Conditional Judgment. Equivalent to the while statement, the loop condition expression is used to determine whether the loop will continue or not.

Expression 3: Loop variable modifier. Equivalent to i++

（1）各个表达式的含义如下：

表达式1：循环变量初始化。等价于while循环之前的初始化工作，在整个for循环里只执行1次。

表达式2：循环条件判断。等价于while语句的循环条件表达式，用来判断循环是否继续进行。

表达式3：循环变量调制器。等价于while语句花括号里循环体语句若只有一条

in the curly braces of a while statement; its function is to add value to the loop variable so that the for loop tends to end, and it is executed after the execution of the loop body in sequence.

的i++；它的作用就是使循环变量增值，使for循环趋于结束，它是在执行完循环体后顺次执行的。

(2) The two semicolons “;” of the for loop are part of the grammar and cannot be omitted.

（2）for循环的两个分号“;”是语法的一部分，不能省略。

(3) If there is only one statement in the loop body statement, then {} can be omitted, otherwise it can not be omitted, this point is the same as while.

（3）语句，则{}可省略，否则不能省略，这点与while语句相同。

The Execution Flow of the while Statement while 语句的执行流程

We rewrite the problem of finding the cumulative sum of 1 to 100 from a while statement to a for statement:

我们将求解1~100累加和问题，从while语句改写成for语句来实现：

```
for(i=0,sum=0;i<=100;i++)
{
    sum=sum+i;
}
```

The execution logic of the for statement is the same as that of the while statement. We give the order and process of execution of for as shown in Figure 4-8.

for语句的执行逻辑和while语句的执行逻辑是一致的。我们给出for的执行顺序和过程，如图4-8所示。

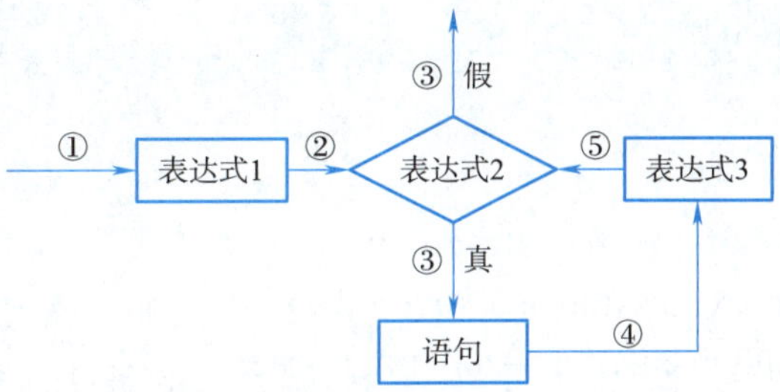

图4-8 for循环的执行顺序

First execute expression 1, then unconditionally execute expression 2, if the value of conditional expression 2 is false, then jump out of the for loop, if it is true, then execute the loop body statement, and finally execute expression 3. this is one loop of the for loop. Then it continues to execute in the order of ⑤, ③(true), ④ until the value of the conditional expression of expression 2 is false, ending the loop.

先执行表达式1，然后无条件执行表达式2，若条件表达式2的值为假则跳出for循环，若为真则执行循环体语句，最后执行表达式3。这就是for循环的一次循环。然后继续按照⑤、③（真）、④的顺序执行，直到表达式2的条件表达式的值为假，结束循环。

Case 4_5（案例 4_5）

Output all numbers between 100 and 200 that are divisible by 3.

输出100～200之间所有能被3整除的数。

Thinking（解题思路）:

This case can be realized using a for loop, according to the three elements of the loop, first define a loop variable int n,

本案例可以使用for循环来实现，根据循环三个要素，首先定义一个循环变量int n，

representing the data to be judged, so that its initial value is 100; use the variable to represent the loop condition expression, that is, n <= 200, to ensure that the data is transformed from 100 to 200, traversing the judgment of each number; the loop variable adjuster for n ++. Just write the above three elements in the correct place according to the syntax of the for loop.

表示要判定的数据，令其初始值为100；利用变量表示循环条件表达式，即n<=200，保证数据从100变换到200，遍历判断每一个数；循环变量调整器为n++。将上述三个要素根据for循环的语法写在正确的位置即可。

Program Code（程序代码）:

Case4_5.c

```
#include <stdio.h>
int main()
{
    int n;                                    // 循环变量定义
    for(n=100;n<=200;n++)                     // 注意两个分号不能省略
    {
        if(n%3==0)                            // 判断能否被 3 整除的法，注意是两个等号
        printf("%d ",n);                      //%d 后面跟了一个空格，隔开数据
    }
    printf("\n");
    return 0;
}
```

Program Output（运行结果）:

```
102 105 108 111 114 117 120 123 126 129 132 135 138 141 144 147 150 153 156 159 162
165 168 171 174 177 180 183 186 189 192 195 198
```

Case Analysis（案例解析）:

(1) The execution process of the for statement is as follows: first execute n=100, the expression is executed only once, then execute the conditional judgment of n<=200, if it is true, then execute the loop body, if it is false, end the loop. After the execution of the loop body execution n + +, and then the execution of n <= 200 condition judgment, the cycle of judgment conditions, the execution of the loop body and the operation of n + +, until the conditions for the end of the loop for false.

(2) The loop body statement is an if statement that determines whether n is divisible by 3 at this point, and outputs n if the condition holds, or not if it does not.

(3) Changing the for statement in this case to a while statement is essentially just adjusting the position of the expression; note that the assignment of the initial value statement, n=100, should precede the while structure, and that the adjuster, n++, should be within the while loop; each element is indispensable.

（1）for语句的执行流程为：先执行n=100，该表达式只执行一次，再执行n<=200的条件判断，若为真则执行循环体，若为假结束循环。循环体执行完毕后执行n++，再执行n<=200的条件判断，循环往复判断条件，执行循环体以及n++的操作，直到条件为假结束循环。

（2）循环体语句是一条if语句，判断此时n能否被3整除，如果条件成立则输出n，不成立则不输出。

（3）将本案例中的for语句改成while语句，本质上只是调整了表达式的位置，请注意赋初值语句n=100，应在while结构之前，调整器n++应在while循环体内，每个要素均不可或缺。

```
n=100;
while(n<=200)
{
```

```
    if(n%3==0)
        printf("%d",n);
    n++;
}
```

(4) Do not add “;” to the end of a for or while structure. Such as for(n=100;n<=200;n++); or while(n<=200); if followed by a semicolon on behalf of the for or while statement ends early, followed by the {} part of the(loop body statement) is no longer belong to the for or while loop body statement.

(5) The two semicolons within the for statement are formatting and should not be omitted.

（4）for或者while结构的后面不要加“;”。如for(n=100;n<=200;n++);或者while(n<=200);若后面加上分号代表for或while语句提前结束，后面的{}部分（循环体语句）不再属于for或while循环体语句。

（5）for语句内的两个分号是格式，不要省略。

4.3.3 Special Forms of the for Statement（for语句的特殊形式）

In principle, each expression of the for statement is allowed to be defaulted, but the two semicolons still cannot be omitted and belong to the fixed format. We will rewrite the for loop statement of Case 4_5 into an equivalent different form of for loop structure:

The first form: the expression 1 is mentioned in front of the for statement to complete the initial value assignment operation, then the for can omit the expression 1, but the semicolon still can not be omitted.

原则上，for语句的各个表达式都允许省略，但两个分号不可以省略，属于固定格式。我们将案例4_5的for循环语句改写成等价的不同形式的for循环结构：

第一种形式：将表达式1提到了for语句前面完成赋初值操作，则for可省略表达式1，但分号仍不能省略。

```
n=100;
for(;n<=200;n++)
{
    if(n%3==0)
        printf("%d",n);
}
```

The second form: expression 2 of the for statement is null, and a judgment is made inside the loop body.

第二种形式：for语句的表达式2为空，在循环体内部进行了判断。

```
for(n=100; ;n++)                    //两个分号不能省略
{
    if(n>200) break;                //如果条件成立则跳出循环，break为强制结束循环的作用
    if(n%3==0)
        printf("%d ",n);
}
```

The third form: expression 3 is defaulted and its in the loop body statement.

第三种形式：省略了表达式3，其在循环体语句中。

```
for(n=100; n<=200;)
{
    if(n%3==0)
        printf("%d",n);
    n++;
}
```

The fourth form: expressions 1 and 3 are defaulted, expression 1 is written before the for statement, and expression 3 is inside the loop.

第四种形式：省略了表达式1和3，表达式1写在了for语句之前，表达式3在循环体内。

```
n=100;
for(;n<=200;)
{
    if(n%3==0)
        printf("%d",n);
    n++;
}
```

Fifth form: all three expressions are empty, where expression 1 is written before the for statement, and expression 2 and expression 3 are inside the loop.

第五种形式：三个表达式均为空，其中表达式1写在了for语句之前，表达式2和表达式3在循环体内。

```
n=100;
for(;;)                          //所有表达式为空，两个分号不能省略
{
    if(n>200) break;
    if(n%3==0)
        printf("%d",n);
    n++;
}
```

Summary（总结）:

These are the default transformations for each expression in the for statement. The expression can be an empty statement, but the semicolon cannot be omitted. These forms just change the position of the expression, the essence of the loop execution logic is the same, the cycle elements of the implementation of the order and process is consistent, and the three elements are indispensable.

以上是for语句中各表达式缺省的几种变换形式。表达式可以为空语句，但分号不可以省略。这些形式只是变换了表达式位置，本质上循环执行逻辑是一样的，循环要素执行的顺次和流程是相一致的，且三要素不可缺少。

In fact, the execution process of the for statement is fixed, that is, the execution of expression 1 once, expression 2 and expression 3 and the loop body in accordance with the order of Figure 4-8 loop many times, until the end of the expression 2 is false. And the expression as long as it is a legal arbitrary expression can be.

实际上，for语句的执行流程是固定的，即执行表达式1一次，表达式2和表达式3以及循环体按照图4-8所示的顺序循环多次，直到表达式2为假结束。而表达式只要是合法的任意表达式均可。

Expression 1 is executed first and only once, often used to give the initial value of the program, such as the following code, "i = 1, sum = 0;" If more than one variable to give the initial value, you can use the comma "," to separate.

表达式1先执行且只执行一次，往往用于给程序赋予初值，如下列代码中的"i=1,sum=0;"若为多个变量赋予初值，可以使用逗号","隔开。

```
for(i=1,sum=0;i<=100;i++)
{
    sum=sum+i;
}
```

In principle the expression can be any expression, such as rewriting the above code as follows:

原则上，表达式可以是任意表达式，如将上述代码改写如下：

```
i=1;
for(sum=0;i<=100;sum=sum+i)
{
    i++;
}
```

In that particular form expression 1 is any other expression unrelated to the loop variables, the loop variable i is assigned an initial value before the for statement, and expression 1 puts the sum=0 statement. Expression 3 can be any expression unrelated to loop control, expression 3 puts the sum=sum+i statement, and the loop variable adjuster i++ statement is put inside the loop. As long as the loop logic is correct, the expression can be placed in any position and put any expression. It can also be simplified and written as: “for(i=1,sum=0;i<=100;sum=sum+i,i++);” multiple expressions separated by commas.

While arbitrary expressions can be put in principle, care should be taken to make the program readable and maintainable.

在该特殊形式中表达式1是与循环变量无关的其他表达式，循环变量i在for语句之前赋初值，表达式1放的是sum=0语句。表达式3可以是与循环控制无关的任意表达式，表达式3放了sum=sum+i语句，循环变量调整器i++语句放到了循环体内。只要循环逻辑是正确的，表达式可以放到任意位置，放任意表达式。也可以简化写成“for(i=1,sum=0;i<=100;sum=sum+i,i++);”，多个表达式用逗号隔开。

虽然原则上可以放任意表达式，但要注意程序的可读性和可维护性。

4.3.4 Comparison of Three Loop Statements（三种循环语句的比较）

So far, we have learned while, do...while, for statements to implement the loop structure, the three loop structure, although the function is similar, but in the use of scenarios and code structure is different.

while loop: The while loop checks the condition before each iteration and executes the loop body if the condition is true. This means that the loop body may not be executed at all if the initial condition is not met.

do...while loops: A do...while loop is similar to a while loop with one important difference: it executes the loop body at least once and then checks the condition after each iteration. The loop body is executed once even if the initial condition is false.

for loops: for loops are a more structured form of looping that combines initialization, conditional judgment, and post-loop operations. for loops are typically used in scenarios where the number of loops is known or predictable.

Example: Suppose we need to print the numbers 1 through 5, which would be very intuitive using a for loop:

至此，我们学习了while、do...while、for语句实现循环结构，三种循环结构虽然功能相似，但在使用场景和代码结构上有所不同。

while循环：while循环在每次迭代前检查条件，如果条件为真，则执行循环体。如果初始条件就不满足的话，这意味着循环体可能根本不会被执行。

do...while循环：do...while循环与while循环类似，但有一个重要的区别：它至少会执行一次循环体，然后在每次迭代后检查条件。即使初始条件为假，循环体也会执行一次。

for循环：for循环是一种更结构化的循环方式，它结合了初始化、条件判断和循环后操作。for循环通常用于那些循环次数已知或可预见的场景。

示例：假设我们需要打印数字1~5，使用for循环会非常直观：

```
for(int i=1; i<=5; i++)
{
    printf("%d\n",i);
}
```

Use a while loop to achieve the same functionality:

使用while循环实现相同的功能：

```
int i=1;
while(i<=5)
{
    printf("%d\n",i);
    i++;
}
```

Use do...while loops:

使用do...while循环：

```
int i=1;
do
{
    printf("%d\n",i);
    i++;
} while(i<=5);
```

Although these three loops are functionally equivalent, the choice of which one to use usually depends on the individual programming style and the code clarity required in a particular scenario. In actual programming, it is recommended to choose the most appropriate loop structure for the specific situation.

尽管这三种循环在功能上等价，但选择使用哪一种通常取决于个人编程风格和特定场景下的代码清晰度要求。在实际编程中，建议根据具体情况选择最适合的循环结构。

4.4 Loop Control Statements 循环控制语句

4.4节视频

Program statements are generally in order or in accordance with the function of the statement defined by the direction of the execution, such as in the loop statement, if you want to terminate the loop in advance, or skip some statements, you can use the loop control statement, the C language provides a break and continue loop control statements.

程序中的语句一般都是按照顺序或按照语句功能所定义的方向执行的，如在循环语句中，若想提前终止循环，或者跳过某一些语句，则可使用循环控制语句，C语言提供了break和continue循环控制语句。

4.4.1 The break Statement（break语句）

The syntax for using the break statement takes the form:

break语句的使用语法形式为：

```
break;
```

Usage: Used in switch statements and loops(for, while and do...while).

Function: Forces the end of the structure in which it is located.

We used the break statement when we learned about switch statements, and we have a preliminary understanding of it. This section focuses on its use in a loop structure.

使用场合：用于switch语句和循环语句（for、while和do...while）中。

作用：强制结束所在结构。

我们学习switch语句时使用了break语句，对其有了初步的了解。本节主要介绍其在循环结构中的具体用法。

Case 4_6（案例 4_6）

Write a program to simulate the donation process, to realize the acceptance of donations one by one, and calculate the total amount until the total amount of donations reaches 10,000 yuan. Requirements for the program to be able to handle user input, and in reaching the target amount when the end of the donation process, requiring the output of the final total amount of donations.

编写程序模拟捐款过程，实现逐个接受捐款，并计算总额，直至捐款总额达到1万元为止。要求程序能够处理用户的输入，并在达到目标金额时结束捐款过程，要求输出最终捐款总额。

Thinking（解题思路）：

(1) Initialize variables: Initialize a variable to record the total

（1）初始化变量：初始化一个变量来

amount of the current contribution, with the initial value set to 0. Initialize a variable as the target amount, i.e., Ten thousand yuan(or RMB10,000).

(2) Write a loop: Use a loop structure(e.g., a while loop) to model the process of accepting donations one by one. Inside the loop, the inputs are processed and the totals are updated: it is checked whether the updated total of donations meets or exceeds the target amount.

(3) Judge and end the loop: If the total amount of donations reaches or exceeds the target amount, use the break statement to jump out of the loop. Otherwise, continue the next loop and wait for the next donation input.

(4) Output result: After jumping out of the loop, output a message informing the user that the donation campaign has ended. Display the final total amount of donations.

记录当前的捐款总额，初始值设为0。初始化一个变量作为目标金额，即1万元（或10 000元）。

（2）编写循环：使用一个循环结构（如while循环）来模拟逐个接受捐款的过程。在循环内部，处理输入并更新总额：检查更新后的捐款总额是否达到或超过目标金额。

（3）判断并结束循环：如果捐款总额达到或超过目标金额，使用break语句跳出循环。否则，继续下一次循环，等待下一个捐款输入。

（4）输出结果：跳出循环后，输出一条消息，告知用户捐款活动已结束。显示最终的捐款总额。

Program Code（程序代码）:

Case4_6.c

```
#include <stdio.h>
int main()
{
    int donation;                              //每次捐款金额
    int total=0;                               //捐款总额
    const int target=10000;                    //目标金额，设为常量
    printf("捐款活动开始，目标金额:%d元\n", target);
    printf("请输入每次的捐款金额，输入负数结束捐款:\n");
    while(1)                                   //循环接受捐款，直到输入负数或者目标达成结束循环
    {
        scanf("%d", &donation);
        //如果输入负数，结束捐款
        if(donation<0) break;
        //更新捐款总额
        total+=donation;
        //检查是否达到目标金额
        if(total>=target) break;
    }
    //输出结果
    if(total>=target)
        printf("捐款结束、目标达成，最终捐款总额为:%d元\n", total);
    else printf("捐款结束、目标未达成，最终捐款总额为:%d元\n", total);
    return 0;
}
```

Program Output 1（运行结果1）:

```
捐款活动开始，目标金额:10000元
请输入每次的捐款金额，输入负数结束捐款:
100
1000
200
3000
3500
```

```
4000
捐款结束、目标达成，最终捐款总额为：11800 元
```

Program Output 2（运行结果2）：

```
捐款活动开始，目标金额：10000 元
请输入每次的捐款金额，输入负数结束捐款：
1000
2000
500
-10
捐款结束、目标未达成，最终捐款总额为：3500 元
```

Case Analysis（案例解析）：

(1) Line 4 defines a variable as the total amount of the donation, which is used to enter the amount of the donation within the loop.

(2) Line 5 defines the total variable, initialized to 0, to store the total amount of donations.

(3) Line 6 defines a constant variable of type const with an initial value of ¥10,000 for target, which is not allowed to be modified subsequently, representing the total target amount of the contribution.

(4) Lines 7, 8 are two prompt statements to make the interface more user-friendly.

(5) The 9th line is a while statement, the condition is 1, which represents the constant truth, and there is a break inside to end the loop according to the condition, which can avoid the dead loop.

(6) In line 11, data is entered from the keyboard and deposited into the variable donation to simulate the amount of the donation.

(7) Line 13 is an if statement controlled by a break statement, which is executed if the condition is valid and serves to end the loop in which it is placed.

(8) Line 15, if line 13 does not end the loop, the execution of line 15 indicates that the amount of the contribution is a valid data greater than 0, then update the total amount of the total contribution.

(9) Line 17 is an if statement that determines whether the current updated total data has reached the target goal, and if total >= target, then a break statement is executed to end the loop.

(10) Lines 20-22 is an if...else statement, if total >= target condition is established, it means that the loop is to meet the conditions of the end of the target is reached, the total amount of donations output. Otherwise, the loop ends with a negative donation, the target is not reached, and the total amount of donations is output.

（1）第4行定义了一个变量为捐款总额，用于循环内输入捐款金额的变量。

（2）第5行定义了total变量，初始化为0，用于存储捐款总额。

（3）第6行定义了一个const型的常变量，target初始值为10000元，后续该变量不允许修改，代表捐款目标总额。

（4）第7、8行为两条提示语句，让界面更友好。

（5）第9行为while语句，条件为1，代表恒真，内部有break根据条件结束循环，可以避免死循环。

（6）第11行，从键盘输入数据存入到变量donation中，模拟捐款的数额。

（7）第13行是一条if语句，控制的语句为break语句，若条件成立则执行break语句，其作用是结束所在的循环。

（8）第15行，若第13行没有结束循环，执行到15行说明捐款数额为大于0的有效数据，则更新total捐款总额。

（9）第17行为一条if语句，判断当前更新后的total数据有没有达到target目标，若total >= target，则执行break语句，结束循环。

（10）第20~22行为一条if...else语句，若total >= target条件成立，说明循环是满足该条件结束的，目标达成，输出捐款总额。否则是遇到了donation为负数的情况结束循环，目标未达成，输出捐款总额。

Case Summary（案例总结）:

This case simulates the process of donating money, and executes the break statement after certain conditions are met. The break statement is used in loops to force the end of the loop in which it occurs.

本案例模拟捐款的过程，满足一定条件后执行break语句，break语句用于循环，则强制结束所在循环。

4.4.2 The continue Statement（continue语句）

The syntax for using the continue statement is of the form:

continue语句的使用语法形式为：

```
continue;
```

Usage: It is used in looping statements, such as for, while, do...while statements.

Role: end the loop, i.e., only skip the statements that have not been executed after continue in the loop.

使用场合：适用于循环语句中，如for、while、do...while语句。

作用：结束本次循环，即仅跳过本次循环中continue后尚未执行的语句。

Case 4_7（案例 4_7）

Utilize a loop construct in conjunction with the continue statement to iterate through the numbers ranging from 1 to 100, and selectively print only those numbers that are divisible by 5.

利用循环结构结合continue语句，遍历1~100的数字，并选择性地仅打印那些能被5整除的数字。

Thinking（解题思路）:

Iterate through each number from 1 to 100 in a loop, and if a number is divisible by 5, output it; otherwise, skip the current iteration using the continue statement to achieve the functionality of only outputting numbers divisible by 5.

通过循环结构遍历1~100之间的每一个数字，如果某个数字能被5整除，则输出该数字；否则，利用continue语句跳过当前循环迭代，不执行输出操作，以实现仅输出能被5整除的数字的功能。

Program Code（程序代码）:

Case4_7.c

```
1   #include <stdio.h>
2   int main()
3   {
4       int i;
5       for(i=1;i<=100;i++)
6       {
7           if(i%5!=0)
8               continue;
9           printf("%d ",i);
10      }
11      return 0;
12  }
```

Program Output（运行结果）:

```
5 10 15 20 25 30 35 40 45 50 55 60 65 70 75 80 85 90 95 100
```

Case Description（案例解析）:

(1) The continue statement in line 8 only ends the loop. That is, the printf statement after the current loop is not executed, and the statement is skipped. At this time, the loop body ends, and according to the execution process, the i++ operation is executed, and the next loop is carried out.

（1）第8行的continue语句仅结束本次循环。即在本次循环中后面的printf语句不执行，跳过该语句，此时循环体结束，按照执行流程，执行i++操作，进行下一次循环。

(2) By running the result we can see that the data of condition(i%5==0); are output by the compiler.

Notes on the use of the continue statement: First, it can only be used in a loop structure; second, the continue statement only ends this loop.

（2）通过运行结果可见，条件(i%5==0);的数据都被编译器输出。

使用continue语句的注意事项：第一，只能用于循环结构中；第二，continue语句仅结束本次循环。

4.4.3 Difference Between break and continue（break和continue的区别）

From 4.3.1 and 4.3.2, we recognize the break statement and continue statement, in order to deepen the understanding and memory, we make a comparison below, break statement is used in switch statement and loop structure, continue statement is only used in loop structure, can not be used in switch statement. break statement is the end of the current level of the loop, while the continue statement only ends the current loop. The break statement ends the current level of the loop, while the continue statement ends the current loop.

Visualize the difference between the break and continue statements by using a flowchart diagramming method. Take the example of outputting a number divisible by 5 within 100 as an example. Combined with the if statement, if n is not divisible by 5, i.e., the condition is valid, then the continue statement is executed, and the arrow enters the next loop, skipping only this operation of outputting n, as shown in Figure 4-9. Suppose we change the continue statement to a break statement, how does the flowchart change? As shown in Figure 4-10, if the conditional judgment is valid, then the execution of the break statement, this time is directly out of this layer of the loop, the end of the loop.

由4.3.1节和4.3.2节我们认识了break语句和continue语句，为了加深理解和记忆，下面我们做一个对比。break语句用于switch语句和循环结构中，continue语句仅用于循环结构中，不能用于switch语句中。break语句是结束本层循环，而continue语句仅结束本次循环。

通过流程图图示方法，来直观对比break和continue语句的区别。以输出100以内能被5整除的数的例子为例。结合if语句，如果n不能被5整除，即条件成立，则执行continue语句，箭头进入了下一次循环，仅跳过本次输出n的这个操作，如图4-9所示。假设我们将continue语句改为break语句，那么流程图如何变化呢？如图4-10所示，如果条件判断成立，则执行break语句，此时是直接跳出了这层循环，结束了循环。

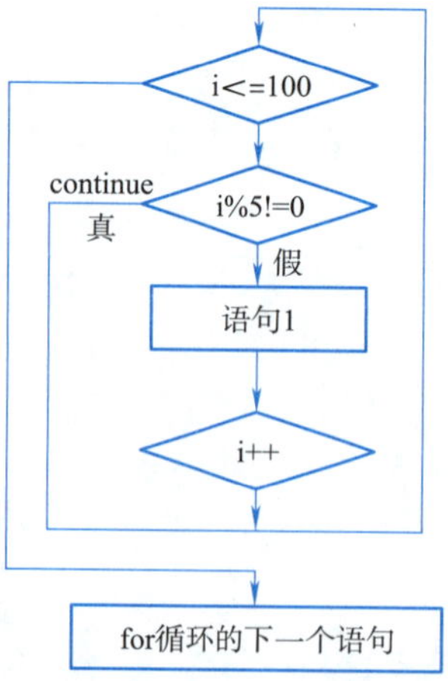

图4-9 continue语句

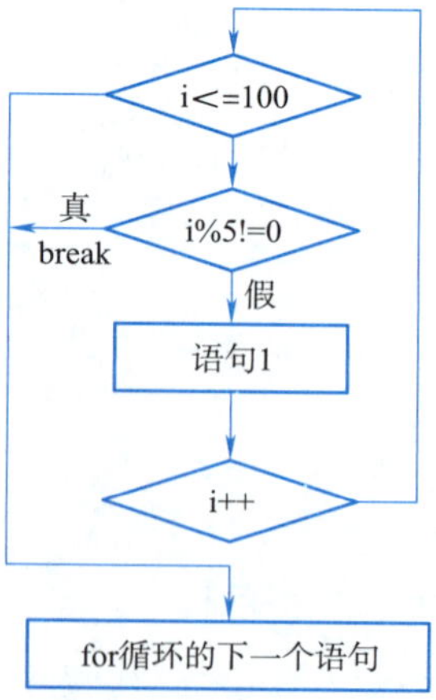

图4-10 break语句

According to the actual situation of the topic to be used flexibly, if you need to jump out of this level of the loop, then use break statement, if only the termination of the cycle of certain statements, then use continue statement.

根据题目的实际情况灵活使用，若需要跳出本层循环则使用break语句，若仅是终止本次循环某些语句，则使用continue语句。

4.5 Nesting of Loops 循环的嵌套

4.5节视频

Nesting of loops is an important concept in programming, which greatly expands the application scope of loop structure and makes the processing of complex problems more efficient. Although beginners may find loop nesting difficult to understand or master, through in-depth understanding and analysis of actual cases, we can gradually master its usage and thus strengthen our programming ability.

循环的嵌套是编程中的一个重要概念，它极大地扩展了循环结构的应用范围，并使得复杂问题的处理变得更为高效。尽管初学者可能觉得循环嵌套难以理解或掌握，但通过对实际案例的深入理解和分析，我们可以逐步掌握其用法，进而强化编程能力。

4.5.1 What Is Nesting of Loops（什么是循环的嵌套）

Nested loops are loops that contain another complete loop structure within the loop, which can be nested in multiple layers, like Russian nesting dolls, and each layer of the nesting doll needs to be intact. The embedded loop can be while, do...while, or for, and any other loop structure we have learned, and different structures can be nested within each other. However, it should be noted that when nesting, the loops must not cross each other and must maintain the integrity of the structure, which is the basic principle of nesting. The following is a simple example to visualize the logic of loop nesting.

循环嵌套指的是一个循环体内包含另一个完整的循环结构，这种嵌套可以多层进行，如同套娃般层层相套，且每层套娃都需完整无缺。内嵌的循环可以是while、do...while或for等任何我们学过的循环结构，不同结构间可相互嵌套。但嵌套时需注意，循环之间不可交叉，必须保持结构的完整性，这是嵌套的基本原则。下面通过一个简单例子，直观地理解循环嵌套的逻辑。

Case 4_8（案例 4_8）

Output 4 lines of "*******", i.e. output:

输出4行"*******"，即输出：

```
*******
*******
*******
*******
```

Thinking（解题思路）:

Printing graphics is one of the occasions for the use of nested loops, in this case you can use a double for loop, that is, the for loop and then nested within a for loop, the outer loop to control the number of lines of 4 lines, the inner loop to control the loop to print 7 *, each line of the output is completed and then output a new line character '\n'. both inside and outside the loop need to define the loop variable, so that i control the outer loop, j control the inner loop.

打印图形是循环嵌套的使用场合之一，在本案例中可以使用双重for循环，即for循环内再嵌套一个for循环，外循环控制行数4行，内循环控制循环打印7个*，每行输出完毕后再输出一个换行符'\n'。内外循环均需要定义循环变量，令i控制外循环，j控制内循环。

Program Code（程序代码）:

Case4_8.c

```
#include <stdio.h>
int main()
```

```
3   {
4       int i,j;                              //循环变量
5       for(i=1;i<=4;i++)                     //控制行的变化，一共 4 行
6       {
7           for(j=1;j<=7;j++)                 //控制列的变换，每列输出一个 *，即一行 7*
8           {
9               printf("*");
10          }
11          printf("\n");                     //一行输出完毕后，输出一个换行符
12      }
13      return 0;
14  }
```

Program Output（运行结果）:

```
*******
*******
*******
*******
```

Case Analysis（案例解析）:

(1) Line 4, first define the inner and outer loop variables, i controls the outer loop changes, j controls the inner loop changes, the outer loop represents the line changes, and the inner loop represents how many *'s are printed.

(2) The for statement in line 5 is an outer loop, and the loop variable i controls the line change.

(3) The for statement in line 7 is an inner loop, controlled by the loop variable j. i is transformed once, j is incremented from 1 to 7, and 7 * are output, and the condition j <= 7 does not hold when j is 8, ending the loop.

(4) According to the execution process of the for loop, the loop variable changes once only after the loop body is executed. For the outer loop, the loop body statement is the inner loop for statement and printf output statement, and this loop body statement is executed before i++ is executed for the next loop.

The specific execution process is described as follows: first execute i=1, and then judge the condition i<=4, at this time the condition is established, the execution of the loop body statement, the first execution of the inner loop(lines 7-10), j from 1 to 8, the end of the loop, the output of the first line of the 7 *, the output of a new line(line 11), and then the program to return to line 5, the execution of the i + + statement, i becomes 2 after i<=4 still holds and continues to the next loop.

(5) The statement that outputs a line break on line 11 should come after an internal for loop that ends, representing the completion of one line of output, followed by a line break.

（1）第4行，首先定义内外循环变量，i控制外循环变化，j控制内循环变化，外循环代表行的变化，内循环代表打印多少个*。

（2）第5行的for语句是外循环，循环变量i控制行的变化。

（3）第7行for语句是内循环，由循环变量j控制，i变换一次，j从1递增到7，输出7个*，j为8时条件j<=7不成立，结束循环。

（4）根据for循环的执行过程，循环体执行完毕后，循环变量才会变化一次，对于外循环来说，循环体语句是内循环for语句和printf输出语句，该循环体语句执行完，才会执行i++，进行下一次循环。

具体执行过程描述如下：首先执行i=1，随后判断条件i<=4，此时条件成立，执行循环体语句，先执行内循环（第7~10行），j从1变换到8，结束循环，输出第一行7个*，输出一个换行符（第11行），随后程序回到第5行，执行i++语句，i变为2后，i<=4仍然成立，继续下一次循环。

（5）第11行输出换行符的语句，应在内部的for循环之后，该for循环结束，代表一行输出完毕，后再换行。

Case Summary（案例总结）:

Nested statements in loops still follow the execution flow of the loop structure. As in this case, the execution flow of the for loop is followed. The variables change once in the outer loop and once in the inner loop.

循环嵌套语句，仍然遵循循环结构的执行流程。如本案例中，遵循的均是for循环的执行流程。外循环变量变化一次，内循环变化一轮。

4.5.2 Examples of Loop Nesting（循环嵌套的举例）

Case 4_9（案例 4_9）

The question asks to output all the prime numbers from 3 to 100.

题目要求输出3~100中的所有素数。

Thinking（解题思路）:

A prime number, or prime number, is a natural number greater than 1, an integer with no factors other than 1 and itself. First, a loop is needed to iterate through each of the numbers from 3 to 100. Second, in the loop within the realization of the current number is a prime number, the loop to determine whether there are other factors in addition to 1 and itself, this judgment process needs to be processed into a loop, the loop can be combined with a break statement, if a factor is found, then the end of the judgment, it can be concluded that it is not a prime number, otherwise it is a prime number. This case should be a double for loop.

素数即为质数，是指一个大于1的自然数，除了1和它本身之外，没有其他因子的整数。首先，需要一个循环遍历3~100中的每一个数。其次，在循环体内实现判断当前的数是否为素数，循环判断除了1和它本身之外是否还有其他因子，这个判断过程需要处理成循环，循环可以结合break语句，若找到一个因子，则结束判断，则可断定其非素数，否则为素数。本案例应为双重for循环。

Program Code（程序代码）:

Case4_9.c

```c
#include <stdio.h>
int main()
{
    int n,i,flag=1;                    //flag 为标志位，初始状态为 1
    for(n=3;n<=100;n++)
    {
        for(i=2;i<=n-1;i++)
        {
            if(n%i==0)
            {
                flag=0; break;         //若有能整除的数据，设 flag 状态为 0，代表非素数
            }
        }
        if(flag==1)                    //若状态非 0，则为素数
        {
            printf("%d ",n);
        }
        flag=1;                        //恢复 flag 的初始状态值，用于下一个数的判断
    }
    return 0;
}
```

Program Output（运行结果）:

```
3 5 7 11 13 17 19 23 29 31 37 41 43 47 53 59 61 67 71 73 79 83 89 97
```

Case Analysis（案例解析）:

(1) Line 4 sets the loop variables n and i, as well as the state flag bit flag, which sets the initial state value to 1. Its role is mainly used to determine whether the state changes after a loop and other structures, which can simplify programming.

(2) Line 5 is an outer loop where the control variable changes from 2 to 100.

(3) Line 7 is an inner loop that controls whether each piece of data for i from 2 to n−1 can divide the number n, i.e., it determines whether there is a factor of it for data other than 1 and itself.

(4) Lines 9 through 12, is an if statement that determines if n is divisible by the current i. If it is divisible, change the status of flag to 0 and end the loop. Because as long as we find a factor can determine that it is not a prime number, the subsequent do not have to judge, break is the role of the end of the level of the loop structure, so the end of the inner loop. If all the data are not divisible, the loop ends normally and the flag=0 statement is not executed.

(5) After the end of the inner loop, line 14 is executed in order, and the value of flag is determined; if it is 1, it is a prime number, otherwise it is a non-prime number.

(6) The reason for the reassignment of flag to 1 in line 18 is that it is used in the judgment of the next piece of data, which is 1 assuming it is a prime number, and if it is a non-prime number, it will be modified to 0 after it goes through an inner loop and a factor is found.Flag bits are often used in programming for judging some kind of change in state.

（1）第4行设置了循环变量n和i，以及状态标志位flag，flag设置了初始状态值为1，它的作用主要用于判断经过循环等结构，状态是否改变的判断，可以简化编程。

（2）第5行为外循环，控制变量从2变化到100。

（3）第7行为内循环，控制i从2到n−1的每个数据能否整除数n，即判断除了1和它本身之外的数据是否存在其因子。

（4）第9行到第12行是一条if语句，判断n是否能被当前的i整除。若能整除改变flag的状态为0，同时结束循环。因为只要找到一个因子则可确定其不是素数，后续不用判断，break的作用是结束其所在的那一层循环结构，故结束内循环。若所有数据都不能整除，则循环正常结束，flag=0语句不会被执行到。

（5）内循环结束之后，按照顺次执行第14行，判断flag的值，若为1则为素数，否则为非素数。

（6）第18行flag重新赋值为1的原因，是用于下个数据的判断，为1假设是素数，若为非素数，在经过内循环，找到一个因子后，会被修改为0。标志位往往用于编程中，用于判断某种状态的变化。

Case Expansion（案例拓展）:

(1)The for(i=2;i<=n−1;i++) statement in line 7 can be simplified to for(i=2;i<=sqrt(n);i++), sqrt is a C standard library function , to find the square root of an integer, the math.h header file should be included before using it. A number greater than the square root of n cannot divide n, so the data can be compared less often and the execution efficiency of the program can be improved.

(2)The flag bit flag is used in the case, if not, how to determine whether it is prime or not?

The case is reimplemented as follows:

（1）第7行的for(i=2;i<=n−1;i++)语句，可以简化为for(i=2;i<=sqrt(n);i++)，sqrt是C标准库函数，求一个整数的平方根，在使用它之前应该包含math.h头文件。大于数n平方根的数不可能整除n，故可以少比较几次数据，提升程序的执行效率。

（2）案例中使用了标志位flag，若不使用flag，如何判断其是否为素数呢？

案例重新实现如下：

```
#include <stdio.h>
#include <math.h>                          //使用 sqrt 要包含该头文件
int main()
{
    int n,i;
```

```
6        for(n=3;n<=100;n++)                    //外循环，遍历 3 到 100 数据
7        {
8            for(i=2;i<=sqrt(n);i++)            //内循环，判断是否其他因子
9            {
10               if(n%i==0) break;              //若有因子，则强制结束
11           }
12           if(i>sqrt(n)) printf("%d ",n);     //判断为素数则输出
13       }
14       return 0;
15   }
```

The if statement on line 9 controls the execution of the break statement. If the condition holds, i.e., there exists a factor that is not prime, the program immediately executes the break statement to force termination of the loop, which terminates abnormally, i.e., i<=sqrt(n). The inverse of this condition is i>sqrt(n), which can be used to determine whether a factor is prime or not, i.e. if(i>sqrt(n)) printf("%d",n).

第9行的if语句控制着break语句的执行。若条件成立，即存在一个因子不是素数，程序会立即执行break语句，强制终止循环，而此时非正常终止，即i<=sqrt(n)。这个条件的取反为i>sqrt(n)，可以用这一条件来判断是否为素数，即if(i>sqrt(n)) printf("%d",n)。

Note（说明）:

The break statement only forces the end of the structure in which it is placed, as in this case, only the inner loop. The outer loop continues to execute normally.

break语句只强制结束其所在的结构，如本案例中，只跳出内循环。外循环仍继续正常执行。

Case 4_10（案例 4_10）

Read the following program and write the results.

阅读下列程序，写出运行结果。

```
#include <stdio.h>
int main()
{
    int i,j,k;
    for(i=1;i<=2;i++)
    {
        for(j=1;j<=3;j++)
        {
            for(k=1;k<=4;k++)
                printf("A");
            printf("\n");
        }
    }
    return 0;
}
```

Program Output（运行结果）:

```
AAAA
AAAA
AAAA
AAAA
AAAA
AAAA
```

Case Analysis（案例解析）:

(1) This example helps us to rationalize the execution logic

（1）该案例可帮助我们理顺嵌套循环

of nested loops. The code contains a triple for loop. i is used as the loop variable in the outer loop, which loops twice; j is used as the loop variable in the second loop, which loops three times; and k is used as the loop variable in the third loop, which loops four times.

的执行逻辑，代码包含一个三重for循环。i作为外循环的循环变量，循环2次；j作为第二重循环的循环变量，循环3次；k作为第三重循环的循环变量，循环4次。

(2) The execution logic still follows the execution flow of the for loop.

（2）执行逻辑仍然遵循for循环的执行流程。

For the outermost loop the execution process is described as follows:

对于最外重循环的执行过程描述如下：

①Execute “i=1”;.

①执行i=1;。

②Judge the condition “i<=2;”, if the condition is true, execute the loop body statement, otherwise end the loop. Note: The loop body statement is a double for loop.

②判断条件i<=2;，若条件为真，执行循环体语句，否则结束循环。说明：循环体语句为一个双重for循环。

③After the double for loop is executed, execute “i++;” go back to step ② until the judgment condition is false then end the loop.

③双重for循环执行完毕后，执行i++;回到步骤②，直到判断条件为假则结束循环。

For the second loop(double for loop) the execution is described as follows:

对于第二重循环（双重for循环）的执行过程描述如下：

①Execute j=1;.

①执行j=1;。

②Judgment condition j<=3;, if the condition is valid, then execute the loop body statement, otherwise end the loop. The loop body statement consists of a for loop and an output statement, which serves to output a line with 4 A’s followed by a new line.

②判断条件j<=3;，若条件成立，则执行循环体语句，否则结束循环。循环体语句由一个for循环和一条输出语句构成，作用是输出一行4个A后换行。

③Execute j++; return to step ②.

③执行j++; 返回步骤②。

According to the conclusion that the outer loop variable changes once and the inner loop variable transforms for one round, i can make the outer loop execute twice, so the inner double for loop executes twice, then j transforms for 2 rounds, and the second double loop within the for loop, j transforms once then k transforms for one round, and each time it outputs one line of A and a newline character. j can output 3 lines of A for each round of transformations, so it outputs 6 lines of A in total.

根据外循环变量变化一次，内循环变量变换一轮的结论，i可以令外循环执行2次，故内部的双重for循环执行了2次，则j变换了2轮，第二重循环内的for循环，j变换一次则k变换一轮，每次输出一行A和换行符。j每一轮的变换可以输出3行A，故一共输出了6行A。

(3) Learners can simulate the execution process of a computer or use a compiler for step-by-step debugging to observe the changes in variables and the output of results.

（3）学习者可以模拟计算机的执行过程，或者使用编译器逐步调试，观察变量的变化和结果的输出。

Case 4_11（案例 4_11）

Use a loop structure to print a multiplication table in the following format.

使用循环结构打印下列格式的九九乘法表。

```
1*1=1
1*2=2  2*2=4
1*3=3  2*3=6  3*3=9
```

```
1*4=4  2*4=8  3*4=12  4*4=16
1*5=5  2*5=10  3*5=15  4*5=20  5*5=25
1*6=6  2*6=12  3*6=18  4*6=24  5*6  30  6*6=36
1*7=7  2*7=14  3*7=21  4*7=28  5*7=35  6*7=42  7*7=49
1*8=8  2*8=16  3*8=24  4*8=32  5*8=40  6*8=48  7*8=56  8*8=64
1*9=9  2*9=18  3*9=27  4*9=36  5*9=45  6*9=54  7*9=63  8*9=72  9*9=81
```

Thinking（解题思路）:

Essentially still a two-dimensional graphics, print graphics is a line-by-line output, generally a double loop, the outer loop to control the transformation of rows, the inner loop to control the transformation of columns. First of all, we need to find the rows and columns of the nine-nine multiplication table law. We observe a total of 9 rows, each row of the formula, the second operand and the number of rows in the same, the first operand from 1 to transform the number of rows. For example, in row 4, 1*4=4 2*4=8 3*4=12 4*4=16, there are 4 equations, i.e., there are 4 columns in total. Row 8 has a total of 8 formulas, that is, 8 columns, so it can be concluded that the maximum column number and the number of rows in the same.

The result of each expression can be calculated by the program, the arithmetic expression can be presented according to the line number and column number, "*" the first operand for the location of the line number, assuming that i represents the line number; the second operand and the location of the same column number, assuming that j represents the column number, then i*j, a total of 9 lines, i transformed from 1 to 9, and each line of the column changes from 1 to where the row number, so j changes from 1 to i. The algorithm flowchart is shown in Figure 4-11.

First execute i=1, judge condition i<=9 is established, execute j=1, if j<=i, then print the multiplication table, j=j+1, print the j-th expression in the i-th row, if j> i, then this line is printed, execute i=i+1, and print the next line. If the condition of i>9 is false, it ends. At this time, the multiplication table is printed.

本质上仍然是一个二维图形，打印图形是逐行输出的，一般是双重循环，外循环控制行的变换，内循环控制列的变换。首先要找到该九九乘法表的行列规律。我们观察到一共9行，每一行的式子中，第2个操作数和所在行号相同，第1个操作数从1变换到所在行号。如第4行，1*4=4 2*4=8 3*4=12 4*4=16，一共有4个式子，即一共有4列。第8行一共有8个式子，即为8列，故可以得出最大列号和所在行号相同。

每个表达式的结果可以由程序算出来的，算术表达式可以根据行号和列号呈现出来，"*"的第1个操作数为所在行号，假设i表示行号；第2个操作数和所在的列号相同，假设j表示列号，则i*j，一共9行，i从1变换到9，每一行的列从1变化到所在行号，故j从1变化到i。算法流程图如图4-11所示。

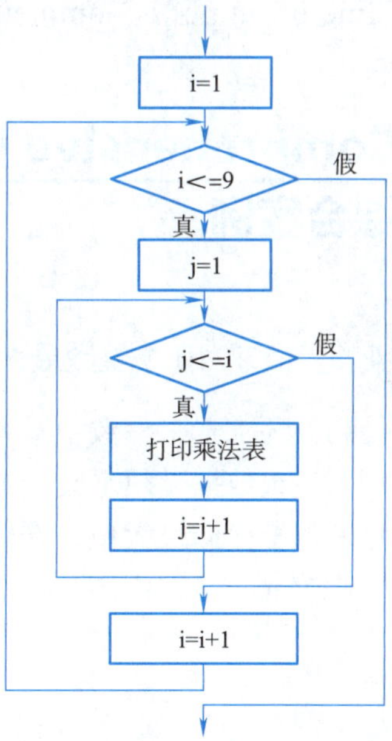

图4-11 打印九九乘法表算法流程图

首先执行i=1，判断条件i<=9成立，执行j=1，若j<=i，则打印乘法表，j=j+1，打印第i行的第j个式子，若j>i,则这一行打印完毕，执行i=i+1，打印下一行，若i>9条件为假，则结束，此时乘法表打印完毕。

Program Code（程序代码）:

Case4_11.c

```
1   #include <stdio.h>
2   int main()
3   {
4       int i,j;
5       for(i=1;i<=9;i++)                              //i 表示行号
6       {
7           for(j=1;j<=i;j++)
8               printf("%d*%d=%d ",j,i,j*i);
9           printf("\n");
10      }
11      return 0;
12  }
```

Case Analysis（案例解析）:

(1) The for loop in line 5 controls the transformation of the row and outputs 9 rows. The for loop in line 7 controls the transformation of the row and column and outputs column i.

(2) line 8 output statement, in accordance with the format of the first display the value of j, and then display the value of i, the product can be obtained by the expression operation, with % d control the result of the output, pay attention to the back of the empty space, separated from the output of the next expression. One line of output is completed by outputting a newline character.

（1）第5行for循环控制行的变化，输出9行。第7行for循环控制该行列的变换，输出i列。

（2）第8行的输出语句，应该按照格式首先显示j的值，然后显示i的值，乘积可以通过表达式运算得到，使用%d控制输出结果，注意后面的空格，用于分隔下一个表达式的输出。一行输出完毕输出一个换行符。

4.6 Comprehensive Cases 综合案例

综合案例 4_6_1 求 10 个整数最大值。

案例描述：使用循环求10个整数中最大值。

解题思路：10个整数数据无须定义10个变量，可以重复使用某个变量如a变量结合循环读入该数据，需要定义一个max变量用于存储最大值，在输入数据时直接比较当前输入的数据与当前的max的大小，若当前输入的数据大则修改max，否则维持原来的max值不变。

程序代码：

Case4_6_1.c

```
#include <stdio.h>
int main()
{
    int a,i=1,max;                  //a 存放数据，i 作为循环变量，max 用来存放最大值
    printf("请输入10个整数:\n");
    scanf("%d",&a);                 //首先输入一个数据
    max=a;                          //给 max 赋值为 a
    while(i<=9)                     //循环输入和比较剩下的 9 个数据
    {
        scanf("%d",&a);
```

```
11            if(max<a) max=a;
12            i++;
13        }
14        printf("max=%d\n",max);                        // 输出最大值
15        return 0;
16    }
```

运行结果：

```
请输入 10 个整数：
1
2 3 4 5 10 7 8 9 6
max=10
```

案例解析：

（1）首先定义变量，本案例中定义了变量a用于存放数据，max用于存放最大值，i作为循环变量，如第4行代码所示，此时3个变量中只有i赋予了初值1，其他两个变量没有赋予初值，是随机值。

（2）max应赋予初值，否则后续使用该随机值与输入的数据比较，将导致结果不准确，故先输入一个数据a，并赋值给max，如第6、7行所示，令第一个数据为当前的最大值。

（3）使用while循环输入并比较剩下的9个数据，i在第4行赋予了初值1，故循环条件为i<=9或i<10，控制循环执行9次。

（4）循环体内，第10行为输入数据a，第11行比较max和a的大小，若a大则修改max值，保证经过循环后，max存放的是10个整数数据中的最大值。这样逐一比较的算法，也常被形象化地称为“打擂台算法”。

（5）第14行为输出数据，将所求得的max值输出。

案例拓展：

（1）在该案例中，max需要赋予初值，否则一个随机值与输入的10个数据进行比较是没有意义的，我们可以赋予max足够小的值，如-999999。保证读入的第一个数据比它大，可以修改max值为第一个数据。代码可以改写为：

```
1    #include <stdio.h>
2    int main()
3    {
4        int a,i=1,max=-999999;            //a 存放数据，i 作为循环变量，max 用来存放最大值
5        printf("请输入 10 个整数:\n");
6        while(i<=10)                      // 循环输入和比较剩下的 9 个数据
7        {
8            scanf("%d",&a);
9            if(max<a) max=a;
10           i++;
11       }
12       printf("max=%d\n",max);           // 输出最大值
13       return 0;
14   }
```

运行结果：

```
请输入 10 个整数：
1 2 3 4 5 10 7 8 9 6
max=10
```

案例解析：在代码中，第4行增加了max的初始化，将原来的第6、7行代码去掉了，因无须先输入一个数据给max赋值了。循环条件需要修改为i<=10，保证有10个数据的输入和比较。

（2）因为有比较精确的次数，故使用for循环更为简洁。代码如下：

```
#include <stdio.h>
```

```
2    int main()
3    {
4        int a,i,max=-999999;               //a 存放数据，i 作为循环变量，max 用来存放最大值
5        printf(" 请输入 10 个整数:\n");
6        for(i=1;i<=10,i++)                 // 循环输入和比较剩下的 9 个数据
7        {
8            scanf("%d",&a);
9            if(max<a) max=a;
10       }
11       printf("max=%d\n",max);            // 输出最大值
12       return 0;
13   }
```

i可在for循环的第一个表达式位置赋值，i++可以放到第三个表达式的位置。其他不变。

（3）若求最小值如何改写呢？代码如下：

```
1    #include <stdio.h>
2    int main()
3    {
4        int a,i,min=999999;
5        printf(" 请输入 10 个整数:\n");
6        for(i=1;i<=10,i++)                 // 循环输入和比较剩下的 9 个数据
7        {
8            scanf("%d",&a);
9            if(min>a) max=a;
10       }
11       printf("min=%d\n",min);            // 输出最大值
12       return 0;
13   }
```

运行结果：

```
请输入 10 个整数:
2 3 4 1 5 6 7 8 9 10
min=1
```

案例解析：第4行的max改为min，注意变量定义尽可能见名知意，初值赋予一个比较大的数，如999999，当遇到输入的第一个整数数据时，比该初始值小，min值会被修改。第9行的小于号改为大于号，若有a比当前的min还小，则修改min值，在循环结束后min中存放的一定是最小值。

（4）若同时求最大值和最小值如何改写代码呢？代码如下：

```
#include <stdio.h>
int main()
{
    int a,i,max=-999999,min=999999;
    printf(" 请输入 10 个整数:\n");
    for(i=1;i<=10,i++)                          // 循环输入和比较剩下的 9 个数据
    {
        scanf("%d",&a);
        if(max<a) max=a;
        if(min>a) min=a;
    }
    printf("max=%d\n min=%d\n",max,min);        // 输出最大值
    return 0;
}
```

运行结果：

```
请输入 10 个整数:
```

```
2 3 4 1 5 6 7 8 9 10
max=10
min=1
```

案例解析：第4行中要对max和min同时初始化；第9行和第10行分别比较max和a，min和a的大小关系，最大值和最小值分别比较，是两条独立的if语句，互不干扰。最后输出max和min的值。

案例总结：通过上述案例，加深对循环结构的理解，拓展了求最大值和最小值问题的思路。

综合案例 4_6_2 水仙花数。

案例描述：要求输出所有的”水仙花数”。所谓”水仙花数”是指一个3位的整数，其各个位的数字立方和等于该数本身。例如，153=1^3+5+3^3。

解题思路：首先水仙花数是一个三位数，可以使用一个循环遍历100~999之间的每一个数字，测试每一个数字是否为水仙花数。水仙花数需要分离所测试的三位数的各个数位，分离操作在前面的章节中介绍过，需要用到“%”和“/”与10进行算术运算，最后使用关系运算符判断各个数位的立方和是否和该三位数相等。

程序代码：

Case4_6_2.c

```
1   #include <stdio.h>
2   int main()
3   {
4       int num;
5       int u,t,h;                          //存放分离出的数位：个位、十位、百位
6       for(num=100;num<=999;num++)         //遍历所有三位数的整数
7       {
8           u=num%10;                       //分离出个位
9           t=num/10%10;                    //分离出十位
10          h=num/100;                      //分离出百位
11          if(u*u*u+t*t*t+h*h*h==num)      //判断是否为水仙花数
12              printf("%d\t",num);
13      }
14      return 0;
15  }
```

运行结果：

```
153     370     371     407
```

案例解析：

（1）第4行num作为循环变量，第5行u,t,h用于存放各个数位。

（2）第6行for循环的写法，可以遍历100~999之间的每一个数。

（3）循环体内测试当前的数num是否为水仙花数。若是则输出。

（4）第8~10行，分离各个数位。第8行对10求余数，可以分离个位，如153%10得到3；第9行num/10去掉个位后再对10求余数，可以分离出十位，如153/10获得15,15%10得到5；第10行可以得到百位，如153/100结果为1。

（5）第11行为if语句，条件为“u*u*u+t*t*t+h*h*h==num”，注意这里是关系表达式，判断是否相等是两个等号，若少写一个等号则会报错，因为这里的左值是一个表达式，赋值语句左侧不允许是个表达式。

（6）在第12行的输出语句中，'\t'为转义字符，编译器会将其替换为一个制表符，用于在输出中创建水平间隔。制表符通常用于对齐文本，但具体的对齐方式取决于终端或控制台的设置以及制表符的宽度。同样可以采用“%5d”或者“%d ”，加上适当位宽或者在%d后面加上空格，达到隔开各个数据的目的。

（7）第11行条件中求各个数位的立方和问题可以采用C语言函数pow实现。pow函数是用于计算幂运

算的，定义在math.h头文件中，使用它需要在程序开头加上#include <math.h>预编译命令。调用格式为“pow(x,y);”，作用是计算x的y次方，故u*u*u+t*t*t+h*h*h==num可以改为：pow(u,3)+pow(t,3)+pow(h,3)==num。

综合案例 4_6_3 求 1000 以内的“完全数”。

案例描述： 要求输出1000以内的所有“完全数”。所谓的完全数是指一个数恰好等于它的真因子之和。如：6的真因子为1、2、3，而6=1+2+3。

解题思路： 根据“完全数的定义”一个数应该有真因子，而1不包含真因子，应从2开始。可以使用循环令循环变量遍历2~1 000内的每一个数，设循环变量为m，在循环内判断当前的数m是否是完全数。首先，求m的所有真因子，计算所有真因子之和，判断该和是否与数m相等。

程序代码：

Case4_6_3.c：

```
1     #include <stdio.h>
2     int main()
3     {    //number 代表当前数字，sumOfFactors 代表因子之和，factor 代表循环中的因子
4         int number, sumOfFactors, factor;
5         for(number=2; number<1000; number++)
6         {
7             sumOfFactors=0;                                    //初始化因子之和为 0
8             for(factor=1; factor<number; factor++)
9             {   //遍历寻找真因子，并计算其和
10                if((number%factor)==0)
11                    sumOfFactors+=factor;                      //累加真因子
12            }
13            if(sumOfFactors==number)
14            {   //判断当前数字是否为完全数
15                printf("%d, its factors are ", number);
16                for(factor=1; factor<number; factor++)         //再次遍历打印真因子
17                    if(number%factor==0) printf("%d ", factor);
18                printf("\n");
19            }
20        }
21        return 0;
22    }
```

运行结果：

```
6,its factors are 1 2 3
28,its factors are 1 2 4 7 14
496,its factors are 1 2 4 8 16 31 62 124 248
```

案例解析：

（1）第4行的number代表数，sumOfFactors代表真因子之和，factor代表循环中的因子。在程序的开头定义所需要的变量。

（2）第5行，for循环语句可以令number遍历2~1000的每一个数，满足求1000以内完全数的要求。

（3）第7~12行是求真因子之和的功能。首先给sumOfFactors赋值0，循环体会被执行多次，每次判断不同的数据，为不同的数据求得真因子之和，故sumOfFactors要清零，否则结果不正确。第8~12行为一条for循环语句，求数number的真因子，factor从1变换到number−1，注意不要写成factor<=number，真因子不包含其本身。若number%factor==0成立，说明factor是number的真因子，加到sumOfFactors中，循环结束后sumOfFactors存放的就是数number的所有真因子之和。

（4）第13~19行，是一条if语句，条件为sumOfFactors == number（注意是两个等号的关系运算符，不要写成一个等号的赋值运算符），判断真因子之和sumOfFactors和数number是否相等，若相同则是完全数，则输出完全数和对应的真因子。

（5）因为真因子没有存储，故要输出真因子还需要再计算一遍，如第16、17行所示。输出一个完全数之后，输出一个换行符，如第18行所示。

Summary 小 结

Through the study of this chapter, we have a deep understanding of the use of loops and the necessity of loops in the program, mastered the while, do...while and for the three major similarities and differences in the use of loops, and in-depth understanding of the logic of loops. At the same time, we have learned the syntax, function and difference of loop control statements break and continue, and can be flexibly applied according to the actual situation. In addition, we also understand the concept, principle and execution process of loop nesting, which provides strong support for programming practice.

通过本章学习，我们深入理解了循环的用途以及程序需要循环的必要性，掌握了while、do...while和for三大循环语句的异同及用法，深入理解了循环逻辑。同时，我们学会了循环控制语句break和continue的语法、作用及区别，并能根据实际情况灵活应用。此外，我们还了解了循环嵌套的概念、原则及执行过程，为编程实践提供了有力支持。

Exercise 习 题

一、单选题

1. 语句while(!a)中的条件!a等价于（　　）。

A. a==0　　B. a!=0　　C. a!=1　　D. a==1

2. 有以下程序段：

```
int x=0,p;
do {
    scanf("%d", &p);
    x++
};while(p!=10 && x<3);
```

此处do...while循环的结束条件是（　　）。

A. p 的值不等于 10 并且 x 的值小于 3

B. p 的值等于 10 并且 x 的值大于等于 3

C. p 的值不等于 10 或者 x 的值小于 3

D. p 的值等于 10 或者 x 的值大于等于 3

3. 有以下程序：

```
char a=10;
do{a++;} while(a<9);
printf("%d", a);
```

则正确的运行结果是（　　）。

A. 11　　B. 26　　C. 10　　D. 20

4. 有以下程序:

```
char a='a';
while(a<'y')
{
    do
    {
        a++;
    }while(a<'A');
}
printf("%c", a);
```

则正确的运行结果是（　　）。

A. a　　B. z　　C. y　　D. 无法运行

5. 有以下程序:

```
char c;
int i;
for(i=0; i<3; i++)
{
    scanf("%c", &c);
    printf("%c", c);
}
```

若输入abcd回车，则正确的运行结果是（　　）。

A. abc　　B. bbcd

C. 输入过多无法执行　　D. ab

6. 有以下程序:

```
int i, a=0;
for(i=0; i<3; i++)
{
    printf("%d ", ++a+i);
}
```

则正确的运行结果是（　　）。

A. 1 3 5　　B. 0 2 4　　C. 0 1 2　　D. 0 0 0

7. 下列说法中正确的是（　　）。

A. 在 switch 语句中一定要使用 break 语句

B. 在 switch 语句中不一定要使用 break 语句

C. break 语句是 switch 语句的一部分

D. break 只能用于 switch 语句中

8. 执行下面的程序后，a的值为（　　）。

```
#include <stdio.h>
int main()
{
    int a,b;
    for(a=1,b=1; a<=100; a++)
    {
        b+=3;
        if(b>=10) break;
    }
    return 0;
}
```

A. 3　　B. 4　　C. 5　　D. 6

9. 执行下列程序后，运行结果为（ ）。

```
int a=10, i;
for(i=0; i<a; i++)printf("%d", --a);
```

A. 98765 B. 987654 C. 109876 D. 1098765

10. 有以下程序:

```
char c='a';
int i;
for(i=0, i=1, c='h'; i<3; i++) {printf("%c", c);}
```

则正确的运行结果是（ ）。

A. hh B. hhh C.. HHH D. HH

二、程序分析题

执行下面的程序后，结束循环时a的值为__________。

```
int main()
{
    int a=1,b=1;
    while(a<=10)
    {
        b+=1;
        a++;
        if(b>=3) break;
    }
    return 0;
}
```

三、编程题

1. 编写一个程序，该程序接收一个正整数n作为输入，并计算1～n之间所有奇数的总和，最后输出结果。

2. 设计一个程序，该程序能够计算出1～20之间所有整数的阶乘之和，即1!+2!+3!+…+20!，并输出结果。(提示：若定义为int型，结果会溢出。请思考如何解决溢出的问题。)

3. 编写一个程序，该程序从键盘接收一个整数输入，并确定该整数是几位数。提示：可以通过连续除以10并计数操作次数来实现。

4. 同构数是指一个数恰好等于其平方数末尾的数字。例如，5是同构数，因为5的平方是25，而25的末尾是5。请编写程序找出1~1 000之间的所有同构数。

5. 编写一个程序，该程序接收一个正整数作为输入，并使用循环结构以逆序方式打印出该数字。例如，如果输入是12345，则输出应为54321。

6. 创建一个程序，该程序接收两个整数m和n（m<n）作为输入，然后列出m~n之间的所有素数。

7. 编写一个程序来读取一行字符，并分别统计其中英文字母、空格、数字和其他字符的数量。

8. 设计一个程序来实现一个简单的猜数字游戏。程序应随机生成一个1~100之间的数字，并根据用户的猜测给出“太大了”、“太小了”或“猜对了”的反馈。

9. 编写一个程序，该程序接收一个正整数n作为输入，并使用循环来打印一个由“*”组成的n行n列的直角三角形。

10. 编写程序，输入整数n，随后打印出斐波那契数列的前n项。斐波那契数列是每个数字都是前两个数字之和的序列，例如：0, 1, 1, 2, 3, 5, 8, 13, 21, 34,…。

Chapter 5

Batch Processing of Data Using Arrays

第 5 章

利用数组批量处理数据

Introduction 主要内容

We have systematically studied the three major structures-sequence, selection, and cycle. In order to further deal with more complex program logic, we will start the learning journey of arrays. It mainly studies the source of arrays, the definition, initialization and reference of one-dimensional arrays, the definition, reference and initialization of two-dimensional arrays, and the processing methods of character arrays and strings.

我们已系统学习了三大结构——顺序、选择和循环。为了进一步处理更复杂的程序逻辑，我们将开启数组的学习之旅。主要学习数组的来源、一维数组的定义、初始化与引用，二维数组的定义、引用和初始化，以及字符数组和字符串的处理方法。

Objectives 学习目标

To cultivate students' ability to define and use arrays in structural types, to flexibly use arrays to process data, and to distinguish one-dimensional arrays, two-dimensional arrays and character arrays, so as to make better use of arrays to solve practical problems.

培养学生构造类型中数组的定义和使用的能力，灵活运用数组来处理数据，区分一维数组、二维数组和字符数组的区别，从而能够更好地利用数组来解决实际问题。

5.1 Definition and Use of One-dimensional Array 一维数组的定义和使用

5.1节视频

In C, we often use arrays to deal with problems, so why do we have arrays? How do you define and use arrays? Next, the source of the array and the definition and usage of the array are introduced.

在C语言中，我们常常用到数组来处理问题。那么，为什么要有数组呢？如何定义和使用数组呢？接下来介绍数组的来源以及数组的定义和使用方法。

5.1.1 Source of Array（数组的来源）

To learn more about arrays, let's first explore

为了深入了解数组，首先我们来探讨

where they come from. Let's start with a more intuitive example to show the advantages of arrays in solving problems. If you want to input and store the scores of three students, you can use float to define three simple variables to represent the scores of three students, named score1, score2 and score3 respectively, and then use the scanf statement to input the values of the three variables. The code is as follows:

数组的来源。使用一个较为直观的示例展示数组解决问题的优势。若要输入3名学生成绩并存储，可用float定义3个简单变量表示3名学生的成绩，分别命名为score1、score2、score3，再使用scanf语句输入三个变量的值。代码如下：

```
float score1;
float score2;
float score3;
scanf("%f%f%f",&score1,&score2,&score3);
```

Please think about if there are 30 students, ask to input and store the scores of these 30 students, how to express it? According to the method just mentioned, 30 simple variables of float type need to be defined, and the scanf statement also needs to write 30 format control characters and 30 variable addresses. The code is as follows:

请思考如果有30个学生呢，要求输入并存储这30个学生成绩，如何表示？按照刚刚的方法需要定义30个float型简单变量，scanf语句也要写30个格式控制符和30个变量地址。代码如下：

```
float score1;
float score2;
float score3;
...
float score30;
scanf("%f%f%f...",&score1,&score2,&score3,...,&score30);
```

So if we use simple variables, there are two disadvantages:

(1) programming is complicated.

(2) The link between the data is not reflected.

It is not difficult to find that the performance data of 30 students have the same attributes and are float-type. In order to solve the above two drawbacks, the C language provides an array processing mechanism to process batch data of the same data type. The code to rewrite the array is as follows:

所以如果我们使用简单变量会有以下两个弊端：

（1）编程较为烦琐。

（2）没有体现出数据之间的联系。

不难发现，30名学生成绩数据实质具有相同的属性，同为float型。为了解决上述两个弊端，C语言提供了数组处理机制，来处理数据类型相同的批量数据。数组改写的代码如下：

```
float score[30];
for(i=0;i<30;i++)
{
    scanf("%f",&score[i]);
}
```

By comparing the code before and after, it is found that duplicate variable definitions can be rewritten into one line of code, and arrays can be used to define a group of variables in batches without naming each variable separately. The longer scanf input statement becomes a for loop control statement, and the code is concise and clear. It can be observed that arrays are used in combination with loops, which is more convenient for us to process data.

通过对比前后代码发现，重复的变量定义可以改写成一行代码，使用数组可以批量定义一组变量，不用为每个变量单独命名。而对比较长的scanf输入语句变成了一条for循环控制的语句，代码简洁清晰。观察可以发现数组结合循环使用，更方便我们处理数据。

5.1.2 Definition of One-dimensional Array（一维数组的定义）

The general definition of a one-dimensional array is:

```
data type array name[constant expression];
```

一维数组的一般定义形式：

```
数据类型 数组名[常量表达式];
```

Syntax Explanation 语法说明

(1) The data type indicates the type of the array element, such as int, float, double, char, and so on.

（1）数据类型表示数组元素的类型，如int、float、double、char等。

(2) The array name is named by the user, following the rules of identifier naming, that is, it is composed of characters, numbers and underscores. Symbols beginning with letters or underscores are named as clearly as possible.

（2）数组名由用户命名，遵循标识符命名规则，即由字符、数字和下划线组成，以字母或下划线开头的符号，命名尽可能见名知意。

(3)[] belongs to the type format of array and cannot be omitted. In principle, an array can be seen as a new type constructed from a basic data type, called a constructed type, and[] is part of the type.

（3）[]属于数组的类型格式，不能省略。数组原则上可以看作由基本数据类型构造出的新类型，成为构造类型，[]为类型的部分。

(4) The constant expression in[] represents the length of the array, that is, the number of elements contained in the array. The C99 standard stipulates that constants or constant expressions must be used instead of variables.

（4）[]内的常量表达式表示数组长度，即数组包含的元素个数。C99标准规定必须使用常量或者常量表达式，不能用变量。

(5) Arrays, like variables, must be defined before being used.

（5）数组和变量一样，必须先定义后使用。

For example, to define a one-dimensional array of integers with a size of 10, it should be expressed as: int a[10];. Where a is the name of the array, 10 is the length of the array, that is, the number of array elements, and the type of all array elements is int.

例如：若要定义一个有10大小的整型一维数组，应表示为int a[10];。其中a是数组名，10是数组长度即数组元素个数，所有数组元素的类型均为int型。

5.1.3 Storage of One-dimensional Array（一维数组的存储）

Each element in the array occupies a fixed amount of memory space, and these spaces are arranged consecutively. This storage makes it very efficient to access the elements of an array because the memory address of any element can be located by simple mathematical operations such as a base address plus an offset. The following is a simple example to illustrate the storage of a one-dimensional array. For the int a[10] array, the storage space is continuous, which is briefly expressed as follows:

数组中的每个元素都占据一段固定的内存空间，并且这些空间是连续排列的。这种存储方式使得访问数组中的元素非常高效，因为可以通过简单的数学运算（如基地址加上偏移量）来定位任何元素的内存地址。下面是一个简单的例子来说明一维数组的存储，对于int a[10]数组，存储空间是连续的，简要表示如下：

a[0]	a[1]	a[2]	a[3]	a[4]	a[5]	a[6]	a[7]	a[8]	a[9]

Note（说明）:

The array name a is a constant, and its value is the first address of the first element a[0] of the array. It cannot be assigned

数组名a是常量，其值为数组首元素a[0]的首地址，在程序中不可以为其赋值，

in the program, that is, it cannot be used as the lvalue of the assignment expression.

即不能作为赋值表达式的左值。

5.1.4 Reference of One-dimensional Array（一维数组的引用）

Representation of array elements:

```
array name[subscript]
```

数组元素的表示形式：

```
数组名[下标]
```

Description 说 明

(1) The representation of an array element is uniquely determined by the array name and the subscript. The subscript operator in C language is[]. Such as a[0]and a[1].

(2) Each element is referenced separately, and no attempt is made to reference it as a whole.

(3) It is assumed that the array length is a symbolic constant N, and the array element subscript ranges from 0 to N−1. The subscript starts from 0, and the total length is N, so the maximum subscript is N-1. If there is an array definition int a[10], the elements contained in the array are a[0], a[1], a[2], a[3], a[4], a[5], a[6], a[7], a[8] and a[9].

(4) When using, pay attention to prevent the subscript from crossing the boundary. The int a[10] array cannot take a[10] element, and the maximum subscript is the length minus 1.

(5) The subscript must be an integer type, not a floating point or other type.

(6) The subscript can be an integer constant, variable, or expression.

(7) Because array subscripts can be variables and discrete integers, arrays are often used more efficiently in combination with loops, which control the change of subscripts to reference each element of the array. A one-dimensional array can use a single loop.

(8) The[] operator is used for multiple purposes. Pay attention to the distinction. Is part of an array type when the array is defined, and is a subscript operator when subsequent array elements are accessed.

（1）数组元素的表示形式用数组名和下标唯一确定，C语言中下标运算符为[]。如：a[0]、a[1]。

（2）每个元素单独引用，不能试图整体引用。

（3）假设数组长度为符号常量N，数组元素下标范围为0~N−1。下标是从0开始的，总长度为N，故最大下标为N−1。如有数组定义int a[10]；则数组中包含的元素有a[0]、a[1]、a[2]、a[3]、a[4]、a[5]、a[6]、a[7]、a[8]、a[9]。

（4）使用时注意防止下标越界。int a[10]数组取不到a[10]元素，最大下标是长度减1。

（5）下标必须是整型类型，不能是浮点型或者其他类型。

（6）下标可以是整型的常量、变量或表达式。

（7）因数组下标可以是变量且是离散的整数，故数组往往结合循环使用更高效，循环变量控制下标的变化，来引用数组各个元素。一维数组使用一重循环即可。

（8）[]运算符一名多用，注意区分。在数组定义时属于数组类型的一部分，在后续数组元素访问时，属于下标运算符。

Case5_1（案例5_1）

Determine whether the following examples of array definitions and array element references are legal.

判断下列关于数组定义和数组元素引用的示例是否合法。

①int n=10; int a[n];

Not legal. Parsing: Array length cannot be a variable, it can only be a constant.

不合法。解析：数组长度不能为变量，其只能为常量。

②int n=5,a[10]; a[n]=20;

Legal. Parsing: First, define an integer array a[10] and an integer variable n, where n is initialized to 5. The second sentence is to access a[n] array element, which accesses a[5] element. Note that the length when defining the array cannot be a variable, but the subscript when accessing the array can be a variable.

合法。解析：首先定义了一个整型数组a[10]和整型变量n，n初始化为5。第二句话是访问a[n]数组元素，访问的是a[5]元素，特别注意定义数组时的长度不可以是变量，但访问数组时的下标可以是变量。

③int a[10]; a[0]=a[5]+a[2*3]+a[10];

Not legal. Array element a[10] is used, the subscript is out of bounds, and the maximum subscript is 9. a[2*3] is legal, the subscript is allowed to be an expression, and the element a[6] is accessed after the operation.

不合法。使用了数组元素a[10]，下标越界，最大下标为9。a[2*3]是合法的，下标允许为表达式，运算后访问的是a[6]元素。

5.1.5 Initialization of One-dimensional Array（一维数组的初始化）

Array initialization is an efficient way to uniformly assign values to array elements at one time when defining an array. It reduces the tedious manual assignment and improves the readability and maintainability of the code. Through initialization, we can easily set the initial value of the array elements, making programming more convenient.

数组初始化是在定义数组时，一次性为数组元素统一赋值的高效方式。它减少了手动赋值的烦琐，提高了代码的可读性和维护性。通过初始化，我们可以方便地设置数组元素的初始值，使得编程更加便捷。

Several Initialization Forms of One-dimensional Array　一维数组几种初始化形式

(1) Assign initial values to all array elements. For example, in an int a[10] array, {} is used to enclose all data, commas are used to separate the data, and the values are assigned to elements a[0] to a[9] in turn, specifically expressed as: int a[10] = { 0,1,2,3,4,5,6,7,8,9 };

(2) Assign a value to part of the array elements. If the number of values provided during initialization is less than the size of the array, other elements will be automatically completed by the system. The numeric data system will automatically complete 0, and the character data system will complete '\0', such as int a[10] = { 0,1,2,3,4 }; only the first five data are assigned, and the system automatically adds 0 for the last five data, which is equivalent to int a[10] = { 0,1,2,3,4,0,0,0,0,0 };.

(3) Assign the initial value to all array elements, and omit the length. If the initial value is assigned to all array elements, the array length can be omitted, and the compiler system can calculate the array length by counting the number of elements. For example, int a[5] = { 1,2,3,4,5 }; it can be written as int a[] = { 1, 2, 3, 4, 5 }; the latter determines that the length of the array is 5 by counting the number of assignments.

（1）对全部数组元素赋初值。如int a[10]数组，用{}将各个数据括起来，数据间用逗号隔开，值将依次赋给a[0]到a[9]元素，具体表示为：int a[10]={0,1,2,3,4,5,6,7,8,9};

（2）对部分数组元素赋值。如果初始化时提供的值的数量少于数组的大小，其他元素由系统自动补全，数值型数据系统自动补0，字符型数据系统补'\0'，如int a[10]={0,1,2,3,4};只赋了前五个数据，后五个系统自动补0，相当于int a[10]={0,1,2,3,4,0,0,0,0,0};。

（3）对全部数组元素赋初值，可省略长度。若对全部数组元素赋初值，可省略数组长度，编译系统可以通过统计元素个数来计算出数组长度。如int a[5]={1,2,3,4,5};可写为int a[]={1,2,3,4,5};后者通过统计赋值的个数确定数组的长度为5。

Case 5_2（案例 5_2）

Assign 10 array elements as 0,1,2,3,4,5,6,7,8,9, and output them in reverse order.

对10个数组元素依次赋值为0,1,2,3,4,5,6,7,8,9，要求按逆序输出。

Thinking（解题思路）:

The array should be defined first and then used, so define the array first. According to the data type in the title, the array type should be int. We know the data, so we can use array initialization to assign values to each element. You can use a reverse order loop to output the values of individual array elements.

数组要先定义后使用，故先定义数组。根据题目中的数据类型确定，数组类型应为int型。我们已知数据故可以使用数组初始化为各个元素赋值。逆序输出可以使用循环倒序输出各个数组元素的值。

Program Code（程序代码）:

Case5_2.c

```
1    #include <stdio.h>
2    int main()
3    {
4        int a[10]={0,1,2,3,4,5,6,7,8,9};          //定义整型一维数组，同时初始化
5        int i;                                    //定义循环变量控制下标，注意下标要求是整数
6        for(i=9;i>=0;i--)                         //控制数组元素逆序输出
7        {
8            printf("%d ",a[i]);                   //每输出一个数后加一个空格
9        }
10       return 0;
11   }
```

Program Output（运行结果）:

```
9 8 7 6 5 4 3 2 1 0
```

Case Analysis（案例解析）:

(1) Initialize while defining the array in line 4.

(2) Thei in line 5 is used as the loop variable of the for loop in line 6 to control the change of the subscript. Note that the subscript should be an integer.

(3) In line 6, the array elements are output in reverse order by using the loop. a[9] is output first, so i is assigned to 9, and then 1 is gradually subtracted. Other elements are output in the loop until i > = 0 is not established, and the loop is ended.

(4) If you do not initialize, you can only assign a value to each element of the array element separately, and you cannot try to assign a value as a whole. For example, you cannot write the following two lines of code:

（1）第4行定义数组的同时进行初始化。

（2）第5行的i作为第6行for循环的循环变量，用来控制下标的变化，注意下标应为整数。

（3）第6行使用循环依次逆序输出数组元素，先输出a[9]，故i赋值为9，后逐渐减1，循环输出其他元素，直到i>=0不成立，结束循环。

（4）若不进行初始化，要为数组元素赋值只能单独为每一个元素赋值，不能试图整体赋值。如不能写成下面两行代码：

```
int a[10];
a[10]={0,1,2,3,4,5,6,7,8,9};                //这个是错误的，不能试图整体引用
```

In the above code, a[10] array is defined first, and then the overall assignment is not allowed. If a[10] appears again without initialization, the computer will regard a[10] as a single element, and the maximum subscript is 9. If the subscript is out of range, an error will be reported.

(5) If you do not initialize and want to assign a value to an array element later, you can combine the loop. In this case, it can be changed to:

在上面代码中，先定义了a[10]数组，后面想整体赋值是不允许的。若没有初始化，后面再出现a[10]，计算机会将a[10]看成一个单个元素，而最大下标为9，下标越界，会报错。

（5）若不进行初始化，后面想为数组元素赋值，可以结合循环。本案例中，可以改为：

```
for(i=0;i<10;i++)
{
    a[i]=i;
}
```

If you want to enter any 10 integers from the keyboard, you can write the code:

若想从键盘输入任意10个整数，可以写成代码：

```
for(i=0;i<10;i++)
{
    scanf("%d",&a[i]);
}
```

Case Summary（案例总结）:

When writing a program to process the same type of sequence data, a one-dimensional array can be used. The array should be defined first and then used. When the data is known, it can be initialized at the same time. Note that the one-dimensional array is often combined with a loop. Because the number of data is determined, the for loop is often used, which is more convenient and efficient.

在编写程序中，处理同类型的序列数据，可以使用一维数组，数组要先定义后使用，在已知数据时可以同时初始化，注意一维数组往往结合一重循环，由于数据个数确定，往往采用for循环，更加便捷高效。

5.2 Two-dimensional Array 二维数组

5.2节视频

In the previous section, we have introduced the source of arrays and the definition, reference and initialization of one-dimensional arrays, and mastered the basic operation methods of arrays through cases. This section will continue to introduce the commonly used two-dimensional arrays.

在上节我们已经介绍了数组的来源和一维数组定义、引用及初始化，通过案例掌握了数组的基本操作方法。本小节将会继续介绍常用到的二维数组。

5.2.1 Definition and Storage Mode of Two-dimensional Array（二维数组定义及存储方式）

Definition form of two-dimensional array:

二维数组的定义形式：

```
data type array name[constant expression]
[constant expression];
```

```
数据类型 数组名 [ 常量表达式 ]
[ 常量表达式 ];
```

E.g.（如）:

```
float a[3][4];
```

Description 说 明

(1) A two-dimensional array has a similar definition form to a one-dimensional array, except that there is one more dimension, and the length of each dimension can only be a constant or a constant expression.

(2) How should we look at or understand two-dimensional arrays?

（1）二维数组和一维数组有相似的定义形式，只是多了一个维度，每一维的长度仍只能为常量或者常量表达式。

（2）我们该如何看待或者如何理解二维数组呢？

Think of it as a special one-dimensional array whose elements are still a one-dimensional array. Take the float a[3][4] two-dimensional array as an example, as shown below.

可以把它看成一种特殊的一维数组，这个数组的元素仍是一个一维数组。以刚刚的float a[3][4]二维数组为例，如下所示：

a[0][0]	a[0][1]	a[0][2]	a[0][3]
a[1][0]	a[1][1]	a[1][2]	a[1][3]
a[2][0]	a[2][1]	a[2][2]	a[2][3]

We treat each row as a whole, that is, as an element, the first row is a[0], the second row is a[1], and the third row is a[2]. Each element is a one-dimensional array containing four elements, such as a[0] element, which contains four elements a[0][0], a[0][1], a[0][2], a[0][3].

A two-dimensional array can also be seen as a matrix, with the first dimension being the number of rows and the second dimension being the number of columns. The above diagram can be seen as a matrix, which has three rows and four columns, so we often call the subscript of a two-dimensional array the row and column subscript.

(3) Two-dimensional arrays are often stored consecutively in memory in row-first order. Two-dimensional arrays, like one-dimensional arrays, are stored sequentially and continuously. Two-dimensional arrays are generally stored in memory in the order of row priority, that is, the first row is stored first, and then the second and third rows are stored.

我们把每一行看成一个整体，即看成一个元素，第一行是a[0],第二行是a[1]，第三行是a[2]。每个元素又是一个包含4个元素的一维数组，如a[0]元素，a[0]元素包含4个元素a[0][0]、a[0][1]、a[0][2]、a[0][3]。

二维数组也可以看成矩阵，第一维是行数，第二维是列数。上述图示可以看成矩阵，这个矩阵有3行4列，所以往往也称二维数组的下标为行列下标。

（3）二维数组往往按照行优先的顺序连续存储在内存中。二维数组与一维数组一样，是按顺次连续存储的，二维数组在内存中一般是按行优先的顺序存储的，即先存第一行，再存第二行、第三行。

5.2.2 Reference of Two-dimensional Array（二维数组的引用）

Reference form of two-dimensional array:

```
array Name[subscript][subscript]
```

二维数组的引用形式：

```
数组名[下标][下标]
```

Description 说 明

(1) Two-dimensional arrays are still defined first and then used. Array elements are also referenced separately and cannot be referenced as a whole.

(2) The row and column subscripts all start from 0, so the subscript range is 1 less than the length. Be careful not to generate subscript crossing.

(3) Because there are two subscripts, it is often used in combination with the double loop, the outer loop controls the change of the first dimension subscript, and the inner loop controls the change of the second dimension subscript. For example:

（1）二维数组仍是先定义后使用，数组元素同样是单独引用，不可整体引用。

（2）行列下标均从0开始，故下标范围均比长度少1，注意不要产生下标越界。

（3）因为有两个下标，故往往结合双重循环使用，外循环控制第一维下标的变化，内循环控制第二维下标的变化。举例：

```
int a[2][3];
for(i=0;i<2;i++)
```

```
    {
        for(j=0;j<3;j++)
        {
            printf("%d,%d\n",a[i][j]);
        }
    }
```

From the code, we can see that the first loop is for(i= 0; i< 2; i++), and the second loop is for(j = 0; j< 3; j++). The first loop is used to control the rows of the two-dimensional array, and the second loop is used to control the columns of the two-dimensional array. And the loop condition in the for loop is the "<" array subscript instead of the "<=" array subscript, because the array subscript starts from 0, and the maximum subscript is the length minus 1. Be careful not to generate an out-of-range subscript.

从代码中我们可以看到第一层循环为for(i=0;i<2;i++)，第二层循环为for(j=0;j<3;j++)，其中第一层循环用于控制二维数组的行，第二层循环用于控制二维数组的列。且在for循环中的循环条件都是"<"数组下标而不是"<="数组下标，因为数组下标都是从0开始的，最大下标为长度减1，注意不要产生下标越界。

5.2.3 Initialization of Two-dimensional Array（二维数组的初始化）

In principle, a two-dimensional array should be initialized on separate lines, with curly braces separating each line.

二维数组原则上要分行进行初始化，用花括号隔开每一行。

Several Forms of Two-dimensional Array Initialization 二维数组初始化几种形式

(1) Assign an initial value to a two-dimensional array by line, and use curly brackets {} to separate different lines:

（1）分行给二维数组赋初值，使用花括号{}分隔不同行：

```
int a[3][4]={{1,2,3,4},{5,6,7,8},{9,10,11,12}};
```

Analysis: Integer array a, with 3 rows and 4 columns, three rows separated by curly braces, and rows separated by ",". Assign values to the four elements of each row. The first curly bracket { 1,2,3,4 } assigns values to the four elements in row 0, a[0][0], a[0][1], a[0][2], a[0][3]; { 5, 6, 7, 8 } assigns values to row 1, a[1][0], a[1][1], a[1][2], a[1][3], and so on.

(2) Partial assignment is available, and the system automatically adds values, such as 0 for numerical type and '\0' for character type.

解析：整型数组a，有3行4列，用花括号隔开三行，行和行之间用","隔开。分别为每行的4个元素赋值。第一个花括号{1,2,3,4}，给第0行4个元素，即a[0][0]、a[0][1]、a[0][2]、a[0][3]；{5,6,7,8}给第1行赋值，即a[1][0]、a[1][1]、a[1][2]、a[1][3]，依此类推。

（2）可部分赋值，系统自动补值，数值型补0、字符型补'\0'。

```
int a[3][4]={{1},{5},{9}};
```

Equivalent to（等价于）:

```
int a[3][4]={{1,0,0,0},{5,0,0,0},{9,0,0,0}};
```

Analysis: Two-dimensional arrays can also be partially assigned, and the rest are automatically supplemented by the system. For example, in the second example above, only the first element of each row, that is, the element with a subscript of 0, is assigned, and other elements will be automatically supplemented with 0.

解析：二维数组也可以部分赋值，剩下的系统自动补充，如上面第二个例子只为每行的首元素，即下标为0的元素赋值了，其他的系统会自动补0。

(3) In the case of assignment of all elements, the first dimension can be defaulted, and the second dimension cannot.

（3）全部元素赋值情况下，第一维可省略，第二维不可以。

```
int a[3][4]={1,2,3,4,5,6,7,8,9,10,11,12};
```

Equivalent to（等价于）:

```
int a[][4]={1,2,3,4,5,6,7,8,9,10,11,12};
```

Parsing: If all elements are assigned, the first dimension can be defaulted, but the length of the second dimension cannot be omitted. For example, if all 12 elements are assigned, the {} of each line can be removed, and the first dimension can be omitted. The compilation system can determine the number of lines according to the length of the second dimension and the total number of data.

解析：如果全部元素赋值的情况下，第一维可以省略，但第二维长度不可省略，如这个例子12个元素都赋值了，则可以去掉每行的{}，同时第一维可以省略，编译系统可以根据第二维长度和数据总个数确定行数。

Consider another exampl:

再看一个例子：

```
int a[][4] = { { 0,0,3 }, {}, { 0,10 } };             //This is legal.（这种写法合法。）
```

Analysis: If not all values are assigned, but after using {} to separate lines, the number of lines can also be determined, and the length of the first dimension can also be omitted. You can determine that the first dimension has a length of 3 based on {} that the system has 3 rows.

解析：如果未全部赋值，但使用{}隔开行后，也可以确定行数，同样可以省略第一维长度。可以根据{}判定系统有3行，确定第一维长度为3。

5.2.4 Application Example of Two-dimensional Array（二维数组应用举例）

Case 5_3（案例 5_3）

Transpose a known matrix to form a new matrix, as shown in the following data, and transfer ***A*** matrix to the ***B*** matrix.

将一个已知的矩阵转置，形成一个新矩阵，如下列数据所示，将***A***矩阵转至成***B***矩阵。

$$\boldsymbol{A}=\begin{pmatrix}1 & 2 & 3\\ 4 & 5 & 6\end{pmatrix} \qquad \boldsymbol{B}=\begin{pmatrix}1 & 4\\ 2 & 5\\ 3 & 6\end{pmatrix}$$

Thinking（解题思路）:

First of all, for the matrix ***A***, it can be stored by the integer two-dimensional array a[2][3]. For the matrix ***B*** after transposition, it can be represented by defining the array b[3][2]. The transposition operation can be completed by accessing a[i][j] and storing it in B[j][i] element by element. Accessing each element needs to be combined with a double for loop to implement the program.

首先对于矩阵***A***，可用整型二维数组a[2][3]存储，对于转置后的矩阵***B***，可定义数组b[3][2]来表示，逐个元素访问将a[i][j]存放到b[j][i]中，即可完成转置运算，访问每个元素需要结合双重for循环来实现程序。

Program Code（程序代码）:

Case5_3.c

```
#include <stdio.h>
int main()
{
    int a[2][3]={{1,2,3},{4,5,6}};          //定义数组同时初始化
    int b[3][2],i,j;                        //定义二维数组，及循环变量i，j
    printf("array a:\n");                   //输出提示信息
    for(i=0;i<=1;i++)                       //循环控制二维数组的两个下标变化
```

```
8      {
9          for(j=0;j<=2;j++)
10         {
11             printf("%5d",a[i][j]);              //输出查看
12             b[j][i]=a[i][j];                    //转置运算
13         }
14         printf("\n");                           //输出一行后换行
15     }
16     printf("array b:\n");
17     for(i=0;i<=2;i++)                           //输出转置后数组b
18     {
19         for(j=0;j<=1;j++)
20             printf("%5d",b[i][j]);
21         printf("\n");
22     }
23     return 0;
24 }
```

Program Output（运行结果）：

```
array a:
    1    2    3
    4    5    6
array b:
    1    4
    2    5
    3    6
```

Case Analysis（案例解析）：

(1) Two-dimensional array a stores the matrix to be transposed, and two-dimensional array b stores the matrix after transposition. Array a is initialized to represent the known matrix to be transposed.

(2) Two-dimensional arrays should use double loops, with the outer loop controlling the change of the first dimension subscript and the inner loop controlling the change of the second dimension subscript, such as the for loop in lines 7-9.

(3) In lines 11-12, the elements of the two-dimensional array a are output first, and then the transpose operation in line 12 is performed. Both lines are governed by the inner loop for statement, and {} is added. After a line is processed, a newline character is output, that is, it is output according to the format of the matrix.

(4) After the double loop processing, the two-dimensional array b stores the transposed data.

(5) Lines 17-22 output the transposed two-dimensional array b. Note that {} is not added in line 20 because there is only one statement. The loop body of for can be omitted, but pay attention to indentation to reflect hierarchy and standardization.

（1）二维数组a存放待转置的矩阵，二维数组b存放转置后的矩阵。数组a进行了初始化，代表已知待转置的矩阵。

（2）二维数组应使用双重循环，外循环控制第一维下标变化，内循环控制第二维下标变化，如第7~9行for循环的写法。

（3）第11、12行先输出二维数组a的元素，再进行第12行的转置运算，这两行均归内循环for语句所管辖，要加上{}。一行处理完毕后输出一个换行符，即按照矩阵的格式输出。

（4）经过双重循环处理后，二维数组b存放的就是转置后的数据。

（5）第17~22行输出转置后的二维数组b。注意第20行没有加{}，是因为只有一条语句，for的循环体可以省略，但要注意缩进，体现层次和规范性。

5.3 Character Arrays and Strings 字符数组及字符串

5.3节视频

So far, we have mastered the basic concepts and applications of one-dimensional and two-dimensional arrays. Next, we'll focus on arrays of character data. Although there is a word difference between characters and strings, the storage methods and uses are quite different. Characters are stored in ASCII code and occupy only one byte; A string is a sequence of multiple characters and is widely used. Therefore, this chapter will focus on the knowledge of character arrays and strings.

至今，我们已经掌握了一维和二维数组的基本概念和应用。接下来，我们将聚焦于字符型数据的数组。字符与字符串虽一字之差，但存储方式和用途却大相径庭。字符以ASCII码形式存储，仅占一个字节；而字符串则是由多个字符组成的序列，应用广泛。因此，本章节将重点探讨字符数组与字符串的相关知识。

5.3.1 Character Array（字符数组）

If the type of the array is character, it is called a character array. It is used to store character data, each element occupies 1 byte and stores one character. For example, if a char type one-dimensional array c with 10 elements is defined, the definition format is: char c[10];.

若数组的类型为字符型，则称为字符数组。其用于存放字符型数据，每个元素占1字节，存放一个字符。如定义一个char型有10个元素的一维数组c，定义格式为：char c[10];。

The elements of the array are c[0] to c[9]. If you want to assign a value to each element, you can assign a value as follows:

其数组元素为c[0]~c[9]，若想为其各个元素赋值，则可赋值如下：

```
c[0]='I';c[1]=' ';c[2]='a';c[3]='m';c[4]=' ';c[5]='h'; c[6]='a';c[7]='p'; c[8]='p'; c[9]='y';
```

Each element is referenced separately and stores a single character. Note that a single character constant needs to be enclosed in single quotes. A brief description of the storage space is as follows:

每个元素单独引用，存放单个字符，需要注意的是单个字符常量需要加上单引号。存储空间简要描述如下：

I		a	m		h	a	p	p	y

In the above example, the c array is not initialized, and then it needs to be assigned one element after another, which is a little troublesome. If the initial value is known, the initialization can be used, and the operation is simpler.

上述例子中c数组未初始化，后需要一个个元素赋值，稍显麻烦。若已知初值的情况下，可以使用初始化，使操作更为简便。

Character Array Initialization 字符数组初始化

In the above example, it can be changed to the initialization form:

上述例子中，可改为初始化形式：

```
char c[10] = {'I','', 'a', 'm', ' ', ' H', 'a', 'p', 'p', 'y'};
```

If "China" is assigned, there are only 5 characters, which belong to partial initialization, such as: char c[10] = {'C', '', 'h', 'i', 'n', 'a' }; The system automatically adds '\0' to the remaining unassigned array elements. The storage space is shown below.

如若赋值 "China"，只有5个字符，属于部分初始化，形如：char c[10]={'C','','h','i','n','a'}; 系统为剩下的未赋值的数组元素自动补'\0'，存储空间如下所示。

C	h	i	n	a	\0	\0	\0	\0	\0

For a two-dimensional array, separate each row with {}. Example:

对于二维数组，用{}隔开每一行。举例：

```
char d[5][5] ={{' ',' ', '*' }, { ' ','*',' ', '*' }, { '*', ' ', ' ', ' ', '*' },
{' ','*', ' ', '*' }, {' ', ' ','*'}};
```

Note that there is a single space in ‘’, which represents a single space.

注意' '里面有一个空格，代表是单个空格。

Case 5_4（案例 5_4）

Output a known string: I Love China.

输出一个已知的字符串：I Love China.

Thinking（解题思路）：

First define a one-dimensional character array and its initial “I Love China.”, and then use the loop one by one output character array elements.

先定义一个一维字符数组并对其初始化为“I Love China.”，接着用循环逐个输出字符数组元素。

Program Code（程序代码）：

Case5_4.c

```
1    #include <stdio.h>
2    int main()
3    {
4        char c[13]={'I',' ','L','o','v','e',' ', 'C', 'h', 'i', 'n', 'a', '.'};
                                              //逐个字符初始化
5        int i;                               //作为循环变量
6        for(i=0;i<13;i++)                    //逐个元素输出
7            printf("%c",c[i]);               //使用%c控制单个字符的输出
8        return 0;
9    }
```

Program Output（运行结果）：

```
I Love China.
```

Case Analysis（案例解析）：

(1) Line 4 defines a one-dimensional array and initializes it.

(2) Line 5 defines a loop variable i;

(3) Lines 6 and 7 use a for loop to output the char array elements one by one, using the % c control.

（1）第4行先定义一个一维数组并对其进行初始化。

（2）第5行定义一个循环变量i。

（3）第6、7行运用一重for循环逐一输出char型数组元素，使用%c控制。

5.3.2 String（字符串）

In C, a string constant is a special literal used to represent an immutable sequence of characters. String constants are enclosed in double quotes in the source code, such as "Hello, World!.". String constants are stored in the read-only data section of the program at compile time and cannot be modified. The system automatically adds '\0' to the string as the end marker. Use character arrays to store or manipulate strings in your program.

在C语言中，字符串常量是一种特殊的字面量，用于表示不可变的字符序列。字符串常量在源代码中以双引号括起来，如"Hello, World!"。字符串常量在编译时会被存储在程序的只读数据段中，因此不能被修改。系统会为字符串自动添加'\0'作为结束标志。在程序中存储或处理字符串使用字符数组。

1. String Initialization Character Array（字符串初始化字符数组）

We can simplify initialization by using string constants to

我们可以使用字符串常量初始化一个

initialize a one-dimensional array of characters. For char c[10] = {'I', ' ', 'a','m ',' ',' h ',' a ',' p ',' p ', ' y '}; can be rewritten as char c[] = { "I am happy"};.

一维字符数组，进而简化初始化操作。对于char c[10]={'I',' ','a', 'm', ' ', 'h', 'a', 'p', 'p', 'y'};可以改写为char c[]={"I am happy"};。

Where, if a string is used to initialize a character array, {} can be omitted, so it is further abbreviated as char c[] = "I am happy";. Because initialization is used, the length of the character array definition is omitted. We see a total of 10 characters. Note that a space is a valid character, and the string system will add a '\0' ending flag to it, which also needs to occupy a storage space. Therefore, the actual length of the array will be converted to 11. Equivalent to char c[11] = { "I am happy" };。

其中，若使用字符串初始化字符数组，{}可以省略，故进一步简写为char c[]="I am happy";。因为使用了初始化，故字符数组定义时的长度省略了，我们看到的字符一共10个，注意空格算一个有效的字符，而字符串系统会为其加上'\0'结尾标志，也需要占一个存储空间，故该数组的实际长度会换算为11，等价于char c[11]={"I am happy"};。

In case of partial initialization, the string initialization character array form is: char c[10] = "China";. The storage space is represented as follows:

若是部分初始化，字符串初始化字符数组形式为：char c[10]="China";。存储空间表示如下：

C	h	i	n	a	\0	\0	\0	\0	\0

2. Input and Output of Strings（字符串的输入和输出）

When using scanf and printf to input and output strings, you can use % c with a for loop to output individual characters one by one, or you can use % s to input or output as a whole.

使用scanf和printf输入输出字符串时，可以使用%c结合for循环单个字符逐一输出，也可以使用%s整体输入或输出。

(1) Operation method of printf output string print.

（1）输出字符串的操作方法。

For the char c[10] = "China"; array, if the stored character string is output, % c can be used to control the character output one by one, and the code is as follows:

对于char c[10]="China";数组，若输出其存储的字符串，可以使用%c逐一控制字符输出，代码如下：

```
int i;
for(i=0;c[i]!='\0';i++)
{
    printf("%c",c[i]);
}
```

The output is: China

输出结果为：China

The C language provides a more simplified format control for output strings, namely %s.

C语言提供了一个更为简化的输出字符串的格式控制符，即%s。

The above code can be rewritten as:

上述代码可以改写为：

```
printf("%s",c);
```

Instructions for %s　%s 使用说明

①%s is used for printf output to control the output of the string. The output table lists the first address of the string. Since the array name is the first address, the array name is generally placed.

①%s用于printf输出控制字符串的输出，输出表列放字符串的首地址，因数组名即为首地址，故一般放数组名。

②Output the characters one by one until the '\0' end flag is encountered. Note that '\0' is not output, which is the end flag.

②逐一字符输出，直至遇到'\0'结束标志为止，注意'\0'不输出，为结束标志。

For the above case, the result shows China. When the system encounters the first '\0', it will end the output string, and the following will not be output.

对于上述案例，结果显示China，系统遇到第一个'\0'则结束输出字符串，后面的不再输出。

(2) Operation method of scanf input string scanf.

（2）输入字符串的操作方法。

Suppose there is an array char c[5];. Input a string from the keyboard and assign it to the array. There are the following methods:

假设有数组char c[5];，从键盘输入一个字符串赋值给该数组，有如下方法：

①Use %c to read in character by character: for(i=0; i<5;i++) scanf("% c", & c[i]);

①使用%c逐个字符读入：for(i=0; i<5;i++) scanf("%c",&c[i]);

②Use the entire string input using %s:

②使用%s整个字符串输入：

```
char c[6];
scanf("%s",c);
```

Enter from the keyboard: China

从键盘输入：China

Storage space is:

存储空间为：

C	h	i	n	a	\0

Note（说明）：

When %s is used, the system will automatically add the '\0' end flag at the end of the string, so the storage space needs one more end flag. In addition, the array name is listed in the address table of scanf without adding the & symbol, because the array name itself is the address of the first element of the array, which meets the requirements of the address table column of scanf.

在使用%s时，系统会自动在字符串末尾加上'\0'结束标志，故存储空间需要多一个结束标志的存储空间。另外，scanf的地址表列放数组名，不用加&符号，因为数组名本身就是地址，是数组首元素的地址，符合scanf的地址表列的要求。

③If you use the %s format to control multiple arrays to input multiple strings, separate the multiple strings with a space or a carriage return. If there is the following code:

③若使用%s格式控制多个数组输入多个字符串，输入时以空格或者回车隔开多个字符串。如果有如下代码：

```
char str1[5],str2[5],str3[5];
scanf("%s%s%s", str1, str2, str3);
```

Keyboarding:

键盘输入：

```
How are you?
```

The three character arrays are stored as follows:

则三个字符数组的存储情况如下：

H	o	w	\0		

a	r	e	\0		

y	o	u	?	\0	

When a space or carriage return is encountered, it represents the end of string input. According to the input format, str1stores How, str2 stores are, str3 stores you? , and a '\0' is added at the end of each string as the end flag.

遇到空格或者回车则代表字符串输入结束，根据输入格式str1存放How，str2存放are，str3存放you?，在每个字符串末尾补了一个'\0'作为结尾标志。

In the above two ways of inputting character strings, % c is

上述两种输入字符串的方式，%c是逐

to input characters one by one, which can effectively read in blank symbols such as space, carriage return or【Tab】key, but %s is to use these blank symbols as separators, so it cannot be read in effectively.

一输入字符，可以有效读入空格、回车或者【Tab】等空白符号，但是%s输入方式，是将这些空白符号作为隔开符，故不能有效读入。

5.3.3 String Handling Functions（字符串处理函数）

Strings are often used, but C does not have a string type, so the processing is more complex, so the C function library provides many string processing functions, to use these functions, you need to include the string.h header file, that is, add “# include < string. h >” at the beginning of the program. This section describes the common string processing functions strlen, strcat, strcpy, and strcmp.

字符串是经常被使用的，但C没有字符串类型，处理比较复杂，故C函数库提供了许多字符串处理函数，要使用这些函数需要包含string.h头文件，即在程序的开头加上“#include <string.h>”，本节介绍常用的字符串处理函数strlen、strcat、strcpy、strcmp函数。

1. strlen Function（strlen函数）

Function: calculate the actual length of the string.

Call form: strlen(character array)

Return value: an integer that is the actual length of the string.

If a one-dimensional array is defined to store the "China" string, the specific code is as follows:

功能：计算字符串实际长度。

调用形式：strlen(字符数组)

返回值：一个整数，即字符串的实际长度。

如定义了一个一维数组，存放"China"字符串，具体代码如下：

```
char str[10]="China";
printf("%d", strlen(str));
```

Program output（运行结果）:

```
5
```

Analysis: Use the strlen(str) function to find the actual length of the string, 5, and print the result using printf. Note that '\0' takes up storage space when it is stored, but it is not included in the string length when calculating the string size. For example, the "China" string requires 6 bytes of storage space, but the string size is 5.

解析：使用strlen(str)函数求得字符串的实际长度5，并使用printf输出这个结果。注意'\0'存储的时候占存储空间，但计算字符串大小时，不计入字符串长度中。如"China"字符串需要6个字节的存储空间，但字符串大小为5。

2. strcat Function（strcat函数）

Function: Concatenate two strings.

Call form: strcat(character array 1, character array 2);

Call result: concatenate string 2 to the end of string 1 and place the result in character array 1.

For example, if there are two character arrays str1 and str2, the definitions are as follows: char str1[30] = "People"; char str2[30] = "China";

If you want the string stored in str2 to be connected to the string in str1 and stored in str1, the calling format is:

功能：连接两个字符串。

调用形式：strcat(字符数组1,字符数组2);

调用结果：把字符串2连接到字符串1的后面，结果放在字符数组1中。

举例，若有两个字符数组str1和str2，定义如下：char str1[30]="People";char str2[30]="China";

若想将str2存储的字符串连接到str1字符串后面，并存储到str1中，调用格式为：

```
strcat(str1,str2);
```

Note（说明）:

strcat only needs to use the array names of two arrays. After calling, the "China" string will be spliced behind the "People" string to form a new string "People China" and stored in the str1 array. Note that the str1 array should be large enough, otherwise it will overflow. The result can be viewed using printf output, printf("%s" , str1);.

strcat只需使用两个数组的数组名即可，调用后"China"字符串会拼接到"People"字符串后面，形成新的字符串"PeopleChina"存放到str1数组中，注意str1数组应足够大，否则会产生溢出。可使用printf输出查看结果，printf("%s",str1);。

The complete code is as follows（完整的代码如下）:

```
#include <stdio.h>
int main()
{
    char str1[30]="People";
    char str2[30]="China";
    strcat(str1,str2);                    //注意使用方法
    printf("%s",str1);
    return 0;
}
```

Program output（运行结果）:

```
PeopleChina
```

3. strcpy Function（strcpy 函数）

Function: Duplicate(copy) strings.

Call form: strcpy(character array 1, character array 2)

Call result: copy that string in character array 2 to character array 1.

功能：复制（拷贝）字符串。

调用形式：strcpy(字符数组1,字符数组2)

调用结果：将字符数组2中字符串复制到字符数组1中。

For example（举例）:

```
char str1[10],str2[]="China";
strcpy(str1,str2);
```

Program output（运行结果）:

```
China
```

There are str1 array and str2 array storing China string in the above code. After calling the strcpy function, the value of str1 is the value of str2. Note that the space for str1 should be large enough to hold the copied values.

上述代码中有str1数组和存放China字符串的str2数组，调用strcpy函数后，str1的值就是str2的值。需要注意的是str1的空间要足够大用来存放复制的值。

4. strcmp Function（strcmp 函数）

Function: Compare string sizes.

Call form: strcmp(string 1, string 2)

Call result: return a positive integer if string 1 is greater than string 2, return a negative integer if string 1 is less than string 2, and return 0 if the two strings are equal.

功能：比较字符串大小。

调用形式：strcmp(字符串1,字符串2)

调用结果：若字符串1大于字符串2则返回正整数；若字符串1小于字符串2则返回负整数；若两个字符串相等，则返回0。

String Comparison Rule 字符串比较规则

The rule of character string comparison in C language is based on the ASCII values of characters. The specific rule is to compare the ASCII values of two character strings one by one. If the ASCII values of a pair of characters are not equal at a certain position, that is, to stop and return the result when the first pair of unequal characters is encountered. The size of the first pair of unequal characters is the size result of the whole character string. If all characters are equal, including length, the string is equal.

C语言中字符串比较的规则是基于字符的ASCII值进行的，具体规则为：逐个字符比较两个字符串的ASCII值，若在某个位置某对字符的ASCII值不相等，即遇到第一对不相等的字符则停止并返回结果，该第一对不相等的字符大小即为整个字符串的大小结果。如果所有字符都相等（包括长度），则字符串相等。

(1) If all characters are the same, the string is equal, such as: strcmp("abc", "abc");//The return value of the function is 0.

（1）若全部字符相同，字符串相等，如：strcmp("abc", "abc"); //函数返回值为0。

Both strings encounter the end of '\0' character by character from left to right. All characters are the same, so they are equal.

自左至右逐个字符相比，两个字符串均遇到了'\0'结束，全部字符都相同，故相等。

(2) If different characters appear, when the different characters are encountered for the first time, the size of the pair of characters is the comparison result of the string, such as strcmp("abc", "acc"); If the first pair of characters is the same as 'a', the comparison continues. If the second pair of characters is 'B'<'c', that is, the ASCII value of 'b' is less than the ASCII value of 'c', the return value of the function is a negative integer, and the comparison ends. At this time, the "abc" string is less than the "acc" string.

（2）若出现不相同的字符，当第一次遇到不相同字符时终止，该对字符大小就是字符串的比较结果，如：strcmp("abc", "acc")；第一对为字符'a'相同，继续往后比较，第二对字符'b'<'c'，即'b'的ASCII值小于'c'的ASCII值，函数返回值为负整数，结束比较，此时"abc"字符串小于"acc"字符串。

Example: strcmp("adc", "acc"); The second pair of characters 'd'>'c', so the function returns a positive integer.

举例：strcmp("adc", "acc"); 第2对字符'd'>'c'，故函数返回值为正整数。

Example: strcmp("ac", "acc"); The first string encounters the end of '\0'. The longer string is larger. It can also be understood that the first pair of unequal characters is '\0'<'c'.

举例：strcmp("ac", "acc"); 第1个字符串遇到'\0'结束，长的那个字符串较大。也可以理解成第一对不相等的字符为'\0'<'c'。

Precautions for Using String 字符串使用注意事项

Strings cannot directly use relational operators to compare size relationships, such as:

字符串不可以直接使用关系运算符比较大小关系，如：

```
if(str1>str2) printf("yes");                    // 错误
```

To use strcmp for comparison, change to:

要使用strcmp来进行比较，应改为：

```
if(strcmp(str1,str2)>0) printf("yes");
```

Note（说明）:

You must use strcmp to determine the size of the string by judging the sign of its return value.

必须使用strcmp进行比较，通过判断其返回值正负号，来确定字符串的大小关系。

5.4 Comprehensive Cases 综合案例

综合案例 5_4_1 交换一维数组最大值和最小值。

案例描述： 定义一个一维数组，交换数组中的最大值和最小值并输出查看数组数据。

解题思路： 在题目没有具体指定一维数组的数据类型的情况下，我们可以将其视为整型数组进行处理。我们的任务是交换数组中的最大值和最小值，首先需要确定数组中的最大值和最小值，这一步骤可以采用“打擂台”算法来完成，同时标记最大值和最小值对应的数组下标，接下来交换它们的值来实现最终的目标。

程序代码：

Case5_4_1.c

```
1   #include <stdio.h>
2   #define N 10                                      // 定义符号常量
3   int main()
4   {
5       int a[N];                                     // 定义整型数组，长度为 10
6       int max,min,i,m,n;                            // 定义程序所需变量
7       // 输入一维数组元素值
8       printf(" 输入 10 个数组元素 :\n");
9       for(i=0;i<N;i++) scanf("%d",&a[i]);
10      // 打擂台算法求最大值和最小值
11      max=min=a[0];                                 // 假设 a[0] 为当前的最大值、最小值
12      m=n=0;                                        //m 标记最大值下标，n 标记最小值下标
13      for(i=1;i<N;i++)
14      {
15          if(max<a[i]) {max=a[i]; m=i;}             // 打擂台算法
16          if(min>a[i]) {min=a[i]; n=i;}             // 打擂台算法
17      }
18      // 交换数组元素
19      a[m]=min;
20      a[n]=max;
21      printf(" 输出交换后的数组元素 : \n");
22      for(i=0;i<N;i++) printf("%d", a[i]);
23      return 0;
24  }
```

运行结果1：

```
输入 10 个数组元素:
1 2 3 4 5 6 7 8 9 10
输出交换后的数组元素:
10 2 3 4 5 6 7 8 9 1
```

运行结果2：

```
输入 10 个数组元素:
2 3 4 1 5 6 7 8 10 9
输出交换后的数组元素:
2 3 4 10 5 6 7 8 1 9
```

案例解析：

（1）第2行定义了符号常量N，作为第5行数组定义时的长度，因数组要求必须是常量，使用符号常量

的好处是，可以用N代替10，若想修改为其他数值，可以一改全改。

（2）第5行定义了一个整型数组，长度为10，第6行定义了所需要的变量，max用来存放最大值，min用来存放最小值，m用于存放最大值下标，n用于存放最小值下标，i作为循环变量控制一维数组下标的变化。

（3）第9行使用for循环依次读入数组元素。

（4）第11行令max和min的初值为a[0]元素的值，同时使用了m、n标记最大值和最小值下标，此时均为0。

（5）第13~17行为打擂台算法实现，和剩下的元素进行比较，求得最大值max及最大值下标m、最小值min和最小值下标n。

（6）第19、20行交换了最大值和最小值。最后输出查看结果。

案例拓展及注意事项：

（1）max和min赋予初值的同时，m和n也要赋予初值，否则结果同样不准确；同样地，这4个变量在“打擂台”比较过程中，均要修改。

（2）交换最大值和最小值时，不能交换max和min变量，因max和min与数组是属于不同的存储空间，若交换max和min值，并不能交换数组元素。max和min只是用于辅助存储当前的最大值和最小值，辅助“打擂台”算法的完成。

（3）程序可以优化，可以不定义max和min，只需要用下标标记即可。代码如下：

```
1   #include <stdio.h>
2   #define N 10
3   int main()
4   {
5       int a[N];
6       int i,m,n,temp;
7       // 输入一维数组元素值
8       printf(" 输入 10 个数组元素 :\n");
9       for(i=0;i<10;i++) scanf("%d",&a[i]);
10      m=n=0;                          // 赋予最大值下标和最小值下标初值 0，假设为 a[0]
11      for(i=1;i<10;i++)
12      {
13          if(a[m]<a[i]) m=i;
14          if(a[n]>a[i]) n=i;
15      }
16      // 交换数组元素
17      temp=a[m];
18      a[m]=a[n];
19      a[n]=temp;
20      printf(" 输出交换后的数组元素:\n");
21      for(i=0;i<10;i++) printf("%d ",a[i]);
22      return 0;
23  }
```

解析：第10行，赋予最大值和最小值下标为0，假设a[0]为最大值和最小值。第13行比较a[m]和a[i]的大小，若a[i]大，则修改m为i，令m始终标记最大值下标，对应最大值为a[m]。第14行的最小值下标同样的处理过程，不再赘述。在第17行和第19行为交换最大值和最小值，交换最大值a[m]和最小值a[n]，temp作为辅助变量辅助交换过程。

综合案例 5_4_2 矩阵加法运算。

案例描述： 定义两个矩阵，将两个矩阵相加，输出相加后的新矩阵。

解题思路： 矩阵可使用二维数组表示，矩阵相加转换为二维数组对应数组元素的相加。为了更具有一般性，我们将矩阵处理成float型。

程序代码：

Case5_4_2.c

```
1    #include <stdio.h>
2    #define N 3
3    int main()
4    {
5        float a[N][N],b[N][N];                  // 定义两个数组
6        int i,j;                                // 注意下标为整数，不要和数组一样，定义成flaot型
7        printf("读入数组a:\n");
8        for(i=0;i<N;i++)                        // 输入a数组
9            for(j=0;j<N;j++)
10               scanf("%f",&a[i][j]);
11       printf("读入数组b:\n");
12       for(i=0;i<N;i++)                        // 输入b数组
13           for(j=0;j<N;j++)
14               scanf("%f",&b[i][j]);
15       printf("求得的加法矩阵:\n");
16       for(i=0;i<N;i++)
17       {
18           for(j=0;j<N;j++)
19           {
20               printf("%.2f\t",a[i][j]+b[i][j]);
21           }
22           printf("\n");                       // 按照矩阵格式输出显示，一行后换行
23       }
24       return 0;
25   }
```

运行结果：

```
输入数组a:
1.1   2   1.2
2    1   2.2
3    3    3
输入数组b:
1.1  1.1  1.1
2    2    2
1    1    1
求得的加法矩阵:
2.20     3.10     2.30
4.00     3.00     4.20
4.00     4.00     4.00
```

案例解析：

（1）第5行定义了两个二维数组，用于存放两个3*3的矩阵。当然可以在定义的同时初始化，本案例为了更有普适性，从键盘读入任意两个矩阵。

（2）特别注意下标i和j务必定义成整数，无论数组是何类型，下标要求为整数类型。

（3）第8~10行和第12~14行为两个并列的双重for循环，一个控制二维数组a输入，一个控制二维数组b的输入，注意二维数组是逐个元素输入的，使用双重循环分别控制行列下标。

（4）第16~23行为一个双重for循环，控制a数组和b数组对应行列值的元素相加，并输出，按照矩阵的格式输出查看，输出一行后需换行。

（5）第20行使用了%.2f令结果保留小数点后2位输出显示，后'\t'为制表符，可以令矩阵对齐显示，格式更为整洁。

综合案例 5_4_3 打印杨辉三角形。

案例描述：杨辉三角是中国数学史上的一个伟大成就，最早由中国南宋数学家杨辉在1261年所著的《详解九章算法》一书中出现，而欧洲数学家帕斯卡在1654年也独立发现了这一规律。这一发现中国领先近400年，体现了中国人的无穷智慧。

要求编写程序打印如下的杨辉三角形：

```
1
1  1
1  2  1
1  3  3   1
1  4  6   4   1
1  5  10  10  5  1
```

解题思路：这是典型的图形打印问题，适合通过循环结构解决。杨辉三角是一个下三角形，其第0列和对角线元素均为1，其余元素是前一行的前一列和同列元素之和。二维数组能完美映射其结构，方便访问和修改元素。通过正确初始化与更新二维数组，可轻松生成并打印杨辉三角。

程序代码：

Case5_4_3.c

```
1   #include <stdio.h>
2   #define N 6                                  //定义符号常量
3   int main()
4   {
5       int a[N][N],i,j;                         //定义二维数组和行列下标
6       for(i=0;i<N;i++)
7       {
8           a[i][0]=1;
9           a[i][i]=1;
10      }
11      for(i=2;i<N;i++)                         //从第 2 行开始计算
12      {
13          for(j=1;j<i;j++)
14          {
15              a[i][j]=a[i-1][j]+a[i-1][j-1];
16          }
17      }
18      printf("杨辉三角形为:\n");
19      for(i=0;i<N;i++)
20      {
21          for(j=0;j<=i;j++)
22              printf("%3d",a[i][j]);
23          printf("\n");
24      }
25      return 0;
26  }
```

案例解析：

（1）第2行定义了符号常量N，题目要求打印6行，故定义为6。

（2）第5行定义了一个二维数组和行列下标循环变量。

（3）第6~10行将二维数组a的第0列和对角线上面的元素a[i][i]赋值为1。

（4）第11~17行，计算其他元素，规律为其上一行的同一列和前一列数据之和，如要计算第i行和第j列的数据a[i][j]，其前一行为i-1，前一行同列数据为a[i-1][j]，前一行的前一列为a[i-1][j-1]，即a[i][j]=a[i-1][j]+a[i-1][j-1]。

（5）第19~24行按要求打印二维数组，即可输出杨辉三角形。其中，第21行条件设为j<=i，即可打印下三角。

综合案例 5_4_4 选择排序。

案例描述：排序在解决实际问题中是常规操作，本案例要求对10个数据从大到小排序。

解题思路：选择排序法是一种简单直观的排序算法。它的工作原理是：每一次从未排序的数据元素中选出最大（或最小）的元素，存放在序列的起始位置，然后，再从剩余未排序元素中继续寻找最大（或最小）的元素，然后放到已排序序列的末尾。依此类推，直到全部待排序的数据元素排完。

我们通过图示法来直观理解和分析选择法排序的具体过程。假设待排序序列为一个整型一维数组a[10]={0,4,2,3,5,1,7,9,8,6},现要求从大到小排序，则需要每次在未排序的序列中找到最大值排序到有序序列的末尾。图示中用k来标记未排序序列的最大值下标。

第1轮（Round1）：此时所有数据均为未排序序列，令k标记未排序序列的第一个元素，即k为0，接下来a[k]与剩下的元素依次比较，若遇到某个元素比当前的a[k]大，则修改k的值为该元素的下标，如a[1]比当前的a[k]大，则修改k为1，继续比较后续的元素与当前的a[k]即a[1]比较，若有元素比当前大，则修改k的值，k依次被修改为1、4、6、7，最终k值为7，代表a[7]为当前的最大值。接下来交换a[0]和a[k]（a[7]）的值，则完成第一轮选择排序，如图5-1所示。

第2轮（Round2）：此时有序序列为a[0]，未排序序列为a[1]~a[9]。设定k=1作为起始未排序元素的下标，依次比较a[k]与其后元素，若遇更大值则更新k，则继续比较当前的a[k]和后续元素的值，若有大者，则继续更新k。遍历后，k标记了未排序序列中的最大值a[8]的下标8。交换a[1]与a[k]（a[8]）的值，完成第2轮选择排序，如图5-2所示。

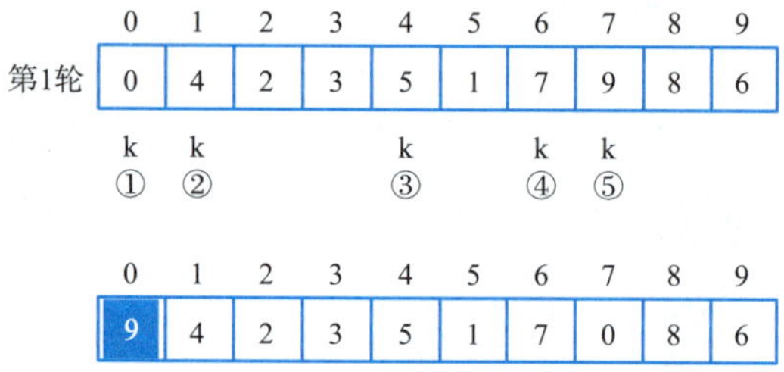

图5-1 第1轮选择排序

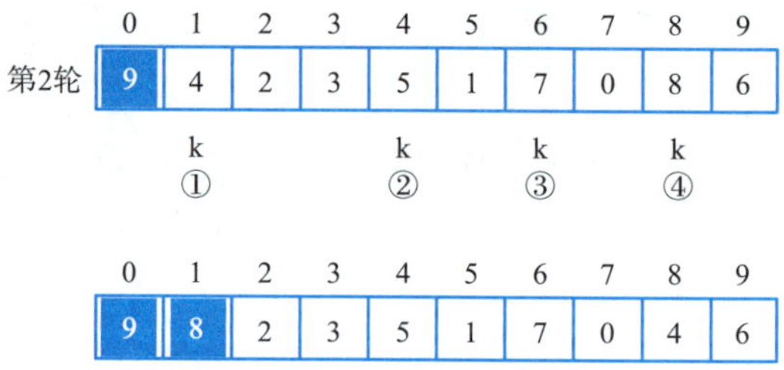

图5-2 第2轮选择排序

第3轮（Round3）：此时有序序列为a[0]~a[1]，未排序序列为a[2]~a[9]。设定k=2作为起始未排序元素的下标，依次比较a[k]与其后元素，若遇更大值则更新k，如a[3]比a[k]（a[2]）大，则令k=3。继续比较当前的a[k]（a[3]）和后续元素的值，若有大者，则继续更新k。遍历后，k标记了未排序序列中的最大值a[6]的下标6。交换a[2]与a[k]（a[6]）的值，完成第3轮选择排序，如图5-3所示。

第4轮（Round4）：此时有序序列为a[0]~a[2]，未排序序列为a[3]~a[9]。设定k=3作为起始未排序元素的下标，继续上述比较和更新k的过程。遍历后，k标记了未排序序列中的最大值a[9]的下标9。交换a[3]与a[k]（a[9]）的值，完成第4轮选择排序，如图5-4所示。

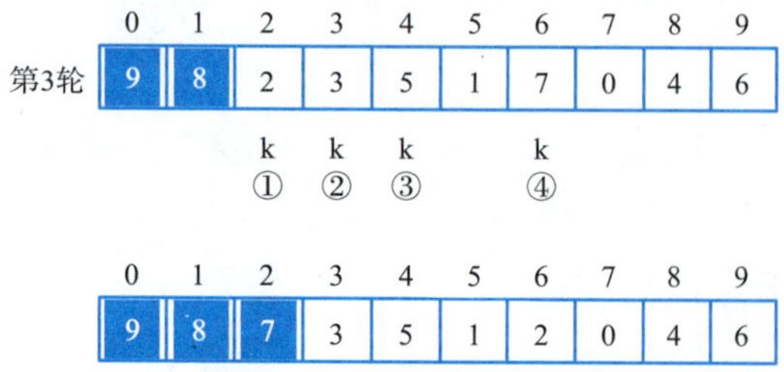

图5-3 第3轮选择排序

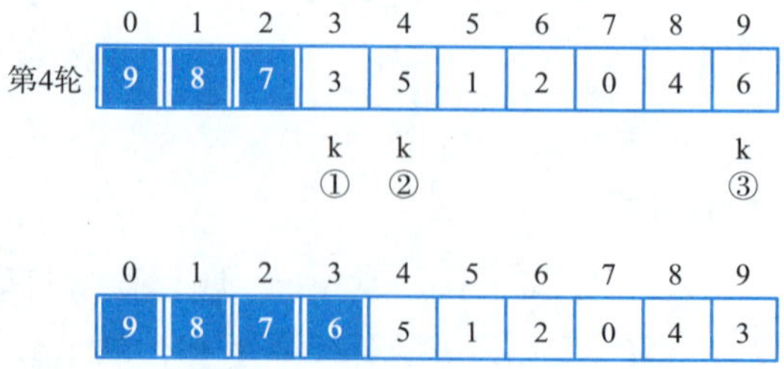

图5-4 第4轮选择排序

第5轮（Round5）：此时有序序列为a[0]~a[3]，未排序序列为a[4]~a[9]。设定k=4作为起始未排序元素的下标，继续上述比较和更新k的过程。遍历后，k标记了未排序序列中的最大值a[4]的下标4。交换a[4]与a[k]（a[4]）的值，完成第5轮选择排序，如图5-5所示。

第6轮（Round6）：此时有序序列为a[0]~a[4]，未排序序列为a[5]~a[9]。设定k=5作为起始未排序元素的下标，继续上述比较和更新k的过程。遍历后，k标记了未排序序列中的最大值a[8]的下标8。交换a[5]与a[k]（a[8]）的值，完成第6轮选择排序，如图5-6所示。

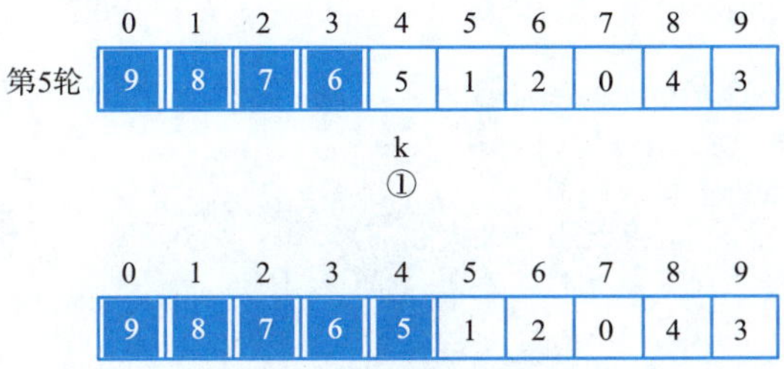

图5-5 第5轮选择排序

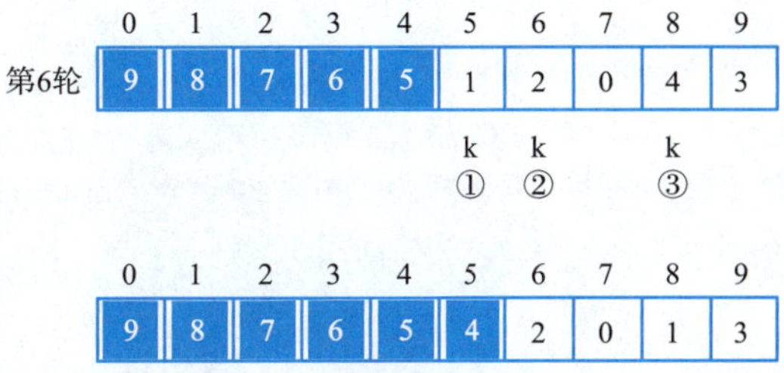

图5-6 第6轮选择排序

第7轮（Round7）：此时有序序列为a[0]~a[5]，未排序序列为a[6]~a[9]。设定k=6作为起始未排序元素的下标，继续上述比较和更新k的过程。遍历后，k标记了未排序序列中的最大值a[9]的下标9。交换a[6]与a[k]（a[9]）的值，完成第7轮选择排序，如图5-7所示。

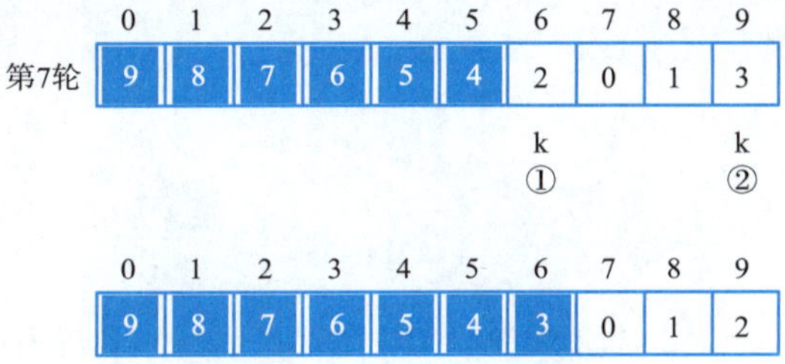

图5-7 第7轮选择排序

第8轮（Round8）：此时有序序列为a[0]~a[6]，未排序序列为a[7]~a[9]。设定k=7作为起始未排序元素的下标，继续上述比较和更新k的过程。遍历后，k标记了未排序序列中的最大值a[9]的下标9。交换a[7]与a[k]（a[9]）的值，完成第8轮选择排序，如图5-8所示。

第9轮（Round9）：此时有序序列为a[0]~a[7]，未排序序列为a[8]~a[9]。设定k=8作为起始未排序元素的下标，继续上述比较和更新k的过程。遍历后，k标记了未排序序列中的最大值a[8]的下标8。交换a[8]与a[k]（a[8]）的值，完成第9轮选择排序，如图5-9所示。

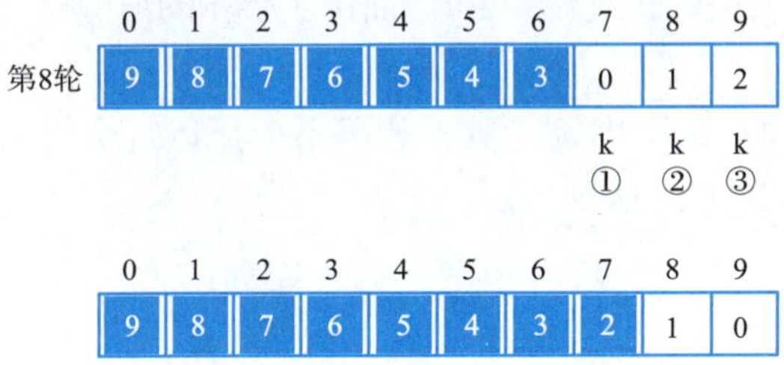

图5-8 第8轮选择排序

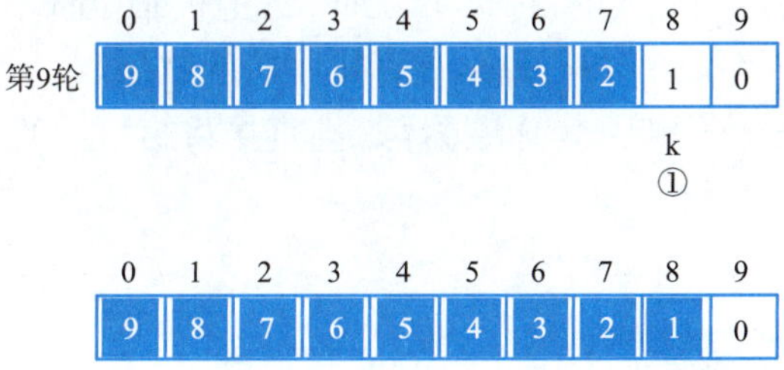

图5-9 第9轮选择排序

最后一个元素自然有序，故10个元素，只需要进行9轮排序即可，每一轮排序确定一个元素在其最终的排序位置。从上述描述中我们可以观察到，最终排序位置实际上为有序序列的末尾、未排序序列的第一个元素，故可设一个变量i标记该位置，其值从0变化到8。概括地讲，假设有N个元素，则只需要进行N-1轮排序，每次排序i从0取到N-2即可，故循环条件可以写为i<N-1，既可控制i的变换又可执行N-1轮排序过程。

程序代码：

Case5_4_4.c

```
#include <stdio.h>
#define N 10                    // 定义符号常量
```

```
3    int main()
4    {
5        int a[N]={0,4,2,3,5,1,7,9,8,6};
6        int i,j,temp,k;               //i 和 j 为循环变量，temp 辅助交换，k 标记当前最小值下标
7        for(i=0;i<N-1;i++)            //N-1 轮，待排序的第 i 个位置
8        {
9            k=i;                      //标记初始的最小值下标为 i
10           for(j=i+1;j<N;j++)        //for 循环第一条 if 语句
11               if(a[k]<a[j])         //若当前的 a[k] 比 a[j] 小
12                   k=j;              //将 j 赋值给 k，k 存放当前最大值下标
13           temp=a[k];                //以下三行作用是交换 a[i] 和 a[k]
14           a[k]=a[i];
15           a[i]=temp;
16       }
17       printf("排序后:\n");
18       for(i=0;i<N;i++)              //输出排序后结果
19         printf("%-3d",a[i]);
20       return 0;
21   }
```

运行结果：

```
排序后:
9  8  7  6  5  4  3  2  1  0
```

案例解析：

（1）第5行定义了一个有N个元素的一维数组，并初始化。这个序列为待排序的序列。

（2）第7行为外循环，控制排序过程进行N−1轮，本案例中有10个数据，应进行9轮排序，每轮排序过程选择一个最大的数放到有序序列的末尾。

（3）第9行的k=i表示每轮排序标记未排序序列的第一个元素。通过打擂台算法与剩下的未排序序列依次比较，找到当前未排序序列的最大值下标，参考第10~12行。

（4）第13、14行为交换a[i]和a[k]的操作过程，将每轮未排序序列的最大值放到a[i]的位置，使用交换可以实现。交换过程需要使用一个辅助变量temp，先将a[k]的值存到temp中，再将a[k]重新赋值为a[i]，此时a[k]的值修改为a[i]的值，a[i]再赋值为temp的值，即可达到交换的目的。若无辅助变量，直接执行a[k]=a[i];a[i]=a[k];并不能达到交换的目的，因为a[k]的值被a[i]覆盖掉了，数据丢失，最终结果a[k]和a[i]均为a[i]的值。

（5）第17~19行输出查看排序后的结果。

（6）若想从小到大排序，则直接将第11行的“>”替换成“<”即可。其他代码不变。运行结果为：

```
排序后:
0  1  2  3  4  5  6  7  8  9
```

综合案例 5_4_5　回文字符串。

案例描述：回文字符串也称“回文串”，是指一个正读和反读都一样的字符串。这种字符串具有中心对称的特性，即从左到右和从右到左读取的结果是相同的。例如，"level""abcdcba""12321""madedam"都是回文串。请编写程序从键盘任意读入一个字符串，判断其是否为回文字符串。

解题思路：首先需要一个足够大的char型一维数组，使用scanf和printf结合%s格式控制符读入和输出字符串。判断是否为回文串，可比较字符串的首尾字符，逐步向内推进，直至中间，若存在不相等的字符则不是回文字符串，否则为回文字符串。

程序代码：

Case5_4_5.c

```
1    #include <stdio.h>
```

```
2   #include <string.h>                    //使用了 strlen 函数，需要包含该头文件
3   int main()
4   {
5       char str[80];                      //定义一个一维数组存储一个字符串
6       printf("请输入字符串:\n");
7       scanf("%s",str);
8       int n;
9       n=strlen(str);
10      int i=0,j=n-1;
11      int flag=1;                        //标志位
12      while(i<j)
13      {
14          if(str[i]==str[j])             //若相等则首尾下标向内推进
15          {
16              i++;j--;
17          }
18          else
19          {
20              flag=0;                    //若不有一对字符不相等，则令 flag 标志位的状态变化
21              break;                     //结束所在的循环
22          }
23      }
24      if(flag) printf("是回文字符串!\n");          //根据标志位判断是否为回文字符串
25      else printf("不是回文字符串! \n");
26      return 0;
27  }
```

运行结果示例：

```
请输入字符串:
madedam
是回文字符串!
```

运行结果示例：

```
请输入字符串:
abc
不是回文字符串!
```

案例解析：

（1）第2行，因为第9行使用了strlen函数求字符串长度，故在程序的开头应包含string.h头文件。

（2）第5行，定义了一个char型的一维数组，用于存储一个字符串，长度应定义得相对足够大。

（3）第6行为提示语句，第7行使用scanf函数输入一个字符串，注意%s的用法，后面直接跟上数组名即可。scanf使用空白字符作为隔开符号，如空格或回车。

（4）第9行使用strlen函数计算字符串的实际长度，存放到变量n中，因为字符串是从键盘任意读入的，长度未知，而我们需要知道具体长度，方便判断首尾字符是否相同。

（5）第10行的i和j两个变量作为字符串的首尾下标，用于后续比较，因下标是从0开始的，故i=0,j=n-1。

（6）第11行的flag是常用的方法，该变量用于作为状态判断，也称标志位，或者形象地称为“哨兵”。用于标志某种状态经过程序结构（如循环）后是否发生改变。本程序是用于标记是否有不相等的字符，一开始假设没有不相等的字符，状态标记为1。

（7）第12~23行为一个while循环，该循环若满足i<j的条件，则不断循环，i和j为字符串的首尾下标，往中间靠拢，若到最中间或者发生i>j的情况，则结束循环，说明已经判断完毕。

（8）第14~22行，即while的循环体部分为一个if...else语句，if条件为判断当前的这对首尾字符是否相等，注意是两个等号，若相等，则i和j继续往中间靠拢，即i++, j--，若不相等则改变flag的状态，flag=0，同时

跳出循环，此时说明不是回文字符串，无须继续判断其他字符。

（9）第24、25行为if...else语句，条件为flag，根据“非零即为真”的原则，若经过while循环后flag仍为1，说明没有不相等的字符，输出其为回文字符串，若flag为0，则说明有不相等的字符，输出不是回文字符串的提示信息。

Summary 小结

(1) In the past, we introduced basic data types such as int, float, double, and char. Arrays are constructed types, which are new data types composed of basic data types according to certain rules. C constructs types such as pointers and structures, which will be learned later. An array is an ordered collection of data of the same type. We will first give its characteristics, and then learn more about it later. The data in an array is called an element, and each element belongs to the same data type. The array name and subscript can uniquely determine the elements in the array. The computer subscript uses[] to represent it, such as a[2]. In general, using arrays in combination with loops is more efficient.

(2) In this chapter, we learned the origins of arrays, their definitions and characteristics, as well as the definition, storage, reference, and initialization of one-dimensional arrays. We asked you to understand the uses of arrays and master the definition, reference, and initialization of one-dimensional arrays.

(3) For two-dimensional arrays, we learned the definition, storage, reference of two-dimensional array elements, and initialization of two-dimensional arrays, and introduced an application example of two-dimensional arrays. In addition, two-dimensional arrays can often be used for matrix operations.

(4) When using character arrays and strings, it is important to note that each element stores a single character, and strings end with a '\0' marker. To master common string processing functions, it is necessary to include the string.h header at the beginning of the program.

（1）以往我们介绍的是基本数据类型如int、float、double、char，数组是构造类型，构造类型是由基本数据类型按一定规律组成的新的数据类型。C构造类型还有指针、结构体等，日后会学习到。数组是相同类型数据的有序集合，我们先给出其特点，后续学习可慢慢体会。数组中的数据称为元素，每一个元素都属于同一个数据类型。数组名和下标可唯一确定数组中的元素，计算机下标使用[]表示，如a[2]。一般情况下，数组结合循环来使用效率会更高。

（2）在本章中我们学习了数组的来源，数组的定义、特点，以及一维数组的定义、存储、引用及初始化，要求大家理解数组的用途，并掌握一维数组的定义、引用和初始化。

（3）关于二维数组，我们学习了二维数组的定义、存储、二维数组元素的引用以及二维数组初始化，并介绍了一个二维数组的应用实例，除此之外，二维数组往往可以用于矩阵运算。

（4）在字符数组和字符串的使用方面，要注意每个元素存放一个字符，字符串以'\0'为结束标志，掌握常用的字符串处理函数，需要在程序开头加上包含string.h命令。

Exercise 习题

一、单选题

1. 下面错误的初始化语句是（　　）。

A. char str[]="hello"；　　B. char str[100]="hello"；

C. char str[]={'h','e','l','l','o'}；　　D. char str[]={'hello'}；

2. 定义了一维int型数组a[10]后，下面错误的引用是（　　）。

A. a[0]=1;　　B. a[10]=2;　　C. a[0]=5*2;　　D. a[1]=a[2]*a[0];

3. 下面的二维数组初始化语句中，正确的是（　　）。

A. float b[2][2]={0.1,0.2,0.3,0.4};　　B. int a[][]={{1,2},{3,4}} ;

C. int a[2][]={ {1,2},{3,4}};　　D. float a[2][2]={0};

4. 引用数组元素时，数组下标可以是（　　）。

A. 整型常量　　B. 整型变量　　C. 整型表达式　　D. 以上均可

5. 定义了一个int型二维数组a[6][7]后，数组元素a[3][4]前的数组元素个数是（　　）。

A. 24　　B. 25　　C. 18　　D. 17

6. 以下对一维数组的定义中不正确的是（　　）。

A. double x[5]={2.0,4.0,6.0,8.0,10.0};　　B. int y[5]={0,1,3,5,7,9};

C. char ch1[ ]={'1', '2', '3', '4', '5'};　　D. char ch2[ ]={'\x10', '\xa', '\x8'};

7. 假定一个int型变量占用2字节，若有定义:int x[10]={0,2,4};，则数组x在内存中所占字节数是（　　）。

A. 3　　B. 6　　C. 10　　D. 20

8. 以下对二维数组的定义中正确的是（　　）。

A. int a[4][]={1，2，3，4，5，6};　　B. int a[][3];

C. int a[][3]= {1，2，3，4，5，6};　　D. int a[][],{{1,2,3},{4,5,6}};

9. 若有定义：int a[2][4];,则引用数组元素正确的是（　　）。

A. a[0][3]　　B. a[0][4]　　C. a[2][2]　　D. a[2][2+1]

10. 若有定义int aa[8];，则不能代表数组元素aa[1]地址的是（　　）。

A. &aa[0]+1　　B. &aa[1]　　C. &aa[0]++　　D. aa+1

二、程序分析题

1. 下列程序执行后的输出结果是什么？

```
#include <stdio.h>
int main()
{
    int a[3][3], *p,i;
    p=&a[0][0];
    for(i=0;i<9;i++)
    p[i]=i+1;
    printf("%d \n",a[1][2]);
    return 0;
}
```

2. 以下程序的输出结果是什么？

```
#include <stdio.h>
int main()
{
    int b[3][3]={0,1,2,0,1,2,0,1,2},i,j,t=0;
    for(i=0;i<3;i++)
        for(j=i;j<=i;j++)
              t=t+b[i][b[j][j]];
    printf("%d\n",t);
}
```

3. 以下程序的输出结果是什么？

```
#include <stdio.h>
int main()
```

```
{
    int m[][3]={1,4,7,2,5,8,3,6,9};
    int i,j,k=2;
    for(i=0;i<3;i++)
    {
        printf("%d ",m[k][i]);
    }
}
```

4. 以下程序的输出结果是什么？

```
#include <stdio.h>
int main()
{
    int a[4][4]={{1,3,5},{2,4,6},{3,5,7}};
    printf("%d%d%d%d\n",a[0][3],a[1][2],a[2][1],a[3][0]);
    return 0;
}
```

5. 以下程序的输出结果是什么？

```
int main()
{
    int a[][]={{1,2,3},{4,5,6}};
    printf("%d", a[1][1]);
    return 0;
}
```

三、编程题

1. 编写一个程序，用于存储10名学生的成绩，并计算平均分、最高分和最低分，最后输出这些统计结果。

2. 设计一个程序，该程序使用一个数组来保存一周七天的名称。用户输入一个1~7之间的数字，程序将输出对应的星期名称。

3. 请编写一个程序，根据给定的包含N个人年龄的数组，统计并归类不同年龄段的人数。年龄段分为0~9岁、10~19岁、20~29岁等，100岁及以上为一个单独段。程序需将统计结果存储在另一个数组中并输出。

4. 模拟校园歌手大赛评分过程，10名评委为3位选手打分，分数范围在0~100之间。每位选手的最终得分为去掉一个最高分和一个最低分后的平均分。程序需使用数组存储评分，并显示每位选手的最终得分。

5. 编写一个程序，从键盘读取一个4行3列的矩阵，计算并输出矩阵中的最大值和最小值及其所在的行和列。如有重复值，需输出所有位置。

6. 编写一个程序，用于在给定的m×n矩阵中找出所有鞍点。鞍点是指在所在行中最小且在所在列中最大，或在所在行中最大且在所在列中最小的元素。

7. 创建一个程序，接收一串字符，统计并输出每个字符出现的频率。程序需使用字符数组来存储输入的字符。

8. 设计一个程序，允许用户输入一个字符串，然后将其逆序存储于数组中，并输出逆序后的字符串。

9. 编写一个程序，接收一个由单词组成的字符串数组，按照字典顺序对这些单词进行排序，并输出排序后的单词列表。

10. 编写一个程序，接收一个整数数组，删除其中的重复元素（保留一个），并输出处理后的数组。

11. 编写一个程序，接收年月日作为输入，计算并输出这一天是该年的第几天。

12. 在不使用strlen函数的情况下，编写一个程序来计算并输出一个字符串的实际长度。

13. 在不使用strcat函数的前提下，编写一个程序来连接两个字符串，并输出结果。

14. 在不使用strcpy函数的前提下，编写一个程序实现字符串复制功能，并输出结果。

15. 在不使用strcmp函数的情况下，编写一个程序来比较两个字符串的大小，并输出结果。

Chapter 6
Pointer

第6章
指　针

Introduction 主要内容

Pointer is the characteristic of C language and the essence of C language. Using it correctly and flexibly can make the program concise, compact and efficient. The tedious redefinition and reassignment can be eliminated. This chapter explains what is a pointer, the definition and use of pointer variables, the type of pointer, the meaning of addition and subtraction of pointers, how pointers reference array elements, and the comparison of different types of pointers.

指针是C语言的特色，是C语言的精华。正确而灵活地运用指针，可以使程序简洁、紧凑、高效，可以省去烦琐的重新定义与重新赋值。本章主要内容讲解了什么是指针、指针变量的定义及使用、指针指向的类型、指针的加减法的意义、指针如何引用数组元素，以及不同类型指针的比较。

Objectives 学习目标

Learning to use pointers correctly is a key step to master the C language. The close relationship between pointers and addresses is the basis for understanding how they work. The flexible definition of different types of pointers according to different requirements can not only make the code more concise, but also significantly improve the execution efficiency of the program. Through in-depth study of the concept and application of pointers, we are able to write beautiful and efficient C language programs.

学会正确运用指针，是掌握C语言的关键一步。指针与地址的紧密关系，是理解其工作原理的基础。根据不同需求，灵活定义不同类型的指针，不仅能够使代码更加简洁，还能显著提升程序的执行效率。通过深入学习指针的概念和应用，我们能够编写出既美观又高效的C语言程序。

6.1 Pointers and Pointer Variables 指针和指针变量

6.1节视频

Pointers store memory address information and direct us to

指针存储了内存地址信息，

access and manipulate data directly. As the essence of C language, the correct use of pointers can make the program more concise, compact and efficient. Although high-level languages may not expose pointers directly, they still have the idea of pointers implicit within them. In the following study, we will discuss the definition and use of pointers in depth to help you master this powerful tool and write more efficient C language programs.

指引我们直接访问和操作数据。作为C语言的精髓，指针的正确运用能让程序更简洁、紧凑且高效。尽管高级语言可能不直接暴露指针，但它们内部依然隐含着指针的思想。在后续的学习中，我们将深入探讨指针的定义、使用，帮助大家掌握这一强大工具，编写出更高效的C语言程序。

6.1.1 What Is a Pointer（什么是指针）

In the C language, different data types determine the size of the storage space occupied by variables. For example, when we define an integer variable(int) or a character variable(char), the system will automatically allocate the corresponding memory space for these variables. Usually, an int variable will occupy 4 bytes, while a char variable will occupy 1 byte. Each byte has a unique number, which we call the address. It's like every classroom in a school has a unique classroom number, through which we can easily find the corresponding class location.

在C语言中，不同的数据类型决定了变量所占用的存储空间大小。例如，当我们定义一个整型变量（int）或字符型变量（char）时，系统会自动为这些变量分配相应的内存空间，通常int型变量会占用4字节，而char型变量则占用1字节。每个字节都有一个唯一的编号，称为地址。这就像学校里的每个课室都有一个唯一的课室号，通过这个课室号，我们可以轻松地找到对应的上课地点。

In memory, each byte has its own address, which is like a "compass" to guide us to a specific storage space. We therefore refer to these addresses figuratively as "pointers". Through the pointer, we can find the storage space of the variable, and then read or modify it.

在内存中，每个字节都有自己的地址，这些地址就像是指引我们找到具体存储空间的"指南针"。因此，我们形象地将这些地址称为"指针"。通过指针，我们能够找到变量的存储空间，进而对其进行读取或修改。

A pointer variable is a special kind of variable that stores the addresses of other variables. Through a pointer variable, we can indirectly access and manipulate the data in the memory space it points to. This way of accessing data indirectly through addresses is called "indirect access". Direct access to memory space through variables is often referred to as "direct access". Compared with "direct access", "indirect access" provides greater flexibility and efficiency.

指针变量则是一种特殊的变量，它的作用是存储其他变量的地址。通过指针变量，我们可以间接地访问和操作它所指向的内存空间中的数据。这种通过地址间接访问数据的方式称为"间接存取"。通过变量直接访问内存空间的方式，常称为"直接存取"。与"直接存取"相比，"间接存取"提供了更大的灵活性和效率。

"Direct access" and "indirect access" are intuitively understood through an example.

下面通过一个示例，直观理解"直接存取"和"间接存取"。

1. Direct Access（直接存取）

In the C language, the storage of variables is very straightforward. When we define and initialize variables, the compiler will allocate the corresponding memory space for these variables, and associate the variable name with the first address of the allocated memory space. For example, there are codes "int i=2, j=6, k;". Each variable takes up 4 bytes of memory. Although these memory cells may not be contiguous in physical memory, we can assume that they are contiguous

在C语言中，变量的存储方式是非常直接的，当我们定义并初始化变量时，编译器会为这些变量分配相应的内存空间，并将变量名与所分配的内存空间首地址相关联。例如，有代码"int i=2,j=6,k;"，每个变量都会占用4字节的内存单元。尽管这些内存单元在物理内存中可能不是连

for the sake of explanation. As shown in Figure 6-1.

Assuming that the first address of variable i is 2000, its address range is 2000 to 2003. Similarly, the first address of variable j is 2004, and the first address of variable k is 2008. Each variable occupies a length of 4 bytes.

When we want to output the value of variable i to the display, the compiler will find the memory space according to the first address(i.e. 2000) corresponding to the variable name i, and read the data in the corresponding memory unit(i.e. 2). Finally, the value is output to the display.

Similarly, if we execute a statement such as "k = i + j", the compiler first finds the memory locations corresponding to the variable names i and j and reads the data in them(that is, 2 and 6). Then, calculate the sum of these two numbers(i.e., 8). Finally, the memory address(i.e., 2008) corresponding to the variable name k is found, and the calculation result 8 is stored in the memory unit corresponding to the address. This direct access to data through variable names is called "direct access".

续的，但为了讲解方便，我们可以假设它们是连续的，如图6-1所示。

假设变量i的首地址为2000，那么它的地址范围就是2000~2003。同样地，变量j的首地址是2004，k的首地址是2008。每个变量都占据4字节的长度。

当我们想要输出变量i的值到显示器上时，编译器会根据变量名i关联对应的首地址（即2000）找到内存空间，读取相应的内存单元中的数据（即2）。最后，将这个值输出到显示器上。

同样地，如果我们执行"k = i + j"这样的语句，编译器会首先通过变量名i和j找到它们对应的内存单元，并读取其中的数据（即2和6）。然后，计算这两个数的和（即8）。最后，通过变量名k找到其对应的内存地址（即2008），并将计算结果8存储到这个地址所对应的内存单元中。这种通过变量名直接存取数据的方式称为"直接存取"。

2. Indirect Access（间接存取）

Suppose a special variable i_ pointer is defined, which stores the first address of i, that is, the statement "i_ pointer = &i"; Store the address 2000 of i in the i_ pointer variable memory space, and then execute the statement "* i_ pointer = 30;". The meaning of this sentence is to access the memory space of i_ pointer, get the address data 2000 in the space, find the storage space of variable i through the address, and store 30 in the storage space of i. This process is called indirect access. As shown in Figure 6-2.

假设定义了一个特殊变量i_pointer，该变量存了i的首地址，即有语句"i_pointer=&i;"将i的地址2000存到i_pointer变量内存空间，接下来执行语句*i_pointer=30;。这句话的含义是通过访问i_pointer的内存空间，取空间内的地址数据2000，通过该地址找到变量i存储空间，将30存入i的存储空间中。这个过程称为间接存取，如图6-2所示。

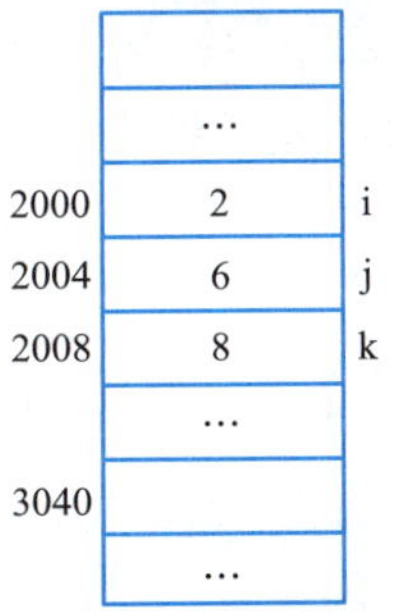

图6-1 变量的存储形式

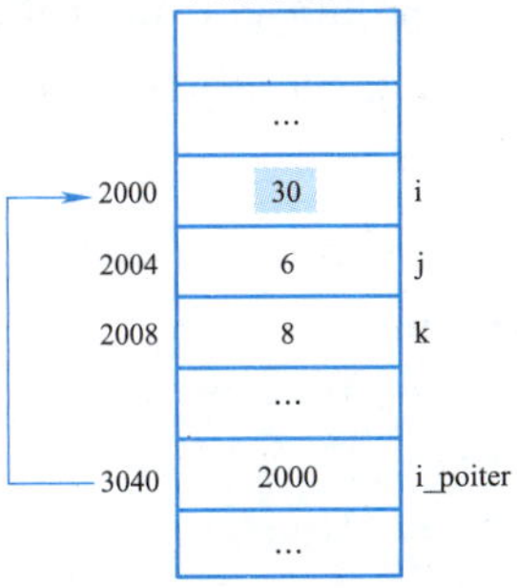

图6-2 指针存储解释

3. Comparison and Analysis（对比与分析）

Above we have introduced two methods of variable access: direct access and indirect access. Data can be accessed directly through variable names, such as "i = 3". You can also

以上我们介绍了变量存取的两种方法：直接存取与间接存取。可以通过变量名直接存取数据，如"i=3"。也可以定义

define a special variable to store the first address of the i variable, such as int*i_pointer = &i;. The meaning of this sentence is to assign the address of i to the variable i_ pointer, and then access the data of I indirectly through the indirect access operation "* i_pointer". The above variable i_pointer stores the address(pointer), which we call the pointer variable visually.

一个特殊的变量存放i变量的首地址，如"int *i_pointer=&i";。这句话的含义是将i的地址赋值给变量i_pointer，再通过间接访问操作*i_pointer来间接访问i的数据。上述的变量i_pointer存放的是地址（指针），称为指针变量。

6.1.2 Definition and Use of Pointer Variables（指针变量的定义与使用）

Pointer variables are designed to store the addresses of other variables, that is, variables that store addresses exclusively. It is worth noting that pointers and pointer variables are two different concepts. A pointer is essentially an address that represents a specific location in memory. It is treated as constant data and points to a fixed point in memory. A pointer variable, on the other hand, is essentially a variable, but the data it stores is an address. In actual use, the two are often confused, and the meaning of the two should be clearly distinguished in strict cases.

指针变量是专为存储其他变量地址而设计的，即专门存储地址的变量。值得注意的是，指针与指针变量是两个不同的概念。指针本质上是一个地址，代表着内存中的具体位置，它被视为常量数据，指向固定的内存点。而指针变量，本质上是一个变量，但其存储的数据是一个地址。在实际使用中两者常被混淆，在严谨场合需明确区分二者的含义。

1. General Form of Pointer Variable Definition（指针变量定义的一般形式）

```
data type *pointer variable name;
```

```
数据类型 * 指针变量名；
```

Description（说明）：

(1) The data type is the "base type" specified for the pointer variable. The type restricts the pointer variable to store only the first address of the "base type" variable.

(2) When defining a pointer variable, "*" indicates that the type is a pointer type, which is part of the pointer type format declaration.

(3) When defining pointer variables, it is recommended to initialize them, or give them valid addresses before using them. The "&" operator is often used to assign the address of a variable to a pointer variable, or to suspend it(assign an initial value of NULL).

Example: Define two pointer variables that point to variables of type int.

（1）数据类型为指针变量指定的"基类型"。该类型限定该指针变量只能存储该"基类型"变量的首地址。

（2）在定义指针变量时，"*"表示类型为指针类型，属于指针类型格式声明的一部分。

（3）定义指针变量的同时建议初始化，或者在使用前一定赋予有效的地址。常使用"&"运算符将某个变量地址为指针变量赋值，或者悬空（赋予初值NULL）。

举例：定义两个指向int型变量的指针变量。

```
int *pointer_1;
int *pointer_2;
```

In this example, two pointer variables, pointer_1 and pointer_2, are defined with the same base type and can be written on the same line as: int*pointer_1,*pointer_2; Note that the second pointer variable is preceded by a *.

在这个例子中，定义了两个指针变量pointer_1和pointer_2，基类型一致，可以写在同一行，写法为：int *pointer_1,*pointer_2; 注意第二个指针变量的前面要加上*。

Assume “int a, b;”. Then “pointer_1 = & a; pointer_2=&b;”. At this point, poiter_1 points to variable a and pointer_2 points to variable b. Special note: The variable pointed to must be int type, consistent with the base type. If it points to float type or other types, an error will be reported.

假设有“int a,b;”，则“pointer_1=&a; pointer_2=&b;”。此时poiter_1指向了变量a,pointer_2指向了变量b。特别注意：所指向的变量必须是int型，与基类型保持一致，若指向float型或者其他类型则报错。

Inspiration and Reflection（启发与思考）:

The pointer variable only stores the first address of a memory space, and its function is to access the data of the space indirectly through the address, so why should the pointer variable be typed? Why is there a base type?

In fact, the pointer itself does not need to be typed, because the size of the space allocated in memory for the pointer variable is fixed, consistent with the number of bytes of the operating system address. The need for typing is to be able to implement indirect access correctly. In order to understand the specific details, we must first master the specific methods of indirect access.

指针变量仅存放一个内存空间的首地址，作用是通过该地址间接访问该空间的数据。那么，为什么要为指针变量区分类型呢？即为什么会有基类型？

实际上，指针本身无须区分类型，因为指针变量在内存中分配的空间大小是固定的，与操作系统地址的字节数保持一致。区分类型的需求是为了能够正确地实现间接访问。为了弄清楚具体细节，首先要掌握间接访问的具体方法。

2. General form of indirect access（间接访问的一般形式）

```
*(pointer variable)
```

```
*（指针变量）
```

Description（说明）:

(1) “*” is a “dereference operator” or an “indirect access operator”. This operator is used to get the value at the memory address pointed to by the pointer variable. In other words, it is used to access the data pointed to by the pointer variable. The dereference operator allows us to read or modify the contents of the memory location pointed to by the pointer.

(2) “*” belongs to a multi-purpose name. If “*” appears in the definition, it is the format of the pointer type, and if it is used alone, it is the “dereference operator” or “indirect access operator”.

(3) In indirect access, we only know the first address of the variable stored in the pointer variable. If we want to access it correctly, we should also know the type of the variable. Only when we know the type, we can know the size of the accessed space and the form of data storage. This information is provided by the “base type”.

（1）“*”为“解引用运算符”或“间接访问运算符”。这个运算符用于获取指针变量所指向的内存地址中的值。换句话说，它用于访问指针变量所指向的数据。通过解引用运算符，我们可以读取或修改指针所指向的内存位置的内容。

（2）“*”属于一名多用。若“*”出现在定义里则作为指针类型的格式，若单独使用则为“解引用运算符”或“间接访问运算符”。

（3）间接访问时，我们只知道指针变量存储的变量的首地址，若想正确访问还应知道变量的类型，知道类型才知道所访问空间的大小和数据存储形式。这个信息由“基类型”提供。

Conclusion（结论）:

Typing a pointer variable, that is, setting the “base type”, is to ensure correct data parsing when accessing the address. Different data types occupy different memory spaces and have specific interpretation methods, so specifying a “base type” ensures that pointer variables are read or written according to the correct data type when dereferenced, preventing memory access errors or data interpretation errors. In short, the “base type” ensures that the pointer variable, when accessed indirectly, correctly interprets the data it points to.

为指针变量区分类型，即设定“基类型”，是为了确保访问该地址时的数据解析正确性。不同的数据类型占用不同的内存空间并有特定的解释方式，因此指定“基类型”能确保指针变量在解引用时按照正确的数据类型进行读取或写入，防止内存访问错误或数据解释错误。简而言之，“基类型”确保指针变量在间接访问时能够正确解释所指向的数据。

Case 6_1（案例 6_1）

For the definition and access of pointer variables, please read the following code to determine whether the code is accurate. If not, please correct it and give the running results.

指针变量的定义与访问，请阅读下列代码，判断代码是否准确。若不准确请修正，并给出运行结果。

Program Code（程序代码）:

Case6_1.c

```
#include <stdio.h>
int main()
{
    int i=3,*p=&i;                          //定义整型数据i，并定义了一个指针变量p指向i

    printf("i=%d\n*p=%d\n",i,*p);          //查看i的值与*p的值
    printf("&i=%d\np=%d\n",&i,p);          //查看&i与p是否相等
    return 0;
}
```

Case Analysis（案例解析）:

The program is syntactically correct, can be executed, and the results are as follows.

该程序经过检查，确认不存在语法问题，并且能够顺利运行。以下是其执行后的结果。

```
i=3
*p=3
&i=6487572
p=6487572
```

We can see that the values of i and *p are equal, both 3. The value of & i is the same as the value of the pointer variable p, which is 6487572, this value is the address, which should always be expressed in hexadecimal, which is not accurate.

When the p=&i statement is executed, p stores the first address of i, which we visualize as p pointing to i. Pay special attention to the difference between p and *p, where p is the pointer variable itself, which stores the first address of i, and *p is the indirect access operation, which accesses the variable it points to, where *p is equivalent to the variable i.

The program has no syntax errors, but using %d for the output address is not accurate enough. Memory addresses are data represented in hex. The %p format specifier should be used for pointers(i.e. addresses). It outputs the memory address pointed to by the pointer in hex and the address needs to be cast to void* to be %p compatible.

我们可以看到i与*p的值是相等的，都是3。&i和指针变量p的值一样都是6487572，这个值为地址，地址往往应该使用十六进制表示，这里不够准确。

当p=&i语句执行后，p存储了i的首地址，我们形象化称为p指向了i。特别注意p和*p的区别，p是指针变量本身，存放了i的首地址，*p是间接访问运算，访问其所指向的变量，这里*p等价于变量i。

程序虽然没有语法错误，但输出地址时使用%d还不够准确。内存地址是用十六进制表示的数据，对于指针（即地址）应该使用%p格式说明符，它能以十六进制形式输出指针所指向的内存地址，并且需要将地址强制转换为void*类型来确保与%p兼容。

Code Improvement（代码改进）:

```
#include <stdio.h>
int main()
{
    int i=3, *p=&i;                         //定义整型数据i，并定义了一个指针变量p指向i
    printf("i=%d\n*p=%d\n",i,*p);          //查看i的值与*p的值
    //使用%p来打印地址，并将地址转换为void*类型
```

```
        printf("&i=%p\np=%p\n",(void*)&i,(void*)p);    //查看&i与p是否相等
        return 0;
    }
```

Program Output（运行结果）:

```
i=3
*p=3
&i=0x[some_address]
p=0x[some_address]
```

Here[some_address] is the hexadecimal representation of the actual memory address, which should be the same because p is initialized to an address pointing to i.

这里的[some_address] 是实际内存地址的十六进制表示，它们应该是相同的，因为p被初始化为指向i的地址。

6.1.3 Precautions for Using Pointer Variable（指针变量使用注意事项）

1. Avoid "Wild Pointers" or "Hanging Pointers"（避免“野指针”或“悬挂指针”）

In C, if a pointer variable is not initialized or assigned, it will contain an undefined value, which could be any memory address. This uninitialized pointer variable is called a "wild pointer" or a "dangling pointer". Using such a pointer variable can cause a number of problems, including but not limited to:

在C语言中，如果指针变量未被初始化或赋值，它将包含一个不确定的值，这个值可能是任何内存地址。这种未初始化的指针变量被称为“野指针”或“悬挂指针”。使用这样的指针变量可能会导致一系列问题，包括但不限于：

Memory access error: Because the address pointed to by the wild pointer is indeterminate, attempting to access or modify memory through that pointer may result in accessing an area of memory that the program should not be accessing, which may cause the program to crash or behave unpredictably.

内存访问错误：由于野指针指向的地址是不确定的，尝试通过该指针访问或修改内存可能会导致访问到程序不应该访问的内存区域，这可能导致程序崩溃或产生不可预测的行为。

Data corruption: Wild pointers may point to other important data structures or variables in the program. Writing to these locations can destroy the original data, resulting in program logic errors or data loss.

数据损坏：野指针可能指向程序中的其他重要数据结构或变量。对这些位置进行写入操作会破坏原有数据，导致程序逻辑错误或数据丢失。

Program crash: If the address pointed to by the wild pointer is an illegal address(for example, a NULL or an area of memory reserved by the operating system), attempting to access memory through that pointer causes the program to crash or throw an exception.

程序崩溃：如果野指针指向的地址是非法地址（如NULL或操作系统保留的内存区域），尝试通过该指针进行内存访问会导致程序崩溃或抛出异常。

Security vulnerabilities: In more complex programs, wild pointers can lead to security vulnerabilities, such as buffer overflows, arbitrary code execution, and so on, allowing attackers to exploit these vulnerabilities to perform malicious operations on the program.

安全漏洞：在更复杂的程序中，野指针可能导致安全漏洞，如缓冲区溢出、任意代码执行等，从而允许攻击者利用这些漏洞对程序进行恶意操作。

To avoid these problems, you should always make sure that you initialize or assign a pointer before using it. A common practice is to initialize the pointer to NULL, indicating that it does not currently point to any valid memory address.

为了避免这些问题，应该始终确保在使用指针之前对其进行初始化或赋值。一种常见的做法是将指针初始化为NULL，表示它当前不指向任何有效的内存地址。

For example, "int*p = NULL; int *pointer_1=NULL,*pointer_2=NULL;". The pointer variable can be reassigned when needed later. If the pointer does not need to point to any variable, it can also be reassigned to NULL.

例如，“int *p=NULL; int *pointer_1=NULL,*pointer_2=NULL;”。后续需要时可以重新对指针变量赋值，若不需要指针指向任何变量，也可以重新赋值NULL。

Case 6_2（案例 6_2）

Read the following code and point out the problem with the program.

阅读下列代码，指出程序的问题。

```
1   #include <stdio.h>
2   int main()
3   {
4       char*p1;                        //定义一个字符型指针
5       char a='c';                     //定义一个字符型变量并赋值
6       *p1=a;                          //指针未有指向就对其赋值，典型的指针使用错误
7       printf("*p1=%c",*p1);           //尝试输出 *p
8       return 0;
9   }
```

Case Analysis（案例解析）:

Although the code compiled without errors, it contained a critical oversight in line 4, where the pointer variable p1 was declared but not properly initialized. Consequently, in line 6, the program utilized an “unassigned pointer” or “dangling pointer”, which is a pointer variable referencing an unpredictable memory location, posing a significant risk and potential for undefined behavior.

虽然代码编译没有错误，但在第4行定义指针变量p1时出现了重大疏忽，没有正确初始化。因此，在第6行，程序使用了一个没有赋值的指针或“野指针”，这是一个指向不可预测的内存位置的指针变量，存在严重的风险和潜在的未定义行为。

Code Improvement（代码改进）:

```
1   #include <stdio.h>
2   int main()
3   {
4       char*p1=NULL;                        //定义一个字符型指针
5       char a='c';                          //定义一个字符型变量并赋值
6       p1=&a;                               //指针指向变量 a
7       printf("*p1=%c\n",*p1);              //查看 *p 的值
8       *p1='a';                             //通过指针修改变量 a 的值
9       printf("a=%c,*p1=%c",a,*p1);         //输出 a 和 *p1 的值
10      return 0;
11  }
```

Program Output（运行结果）:

```
*p1=c
a=A,*p1=A
```

Case Analysis（案例解析）:

(1) It is safer to suspend the pointer definition in line 4 at the same time to avoid the pointer pointing to an uncertain address.

（1）在第4行指针定义的同时悬空，避免指针指向不确定的地址，这样更安全。

(2) The pointer in line 6 is given a valid address, and p1 stores the address of a, pointing to a variable. Line 7 outputs the result of *p1. *p1 indirectly accesses a variable, so it outputs the character 'c'.

（2）第6行指针赋予有效的地址，p1存储a的地址，指向a变量。第7行输出*p1的结果，*p1间接访问的是a变量，故输出字符'c'。

(3) Line 8 accesses the storage space of the variable a through *p1. At this time, *p1 is equivalent to the variable a, and the final effect is to assign the character 'A' to the variable a.

（3）第8行通过*p1访问a变量的存储空间，此时*p1相当于变量a，最终效果是将字符'A'赋值给变量a。

(4) When a and*p1 are output again, they are both 'A' characters.

（4）再输出a和*p1时均为'A'字符。

2. Different Types of Pointer Variables Allocate the Same Amount of Memory Space（不同类型指针变量分配的内存空间大小一样）

In fact, the pointer variable itself is not typed, and any pointer variable can only store one address. For example, two pointer variables p1 and p2 are defined, which point to int and double respectively. The memory size allocated by the compilation system for the pointer variables p1 and p2 is the same. The size of the address of the current compilation system is the same. For example, if the address is 32-bit, all pointer variables are allocated 4 bytes. If the address is 64-bit, 8 bytes are allocated.

实质上指针变量本身是不分类型的，任意一个指针变量只能存放一个地址。如定义了两个指针变量p1、p2，这两个变量分别指向int型和double型。编译系统为指针变量p1、p2分配的内存大小是一样的，当前编译系统地址大小是多少就分配多少，如32位地址，则所有指针变量分配4字节，若地址是64位，则分配8字节。

Case 6_3（案例 6_3）

Show the size of pointer memory space of different data types through code. The following is a simple example code that shows that pointers of different data types occupy the same amount of space in memory.

通过代码展示不同数据类型的指针内存空间大小情况。下面是一个简单的示例代码，展示了不同数据类型的指针在内存中所占的空间大小是一样的。

Program Code（程序代码）:

Case6_3.c

```
#include <stdio.h>
int main()
{
    int *int_ptr;
    char *char_ptr;
    double *double_ptr;
    long long *long_long_ptr;
    void *void_ptr;
    //打印不同类型指针的大小
    printf("Size of int pointer: %zu bytes\n", sizeof(int_ptr));
    printf("Size of char pointer: %zu bytes\n", sizeof(char_ptr));
    printf("Size of double pointer: %zu bytes\n", sizeof(double_ptr));
    printf("Size of long long pointer: %zu bytes\n", sizeof(long_long_ptr));
    printf("Size of void pointer: %zu bytes\n", sizeof(void_ptr));
    return 0;
}
```

Program Output(64-bit system) [运行结果(64位系统)]:

```
Size of int pointer:8 bytes
Size of char pointer:8 bytes
Size of double pointer:8 bytes
Size of long long pointer:8 bytes
Size of void pointer:8 bytes
```

Case Analysis（案例解析）:

In the code, we define multiple types of pointer variables, Include int_ptr(for pointers to integers), char_ptr(for pointers to characters), double_ptr(for pointers to double-precision floating point numbers), long_long_ptr(for pointers to long integers), and void_ ptr(for pointers to general-purpose types, Can point to any

在代码中，我们定义了多种类型的指针变量，包括int_ptr（用于指向整数的指针）、char_ptr（用于指向字符的指针）、double_ptr（用于指向双精度浮点数的指针）、long_long_ptr（用于指向长长整数的

type of data). We then use the sizeof operator to measure how much space each type of pointer occupies in memory and print the result.

When this code is executed, all pointer variables occupy the same amount of memory on both 32-bit and 64-bit operating systems. Specifically, on a 32-bit system, these pointers should all be 4 bytes in size; On 64-bit systems, they are all 8 bytes in size. This fully verifies the fact that pointer variables of different data types occupy the same space in memory.

指针）以及void_ptr（通用类型的指针，可以指向任意类型的数据）。随后，我们利用sizeof运算符来测量每种类型指针在内存中所占的空间大小，并将结果打印出来。

当执行这段代码时，不论是在32位还是64位的操作系统中，所有指针变量所占内存空间的大小都是一致的。具体来说，在32位系统上，这些指针的大小都应该是4字节；而在64位系统上，它们的大小则都是8字节。这充分验证了一个事实，即不同数据类型的指针变量在内存中的空间占用大小是相同的。

6.2 Pointer Reference Array 指针引用数组

6.2节视频

We have introduced the definition and use of pointers and pointer variables in the previous section. Because the storage space of arrays is continuous, we can use the up and down movement of pointer variables to access arrays more efficiently. This section mainly introduces pointer variables pointing to array elements, pointer addition and subtraction operations, and pointer referencing array elements.

上节我们已经介绍了指针和指针变量的定义和使用，由于数组的存储空间是连续的，故可以使用指针变量的上下移动来更为高效地访问数组。本节主要介绍指针变量指向数组元素、指针加减运算、指针引用数组元素。

6.2.1 Pointer Variable Points to Array（指针变量指向数组）

A pointer variable is usually used to point to a single variable of the same type. In this section, the pointer variable points to an array, which often means that the pointer variable points to the first element of the array. So how do you make the pointer point to the first element of the array?

Here is a simple example. Suppose there is an array of int type a[10], and a pointer variable p is defined to point to the first element of the array, namely a[0], as shown in Figure 6-3. The code description is as follows:

指针变量通常用于指向同类型的单个变量，本节所说的指针变量指向数组，往往含义为指针变量指向数组首元素。那么如何令指针指向数组首元素呢？

下面举个简单的例子，假设有一个int型的a[10]数组，定义一个指针变量p指向该数组首元素即a[0]，如图6-3所示，代码描述如下：

```
int a[10]={1,2,3,4,5,6,7,8,9,10};
int *p=a;// 或者 int *p=&a[0]
```

Note（说明）:

The array name a and &a[0] have the same value. The array name is interpreted as a pointer to the first element of the array, that is, its value is the first address of the first element. The pointer variable p points to the first element a[0]. If we output the result of *p, it is the value of a[0].

数组名a和&a[0]的值是一样的，数组名被解释为指向数组首元素的指针，即其值为首元素的首地址。此时指针变量p指向了首元素a[0]。若我们输出*p的结果，则为a[0]的值。

6.2.2 Pointer Addition and Subtraction（指针加减运算）

When the pointer variable points to an array element, addition and subtraction operations are allowed. For example, if the pointer variable adds an integer or subtracts an integer, the autoincrement operator or autodecrement operator can be used, but the accessed data should not exceed the range of the array. If int *p = a;. Then the effect of p++ or ++p is to make the pointer point backward to an element, and if p -- or -- p means that the pointer points forward to an element. Two pointer variables pointing to the same array are allowed to be subtracted. If pointers p1 and p2 point to the same array, p1−p2 represents the number of elements in the difference. The addition of two pointers, such as p1+p2, is meaningless. As shown in Figure 6-4, p1 points to the element of array a[2], and p2 points to the element of array a[5], then p2−p1 = 3, and the difference value indicates that the number of difference elements is 3.

当指针变量指向一个数组元素时，允许加减运算，如令指针变量加一个整数，或者减去一个整数，可以使用自增运算符或者自减运算符，但要保证所访问的数据不能越界，不能超过数组范围。如若有int *p=a;，则p++或者++p的作用是令指针向后指一个元素，若p--或者--p则表示指针向前指一个元素。允许指向同一个数组的两个指针变量进行减法操作，如有指针p1和p2指向同一个数组，p1−p2表示相差的元素个数。两个指针的加法运算如p1+p2是没有任何意义的。如图6-4所示，p1指向了数组a[2]元素，p2指向了该数组a[5]元素，则p2−p1=3，差值表示相差元素的个数为3。

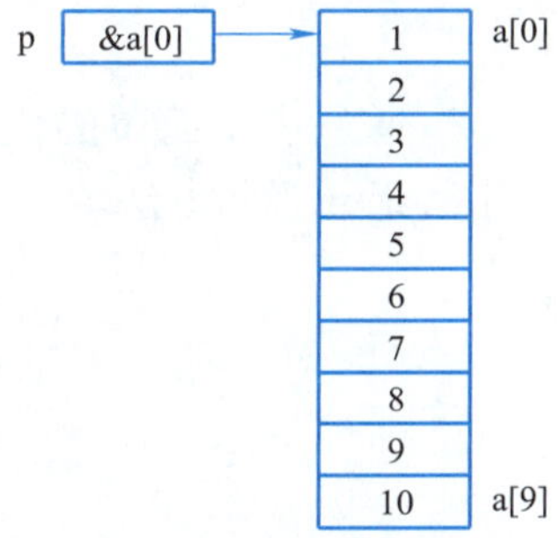

图6-3 指针变量指向数组首元素示意图

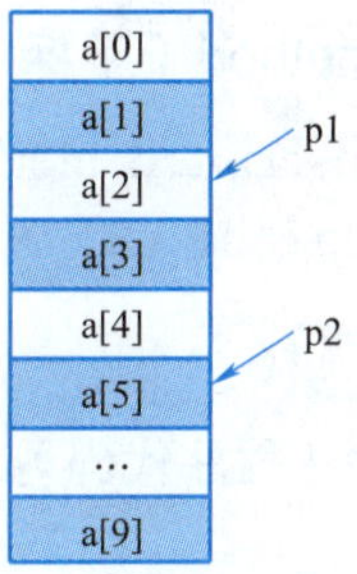

图6-4 指向同一个数组的两个指针变量可以做减法操作

Pointer Access Array Important Note 指针访问数组重要说明

If int*p = a, there are the following important conclusions:

(1) p+i or a+i gets the address pointing to the i+1th element of the array. When you add or subtract a pointer variable or address, you are actually moving the pointer by the size of the data type it points to. For example, if p is a variable that points to an integer(int *p), then p+1 causes p to point to the next integer position, which is the current position plus the number of bytes occupied by an integer. If the array is of type double, with each element taking up 8 bytes, then we need to define double *p to point to this array. At this point, the operation of adding and subtracting i to p actually moves the pointer according to the size of the double type(that is, 8 bytes).

(2) *(p+i) or *(a+i) is equivalent to a[i]. In fact, when we use the subscript operator a[i] in our code, the compilation system automatically converts it to the pointer operation *(a+i). If p and a point to the first address of the same array(i.e., p=a), then *(p+i) and *(a+i) are naturally equivalent to a[i]. In addition, *(p+i) can be further converted into a subscript

若有int *p=a，则有如下重要的结论：

（1）p+i或a+i 得到的是指向数组第i+1个元素的地址。对指针变量或地址进行加减运算时，实际上是按照指针所指向的数据类型的大小来移动指针。例如，如果p是一个指向整数的指针变量（int *p），那么p+1会使p指向下一个整数位置，即当前位置加上一个整数所占的字节数。如果数组是double类型的，每个元素占8字节，那么我们需要定义double *p来指向这个数组。此时，对p进行加减i的操作，实际上是按照double类型的大小（即8字节）来移动指针。

（2）*(p+i)或*(a+i)与a[i] 是等价的。实际上，当我们在代码中使用下标运算符a[i]时，编译系统会自动将其转换为指针运算*(a+i)。如果p和a指向同一个数组的首地址（即p=a），那么*(p+i)和*(a+i)自然也就等

operation p[i], so the access array element a[i] can also be represented by p[i]. Regardless of which representation is used, they are essentially accessing the same memory address, that is, the same memory space.

价于a[i]。此外，*(p+i)还可以进一步转换为下标运算p[i]，因此访问数组元素a[i]也可以用p[i]来表示。无论是使用哪种表示方式，它们本质上都是访问同一个内存地址，即同一片内存空间。

6.2.3 Pointers Referencing Array Elements（指针引用数组元素）

So far, there are two main ways to refer to array elements: one is to use subscripts, and the other is to use pointers. The two methods, while different in form, essentially access the same location in memory where the array is located. The subscript method is intuitive and easy to understand, while the pointer method provides a more low-level memory access method, which is very helpful for in-depth understanding of computer systems and data structures. In the actual programming, we can choose the appropriate method according to the specific needs and scenarios.

至此，引用数组元素主要有两种方法：一种是使用下标法；另一种则是利用指针法。这两种方法虽然形式不同，但本质上都是访问数组在内存中的同一位置。下标法直观易懂，而指针法则提供了更底层的内存访问方式，对于深入理解计算机系统和数据结构很有帮助。在实际编程中，可以根据具体需求和场景选择合适的方法。

1. Subscript method（下标法）

Review the subscripted method of referencing array elements with a simple code.

通过一个简单的代码来回顾下标法引用数组元素的方法。

Case 6_4（案例 6_4）

Program Code（程序代码）:

Case6_4.c

```
1    #include <stdio.h>
2    int main()
3    {
4        int a[10], i;
5        for(i=0;i<10;i++) scanf("%d",&a[i]);
6        for(i=0;i<10;i++) printf("%d ",a[i]);
7        return 0;
8    }
```

Case Analysis（案例解析）:

Line 4 defines the array and the loop variable that controls the subscript change. Line 5 uses a for loop to input the value of each array element, and line 6 uses a for loop to output each element, using a[i] subscript method for array element references.

第4行定义了数组及控制下标变化的循环变量。第5行使用for循环输入每个数组元素的值。第6行使用for循环输出每个元素，数组元素的引用使用了a[i]下标的方法。

2. Pointer Method（指针法）

Case 6_5（案例 6_5）

Use the pointer method to reimplement the code function in Case 6_1, that is, read in 10 elements for the int a[10] array and output these elements.

使用指针法重新实现案例6_1中的代码功能，即为int a[10]数组读入10个元素并输出这些元素。

Thinking（解题思路）:

First, we need to define a pointer variable that points to an int. Next, let the pointer variable point to the first element of the array. The entire array is then iterated by incrementing the pointer variable, with input and output operations on each element. Through the pointer operation, we can access and modify the array elements more flexibly, thus improving the efficiency and readability of the code. The implementation process is as follows:

(1) Define an array a of type int with a length of 10.

(2) Define a pointer p to an int and initialize it to point to the first element of array a.

(3) Using a loop structure, each element of the array a is accessed in turn by the pointer p, and the input value is read using the scanf function.

(4) Using the same loop structure, traverse each element of array a with pointer p, and use the printf function to output the value of each element.

首先，我们需要定义一个指向int类型的指针变量。接着，让这个指针变量指向数组的首元素。随后，通过不断递增指针变量来遍历整个数组，并对每个元素进行输入和输出操作。通过指针操作，我们能够更加灵活地访问和修改数组元素，从而提高了代码的效率和可读性。实现过程如下：

（1）定义一个int类型的数组a，长度为10。

（2）定义一个指向int类型的指针p，并初始化它指向数组a的首元素。

（3）使用循环结构，通过指针p依次访问数组a的每个元素，并使用scanf函数读取输入值。

（4）同样使用循环结构，通过指针p遍历数组a的每个元素，并使用printf函数输出每个元素的值。

Program Code（程序代码）:

Case6_5.c

```
1    #include <stdio.h>
2    int main()
3    {
4        int a[10];
5        int *p=NULL;
6        for(p=a;p<a+10;p++) scanf("%d",p);
7        for(p=a;p<(a+10);p++) printf("%d ",*p);
8        return 0;
9    }
```

Case Analysis（案例解析）:

(1) In line 5, we define a pointer variable named p and initialize it to NULL. This pointer variable p is used to point to an array element of type int. Because p is used to store memory addresses, its base type(or pointer type) is specified as int *, indicating that it points to a variable or array element of type int. Initializing to NULL is a good programming practice that ensures that p does not initially point to any undefined or unexpected memory address.

(2) In line 6, we use the for loop and the previously defined pointer p to override the array element read operation that would otherwise use subscripting. At the beginning of the loop, p is assigned the first address of array a, which points to the first element of the array, a[0]. The loop condition $p < a + 10$ is an address comparison that ensures that p does not point to an address beyond the end of array a. When the condition holds,

（1）在第5行中，我们定义了一个名为p的指针变量，并初始化为NULL。这个指针变量p被用来指向int类型的数组元素。由于p是用来存储内存地址的，因此它的基类型（或称为指针类型）被指定为int*，表示它指向的是一个int类型的变量或数组元素。初始化为NULL是一个良好的编程习惯，它确保了p在初始时不会指向任何未定义或意外的内存地址。

（2）在第6行，我们使用for循环和之前定义的指针p来改写原本使用下标法的数组元素读入操作。循环开始时，p被赋值为数组a的首地址，即指向数组的第一个元素a[0]。循环条件$p < a+10$是一个地址比较，

the scanf function reads the user-entered integer via the pointer p and stores it in the array element currently pointed to by p. The p + + operation then increments the pointer p to point to the next array element(that is, a[1]). This process is repeated until the address pointed to by p reaches or exceeds the end of the array, ensuring that all 10 elements are read in correctly.

它确保p指向的地址不会超出数组a的末尾。当条件成立时，scanf函数通过指针p读取用户输入的整数，并将其存储在p当前指向的数组元素中。然后，p++操作递增指针p，使其指向下一个数组元素（即a[1]）。这个过程重复执行，直到p指向的地址达到或超过数组末尾，从而确保所有10个元素都被正确读入。

(3) Line 7 uses the pointer method to output array elements. Here, with the dereference operation *p, we take the value of the array element currently pointed to by p and print it out using the printf function. Note that p++ cannot be rewritten as a++ here. Because a is an array name, in C, the array name represents the first address of the array, which is a constant pointer whose value(that is, address) is immutable during the execution of the program. Therefore, we cannot increment a. And p is a pointer variable that can be modified, and we can iterate through the elements of the array by incrementing it.

（3）第7行使用指针法输出数组元素。这里，通过解引用操作*p，我们获取了p当前指向的数组元素的值，并使用printf函数将其打印出来。需要注意的是，这里不能将p++改写成a++。因为a是数组名，在C语言中，数组名代表数组的首地址，它是一个常量指针，其值（即地址）在程序执行期间是不可变的。因此，我们不能对a进行递增操作。而p是一个可以修改的指针变量，我们可以通过递增它来遍历数组中的各个元素。

Precautions for Using the Pointer Method 使用指针法注意事项

(1) When performing pointer arithmetic, be sure to ensure that the memory address pointed by the pointer is within the legal range, that is, the address before or after the array limit cannot be exceeded, otherwise it may cause undefined behavior, resulting in program errors or crashes.

（1）在进行指针运算时，务必确保指针指向的内存地址在合法范围内，即不可超出数组界限之前或之后的地址，否则可能引发未定义行为，导致程序出错或崩溃。

(2) During pointer operation, be sure to keep the type of pointer operation matching the data type pointed by the pointer. If the pointer points to data of type char, but the pointer moves with the size of type int, it will cause the pointer offset error and access the wrong memory location.

（2）指针运算时，务必保持指针运算的类型与指针所指向的数据类型相匹配。如果指针指向char类型的数据，却以int类型的大小进行指针移动，将会导致指针偏移错误，进而访问到错误的内存位置。

(3) Be sure to check whether the pointer is NULL before performing pointer operation. A pointer that is NULL means that it does not point to any valid memory address, and a pointer operation or dereference operation at this time will cause the program to crash.

（3）在进行指针运算前，务必检查指针是否为NULL。指针为NULL意味着它未指向任何有效的内存地址，如果此时进行指针运算或解引用操作，将会导致程序崩溃。

(4) When you need to point the pointer to the first element of an array, you should be careful to use the correct syntax. For example, change “int *q = a[0];” to “int *q= & a[0];” or “int *q= a;” to ensure that the pointer correctly points to the first address of the array.

（4）当需要将指针指向数组的首元素时，应注意使用正确的语法。例如，将“int *q=a[0];”更改为“int *q=&a[0];”或者“int *q=a;”，以确保指针正确指向数组的首地址。

(5) When using the scanf function to read the data to the variable pointed by the pointer, since the pointer itself has stored the address of the variable, there is no need to add the address symbol “&”. So, scanf("%d", &p); is wrong, and the correct usage should be scanf("%d", p);.

（5）在使用scanf函数读取数据到指针指向的变量时，由于指针本身已经存储了变量的地址，因此不需要再加取地址符号“&”。所以，scanf(“%d”,&p);是错误的，正确的用法应该是scanf("%d",p);。

3. Comparative Analysis of Efficiency of Subscript Method and Pointer Method（下标法和指针法效率对比分析）

We can access array elements using either the subscript method or the pointer method. However, from an efficiency perspective, the pointer method is generally more efficient. This is because the C compiler system needs to calculate the address of the element by the first address and index of the array every time when dealing with the subscript method. This calculation involves additional multiplication and addition operations. When using the pointer method, we only need to directly access the next element of the array by incrementing the pointer variable p, without recalculating the address each time. Therefore, when dealing with a large amount of data, the pointer method can reduce unnecessary calculations and provide better performance.

无论是使用下标法还是指针法，我们都可以实现对数组元素的访问。但从效率角度来看，指针法通常更为高效。这是因为C编译系统在处理下标法时，每次都需要通过数组的首地址和索引来计算元素的地址。这种计算涉及额外的乘法和加法操作。而使用指针法时，我们只需要通过递增指针变量p来直接访问数组的下一个元素，无须每次重新计算地址。因此，在处理大量数据时，指针法能够减少不必要的计算，从而提供更好的性能。

6.3 Pointer Reference String 指针引用字符串

6.3节视频

In C, strings are stored as arrays of characters, which makes it convenient for us to use pointers to reference and manipulate strings. Each string ends with a null character('\0'), which marks the end of the string. A pointer variable can be used to point to the first character of a string, and then by incrementing the pointer variable, we can access each character in the string one by one, thus traversing the entire string.

在C语言中，字符串是以字符数组的形式存储的，这为我们使用指针来引用和操作字符串提供了便利。每个字符串都以一个空字符（'\0'）作为结尾，这个空字符起着标识字符串结束的作用。指针变量可以被用来指向字符串的首个字符，之后通过递增这个指针变量，我们可以逐个访问字符串中的每个字符，从而实现对整个字符串的遍历。

6.3.1 Storage and Reference of Character Strings（字符串的存储与引用）

Strings in C, whether as variables or constants, are stored in the form of character arrays. But that doesn't mean that all character arrays are strings. For a character array to be treated as a string, you must ensure that it ends with a null character('\0'). If a character array is not properly null-terminated, it may cause undefined behavior or a program crash when using string processing functions.

When we define a character array in our program to store a string, the string will be stored in the variable store, which allows us to modify it at runtime. String constants, on the other hand, are stored in the constant storage area in the form of a special, nameless character array, which ensures that the contents of string constants are immutable during program execution.

In C, a reference to a string can be implemented in two main ways:

(1) The method of storing strings in character arrays allows us to manipulate each character directly through array

字符串在C语言中，无论是作为变量还是常量，都是采用字符数组的形式进行存储的。但并不意味着所有的字符数组都是字符串。一个字符数组要被视为字符串，必须确保它以空字符（'\0'）结尾。如果字符数组没有被正确地以空字符结尾，那么在使用字符串处理函数时可能会导致未定义行为或程序崩溃。

当我们在程序中定义了一个字符数组来存储一个字符串时，这个字符串将被存储在变量存储区中，这允许我们在运行时对其进行修改。而字符串常量则是以一种特殊的、无名的字符数组形式存储在常量存储区，这样的存储方式确保了字符串常量的内容在程序执行期间是不可变的。

在C语言中，字符串的引用可以通过

subscripts. By providing specific array subscripts, we can read, modify, or perform other operations on specific characters in the array. This provides direct control over the contents of the string, allowing us to dynamically modify the string as needed at run time.

(2) A character pointer variable can be used to point to a string constant. In this case, the pointer variable stores the address of the string constant in memory. Each character in a string is accessed indirectly through a pointer variable. This is useful when you need to traverse a string or operate without changing the contents of the string. However, it is important to note that because string constants are stored in the constant store, characters accessed indirectly through pointers are not modifiable.

Each of these two reference methods has its own applicable scenarios, and we should choose the appropriate method to reference and manipulate strings according to our specific needs. When using pointer variables to point to strings, we also need to pay special attention to the initialization and bounds checking of pointers to avoid errors such as pointer out of bounds or null pointer dereference.

两种主要方式实现：

（1）字符数组存放字符串的方法允许我们直接通过数组下标来操作每一个字符。通过提供特定的数组下标，我们可以对数组中的特定字符进行读取、修改或执行其他操作。这种方式提供了对字符串内容的直接控制，使我们能够在运行时根据需要动态地修改字符串。

（2）字符型指针变量可以用于指向一个字符串常量。在这种情况下，指针变量存储的是字符串常量在内存中的地址。通过指针变量间接地访问字符串中的每一个字符。这种方式在需要遍历字符串或在不改变字符串内容的情况下进行操作时非常有用。然而，需要注意的是，由于字符串常量存储在常量存储区，因此通过指针间接访问的字符是不可修改的。

这两种引用方式各有其适用的场景，我们应根据具体的需求选择合适的方法来引用和操作字符串。在使用指针变量指向字符串时，我们还需要特别注意指针的初始化和边界检查，以避免出现指针越界或空指针解引用等错误。

6.3.2 Character Array Reference String（字符数组引用字符串）

First, let's look at the array reference string method, which is the easiest to understand and master. In C, this approach is very intuitive and easy to use. Previously, we covered this method in detail in the chapter on character arrays and strings. Now, let's take a simple example to further deepen our understanding.

首先，我们来看最易理解和掌握的数组引用字符串方法。在C语言中，这种方法非常直观且易于操作。之前，在字符数组与字符串的章节中，我们已经详细介绍了这种方法。现在，通过一个简单的例子来进一步加深理解。

Case 6_6（案例 6_6）

The character array is required to store "I love China!", and the string and the eighth character are output.

要求使用字符数组存放"I love China!"，输出该字符串和第8个字符。

Program Code（程序代码）:

Case6_6.c

```c
#include <stdio.h>
int main()
{
    char string[]={"I love China"};
    printf("%s\n",string);
    printf("%c\n",string[7]);
    return 0;
}
```

Program Output（运行结果）:

```
I love China
C
```

Case Analysis（案例解析）:

(1) In line 4, we define an array of strings of type char and omit the length of the array when initializing. The compiler automatically calculates the length of the array based on the number of characters actually initialized and adds a null character('\0') at the end of the string as an end marker. Therefore, for this string of 13 characters(including the terminator), the compilation system allocates 14 bytes of storage for the string array. It is worth noting that the array name string actually represents the address of the first element of the array.

(2) In line 5, we output the entire string to the display using the printf function and the %s format control. The %s format control tells the printf function to print a string, starting at the supplied address, until a null character('\0') is encountered.

(3) On line 6, we want to output the 8th character in the string(that is, the character with index 7, because the array index starts at 0). We use the array element reference string[7] to access this character alone. Note that when we use a subscript to access an array element, the compiler actually first calculates the address of the corresponding element(string+7), and then accesses the stored character value through this address. Finally, we use the printf function again with the %c format control to print the character.

（1）在第4行中，我们定义了一个char类型的string数组，并在初始化时省略了数组的长度。编译器会根据实际初始化的字符个数自动计算数组的长度，并在字符串末尾添加一个空字符（'\0'）作为结束标志。因此，对于这个包含13个字符（包括结束符）的字符串，编译系统会为string数组分配14字节的存储空间。值得注意的是，数组名string实际上代表了该数组首元素的地址。

（2）在第5行中，我们使用printf函数和%s格式控制符将整个字符串输出到显示器上。%s格式控制符告诉printf函数要输出一个字符串，它会从提供的地址开始，一直打印到遇到空字符（'\0'）为止。

（3）在第6行，我们想要输出字符串中的第8个字符（即索引为7的字符，因为数组索引从0开始）。我们使用数组元素引用的方式string[7]来单独访问这个字符。需要注意的是，当我们使用下标访问数组元素时，实际上编译器会首先计算对应元素的地址（string + 7），然后再通过这个地址来访问存储的字符值。最后，我们再次使用printf函数和%c格式控制符来输出这个字符。

6.3.3 Pointer Variable Reference String（指针变量引用字符串）

The reference method for storing a string in a character array variable store was described in the previous section. Next, we'll look at ways to reference string constants through a character pointer variable.

上小节介绍了将字符串存储于字符数组变量存储区的引用方法。接下来，我们将探讨通过字符型指针变量来引用字符串常量的方法。

Case 6_7（案例 6_7）

Requires manipulating the output of a string constant through a character pointer variable.

要求通过字符型指针变量操作一个字符串常量的输出。

Thinking（解题思路）:

First, define a character pointer variable and initialize it to point to the first character of a string constant. The entire string is then output through the character pointer variable. This approach is very common in C because it provides a flexible and efficient way to work with string constants.

首先，定义一个字符型指针变量，并将其初始化为指向字符串常量的首字符。随后，通过该字符指针变量来输出整个字符串。这种方法在C语言中非常常见，因为它提供了一种灵活且高效的方式来处理字符串常量。

Program Code（程序代码）:

Case6_7.c

```
#include <stdio.h>
```

```
2    int main()
3    {
4        char *string="I love China";           //指针变量指向一个字符串
5        printf("%s\n",string);                 //输出指针所指向的字符串
6        return 0;
7    }
```

Program Output（运行结果）:

```
I love China
```

Case Analysis（案例解析）:

(1) In the fourth line of code, we define a character pointer variable string, which is used to store the first address of the string constant "I love China". In fact, when a string is assigned to a string, the address of the first character of the string, 'I', is assigned to the string. During compilation, the string is stored in the constant area as an anonymous array, and a null character('\0') is automatically added at the end as a closing flag. The string pointer then points to the string.

(2) It should be noted that because the string area pointed to by string is located in the constant area, its content is not modifiable. Any attempt to modify the contents of a string through a string pointer(such as string[2] = 'a'; or*string = 'a';) is an error and will result in a compilation error.

(3) Line 5 is the output statement, and the printf function uses the %s format control character to output the string. Here, we pass in the string variable as a parameter, because string holds the first address of the string constant. The effect of %s is to output characters in sequence, starting at this address, until a null character('\0') is encountered. Therefore, the entire string "I love China" can be correctly output.

（1）在第4行代码中，我们定义了一个字符指针变量string，它用于存放字符串常量"I love China"的首地址。实际上，将字符串赋值给string时，是将字符串的首字符'I'的地址赋给了string。在编译过程中，该字符串以匿名数组的形式存储在常量区，并自动在末尾添加空字符（'\0'）作为结束标志。之后，string指针便指向了这个字符串。

（2）需要注意的是，由于string指向的字符串区域位于常量区，因此其内容是不可修改的。任何尝试通过string指针修改字符串内容的操作（如string[2]='a';或*string = 'a';）都是错误的，并会导致编译错误。

（3）第5行是输出语句，printf函数使用%s格式控制符来输出字符串。在这里，我们传入string变量作为参数，因为string中存放的是字符串常量的首地址。%s的作用是从这个地址开始，依次输出字符，直到遇到空字符（'\0'）为止。因此，可以正确地输出整个字符串"I love China"。

Case Development（案例拓展）:

A character pointer variable is essentially a variable, which means that its value can be modified. We can change the pointer variable to point to another string constant. As shown in the following code:

字符指针变量本质上是一个变量，这意味着它的值是可以修改的。我们可以改变指针变量的指向，使其指向其他字符串常量。如下列代码所示：

```
#include <stdio.h>
int main()
{
    char *string="I love China";
    printf("%s\n",string);
    string="I am student";                //string 指向了新的字符串
    printf("%s\n",string);
    return 0;
}
```

Program Output（运行结果）:

```
I love China
I am student
```

In the above code, line 4 defines a character pointer variable string, which is initialized to point to the first address of the string "I love China". Next, the program outputs the contents of the string, which is "I love China". Then, on line 6, we change the pointer to string to point to another string, the first address of the "I am student". Finally, on line 7, the program again uses string to output the string currently pointed to. Since the string now points to "I am student", the final output is "I am student". This process fully demonstrates the flexibility of character pointer variables in dynamically changing string references.

在上述代码中，第4行定义了一个字符指针变量string，它初始化为指向字符"I love China"的首地址。紧接着，程序输出了这个字符串的内容，即"I love China"。然后，在第6行，我们修改了string的指向，使其转而指向另一个字符串"I am student"的首地址。最后，在第7行，程序再次利用string输出当前指向的字符串。由于此时string已经指向了"I am student"，因此最终的输出结果为"I am student"。这一过程充分展示了字符指针变量在动态改变字符串引用方面的灵活性。

6.3.4 Comparison of Character Array and Character Pointer Variable Reference String Method（字符数组与字符指针变量引用字符串方法的比较）

The essential difference between the two is that the character array deals with string variables, while the pointer variable method deals with string constants.

(1) The character array processing method is to store character strings through several char variables, so several storage spaces need to be allocated. The allocation size is the length of the array. If the character string is stored in a one-dimensional array char a[10], a memory unit with a size of 10 bytes is allocated. A character pointer variable allocates only one pointer variable space. There is a constant store for string constants and a variable store for pointer variables.

(2) The name of the character array method array is the first address. The array name is constant and cannot be assigned. The value of a pointer variable is also the first address, but it is a variable and can be reassigned. For example, char str[14]; str = "I love China";. this is a mistake, because the str array name is a constant and cannot be modified; char *a; a = "I love China"; a ="I am student";. This is correct. A character pointer variable is a variable whose value can be modified.

(3) Using the character array method, the value of the string can be changed, such as char a[] = "House"; a[2]='a';. This description is legal, and the character is modified by the array element a[2], which changes the 'u' to the 'r', and the string is output as "Horse". The string value of the character pointer variable method cannot be changed. For example, char *b ="House"; b[3] = r; is incorrectly written because the string is a constant and is not allowed to be modified.

两者本质区别是字符数组处理的是字符串变量，而指针变量方法处理的是字符串常量。

（1）字符数组处理方法是通过若干char型变量存储字符串，故需要分配若干存储空间，分配大小是数组长度，如字符串存在一维数组char a[10]中，分配10字节的内存单元。而字符指针变量只分配一个指针变量空间。字符串常量存在常量存储区，指针变量存在变量存储区。

（2）字符数组方法数组名为首地址。数组名是常量、不能赋值。指针变量值也为首地址，但其是变量，可以重新赋值。看个对比示例，char str[14];str="I love china";。这个写法错误，因为str数组名是常量，不可以修改；char *a; a="I love China"; a="I am student";。这个正确，字符指针变量是变量，值可以修改。

（3）使用字符数组方法，字符串的值可以改变，如char a[]="House"; a[2]='a';。这个描述是合法的，通过数组元素a[2]修改了字符，将'u'改为'r'，字符串再输出时为"Horse"。而字符指针变量方法字符串值不可以改变。如char *b="House"; b[3]=r;这个写法错误，因为字符串是常量，不允许修改。

6.4 Comprehensive Cases 综合案例

综合案例 6_4_1 利用指针法交换矩阵的最大值和最小值。

案例描述：通过指针方法实现一个矩阵中最大值和最小值的交换，并输出交换后的结果，以验证操作的有效性。

解题思路：首先，我们使用二维数组来存储矩阵中的元素。然后，定义两个指针变量，并利用“打擂台算法”遍历矩阵中的每个元素，找出最大值和最小值所在的位置。接着，让这两个指针变量分别指向最大值和最小值所在的数组元素。最后，通过指针变量的间接访问，交换最大值和最小值元素的值，并输出交换后的矩阵，以查看结果。

程序代码：

Case6_4_1.c

```
#include <stdio.h>
int main()
{
    int a[10][10];                                          //定义足够大的二维数组
    int m,n,i,j,temp;
    int *pmax,*pmin;
    printf("请输入矩阵的行列值:\n");
    scanf("%d%d",&m,&n);
    printf("请输入矩阵元素:\n");
    for(i=0;i<m;i++)
    {
        for(j=0;j<n;j++)
        {
            scanf("%d",&a[i][j]);
        }
    }
    pmax=pmin=&a[0][0];                                     //初始化指针变量值。
    //打擂台方法求最值
    for(i=0;i<m;i++)
    {
        for(j=0;j<n;j++)
        {
            if(*pmax<a[i][j]) pmax=&a[i][j];
            if(*pmin>a[i][j]) pmin=&a[i][j];
        }
    }
    temp=*pmax;
    *pmax=*pmin;
    *pmin=temp;
    printf("处理后的矩阵为:\n");
    for(i=0;i<m;i++)
    {
        for(j=0;j<n;j++)
        {
            printf("%d ",a[i][j]);
        }
        printf("\n");
```

```
38          }
39          return 0;
40      }
```

运行结果：

```
请输入矩阵的行列值：
2 3
请输入矩阵元素：
1 2 3
2 8 4
处理后的矩阵为：
8 2 3
2 1 4
```

案例解析：

（1）第4行定义了一个足够大的10×10整型数组a，这样的做法确保了存储空间尽量充足。在实际使用时，我们可以根据矩阵的实际大小来利用这部分存储空间。

（2）第5行定义了两个变量m和n来分别表示实际矩阵的行数和列数。同时，定义了整型变量i和j作为二维数组a的下标变量，用于遍历数组元素。另外，定义了一个整型变量temp作为交换时的临时存储变量。

（3）第6行定义了两个整型指针变量pmax和pmin，它们分别指向矩阵中的具有最大值和最小值的数组元素。

（4）第10~16行通过嵌套循环结构（双重for循环）读取矩阵的所有元素值。

（5）第17行的语句"pmax=pmin=&a[0][0];"给两个指针变量赋予初始地址，即指向矩阵的第一个元素。

（6）第19~26行实施"打擂台"算法。在比较和更新pmax和pmin指针时，应注意当找到新的最大值或最小值时，应更新指针所指向的地址。特别注意，在比较时应使用*pmax和*pmin来访问指针指向的数据值，不能使用pmax和pmin与数组元素值比较，因pmax和pmin存放的是地址，主要区分pmax、pmin和*pmax、*pmin所操作的存储空间数据不同。

（7）第27~29行进行指针指向的最大值和最小值元素的交换。这里应使用*pmax和*pmin来间接访问并交换它们所指向的值，而不是交换指针本身。

（8）第31~38行通过双重for循环输出矩阵的所有元素。在输出每行后，应使用换行符\n来确保矩阵的格式化输出。

综合案例 6_4_2 指针方法连接两个字符串。

案例描述：从键盘接收两个字符串，并利用指针变量的方式将这两个字符串首尾相接，从而模拟标准库函数strcat的功能。

解题思路：首先，我们定义两个足够大的char型一维数组来分别存储用户输入的两个字符串。接着，我们利用指针遍历第一个字符串，直至找到其末尾的'\0'结束符。随后，我们将第二个字符串的每一个字符依次复制到第一个字符串的末尾，从而实现两个字符串的拼接操作。这种方法无须额外的字符串处理函数，仅通过指针操作即可完成字符串的拼接。

程序代码：

Case6_4_2.c

```
#include <stdio.h>
int main()
{
    char str1[50],str2[20];
    printf("请输入两个字符串:\n");
```

```
6          scanf("%s",str1);                  // 读入一个字符串
7          scanf("%s",str2);                  // 滴入一个字符串
8          // 实现字符串连接功能
9          char *p=str1,*q=str2;
10         while(*p!='\0') p++;
11         while(*q!='\0')
12         {
13             *p=*q;
14             p++;
15             q++;
16         }
17         *p='\0';
18         printf("连接后的字符串为:%s",str1);
19         return 0;
20     }
```

运行结果：

```
请输入两个字符串:
Hello
World!
连接后的字符串为:HelloWorld!
```

案例解析：

（1）在第4行，我们定义了两个字符数组str1和str2。考虑到需要将str2的内容连接到str1的末尾，我们确保str1的长度足够大，以容纳两个字符串的内容以及字符串结束符'\0'。

（2）第5行使用printf函数输出一条提示信息，告知用户输入两个字符串。

（3）第6行利用scanf函数从键盘读入第一个字符串并存储在str1中。

（4）第7行同样使用scanf函数从键盘读入第二个字符串并存储在str2中。

（5）第9行定义了两个char型指针变量p和q，并初始化它们分别指向str1和str2的首地址。通过这种赋值方式，我们使指针变量指向了对应的字符数组。

（6）第10行运用while循环，通过递增指针p来遍历str1，直到找到字符串结束符'\0'。此时，指针p指向str1的末尾位置。

（7）第11~16行再次使用while循环，通过递增指针q来遍历str2中的字符。在循环中，我们将str2中的字符逐个复制到str1的末尾，并同时递增指针p。当指针q指向'\0'时，表示str2中的所有字符都已被复制到str1中，循环结束。

（8）第17行通过*p ='\0'手动在拼接后的字符串末尾添加结束符'\0'。由于之前的循环以'\0'作为终止条件，所以并没有将'\0'复制到str1中。这一步是必要的，因为字符串需要以'\0'作为结束标志，以便于后续的字符串操作和输出。

（9）第18行使用printf函数输出连接后的字符串str1。通过%s格式控制符，可以正确地输出整个字符串。

案例拓展：

介绍几个常用的输入输出字符串函数。

（1）gets、fgets标准库函数。

作用：从标准输入设备（通常是键盘）读入一个字符串，直到遇到换行符或EOF（文件结束符）结束。

在综合案例6_4_2中，第6行可以改为gets(str1);，第7行改为gets(str2);，使用回车符作为分隔符，与scanf的区别是gets可以读入空格作为有效字符。

gets不会检查目标数组的大小，这可能导致缓冲区溢出，引发安全问题。因此在现代C语言编程中，推荐使用fgets函数代替。fgets函数可以从指定的输入流中读取一行，直到达到指定的字符数、遇到换行符或

遇到EOF（文件结束符）。使用fgets可以避免缓冲区溢出的风险。

函数原型：char *fgets(char *str, int n, FILE *stream);

str：指向用于存储读取字符串的字符数组的指针。

n：要读取的最大字符数（包括空字符）。

stream：输入流，通常是stdin（标准输入）。

在综合案例6_4_2中，第6行可以改为：fgets(str1,sizeof(str1),stdin);，第7行可以改为：fgets(str2, sizeof(str2),stdin);，其中stdin是标准输入流，默认关联键盘。

（2）puts标准库函数。

作用：向标准输出（通常是屏幕）写入一个字符串，并在末尾自动添加一个换行符。

函数原型：int puts(const char *str);

str：要输出的字符串。

示例：

```
#include <stdio.h>
int main()
{
    char *message = "Hello, World!";
    puts(message);
    return 0;
}
```

在这个例子中，puts函数将输出字符串"Hello, World!"，并在其后添加一个换行符。

Attention for Using 使用注意事项

（1）使用fgets时，请确保目标数组足够大，以容纳要读取的字符串加上一个空字符（'\0'）。

（2）puts函数在输出字符串后会自动添加换行符，如果不希望有额外的换行符，可使用fputs函数代替。

```
#include <stdio.h>                        // 使用 puts、fputs 包含此头文件
int main()
{
    char str[]="Hello, World!";
    // 使用 puts 输出字符串到标准输出，并自动添加换行符
    puts(str);
    // 使用 fputs 输出字符串到标准输出，不自动添加换行符
    fputs(str, stdout);                   //stdout 是标准输出流，默认关联显示器
    return 0;
}
```

（3）若使用gets、fgets、puts、fputs标准库函数，则应包含头文件<stdio.h>。

（4）fgets和puts是C语言中处理字符串输入输出的安全且常用的函数。在编写涉及字符串处理的代码时，应优先考虑使用这些函数。

综合案例 6_4_3 向有序数列插入一个元素。

案例描述：将一个元素插入已排序的序列中，并确保插入新数据后序列依然保持有序。

解题思路：我们可以采用一维数组来存储有序序列。首先，需要确定待插入元素在序列中的合适位置，这通常涉及遍历数组并与现有元素进行比较。找到插入位置后，我们需要将这个位置及其之后的所有元素依次向后移动一个位置，以便为新元素腾出空间。最后，在空出的位置上插入新元素，从而确保整个序列仍然保持有序。

案例实现：分别给出下标法和指针方法两种处理方法，对比两种实现方法，强化对数组和指针应用的理解。

方法一：下标法。

```
1    #include <stdio.h>
2    #define N 11
3    int main()
4    {
5        int a[N]={5,13,20,24,28,31,45,89,95,100};
6        int num,i;                                        //num 表示待插入的数据，i 控制下标
7        printf(" 请输入待插入的数据 :\n");
8        scanf("%d",&num);
9        for(i=0;i<N-1&&a[i]<num;i++);                     // 循环结束，i 就是待插入的位置
10       int j=N-1;
11       for(;j>i;j--)
12       {
13           a[j]=a[j-1];
14       }
15       a[i]=num;
16       printf(" 输出插入数据后的数组:\n");
17       for(i=0;i<N;i++)
18       {
19           printf("%d ",a[i]);
20       }
21       return 0;
22   }
```

运行结果：

```
请输入待插入的数据:
38
输出插入数据后的数组:
5 13 20 24 28 31 38 45 89 95 100
```

案例解析：

（1）第2行定义了一个符号常量N，其值为11，表示数组的大小。

（2）第5行定义了一个一维数组a，包含11个元素。其中初始化了前10个元素，这些元素已经按照从小到大顺序排列，最后一个元素位置留空，用于后续插入新的数据。

（3）第6行定义了两个变量num和i。其中，num用于存储待插入的数据，而i则作为循环控制变量，用于遍历数组并确定新元素的插入位置。

（4）第7行使用printf函数输出提示信息，提示用户输入待插入的数据。

（5）第8行通过scanf函数从键盘读取用户输入的待插入数据，并将其存储在变量num中。

（6）第9行的for循环用于遍历数组a，并与待插入数据num进行比较。循环的主要目的是找到第一个大于num的数组元素，从而确定num应插入的位置。当循环结束时，变量i的值即为插入位置的下标。

（7）第10行定义了一个整型变量j，并初始化为N-1，即数组的最后一个元素的下标。

（8）第11~14行是一个逆向的for循环，从数组的末尾开始，将下标从j到i（不包括i）的元素依次后移一个位置。这一步骤的目的是为新元素num腾出插入空间。

（9）第15行将待插入的数据num赋值给数组a的第i个元素，即a[i] = num，完成插入操作。

（10）第17~20行使用另一个for循环遍历数组a，并将所有元素输出到控制台，以验证新元素是否已正确插入且序列依然有序。

方法二：指针法。

```
#include <stdio.h>
#define N 11
```

```
3   int main()
4   {
5       int a[N]={5,13,20,24,28,31,45,89,95,100};
6       int *p=a;
7       int num;
8       printf("请输入待插入的数据：\n");
9       scanf("%d",&num);
10      for(;p<a+N-1&&*p<num;p++);
11      int *q=a+N-1;
12      for(;q!=p;q--)
13      {
14          *q=*(q-1);
15      }
16      *p=num;
17      printf("输出插入数据后数组:\n");
18      for(p=a;p<a+N;p++)
19      {
20          printf("%d ",*p);
21      }
22      return 0;
23  }
```

运行结果：

```
请输入待插入的数据:
38
输出插入数据后的数组:
5 13 20 24 28 31 38 45 89 95 100
```

案例解析：

（1）第6行定义了一个整型指针变量p，并将其初始化为指向数组a的首元素，用于后续通过指针操作数组。

（2）第10行利用for循环和指针p，在数组中查找待插入数据num的合适位置。循环过程中，p不断后移，并与num进行比较（注意参与比较的是*p），直到找到第一个大于num的元素位置。此时，p所指向的位置即为应插入num的位置。

（3）第11行定义了一个整型指针q，并将其初始化为指向数组的最后一个元素，用于后续的元素后移操作。

（4）第12~15行利用for循环和指针q，将q指向的元素及其之后的所有元素依次后移一个位置，为插入num腾出空间。循环过程中，q指针逐渐前移，直到q和p指向同一位置时停止循环。

（5）第16行将待插入的数据num插入指针p所指向的位置，完成插入操作。

（6）第18~21行使用for循环，让指针p重新指向数组的首元素，并依次后移，输出每个元素的值，以展示插入后的有序数组。

Summary 小 结

This chapter delves into the fundamental concepts and usage of pointers and pointer variables, offering an in-depth analysis of the mechanism of indirect memory access through pointers. It emphasizes the importance of ensuring that pointer

本章详细讲解了指针和指针变量的基本概念及使用方法，深入剖析了指针间接访问内存的机制。强调在使用指针变量时必须确保其指向正确的内存地址，否则可

variables point to the correct memory address when using them, as incorrect access to random memory areas may lead to system errors. Additionally, it introduces two primary methods for referencing arrays: the subscript method and the pointer method, and clarifies that subscripts are a form of pointer operation, such as a[10] being equivalent to *(a+10). Furthermore, it explains how pointers reference strings, including the specific methods and differences between character arrays storing strings and character pointers pointing to strings.

能访问到随机内存区域，从而引发系统错误。同时，介绍了引用数组的两种主要方法：下标法和指针法，并阐明了下标其实是指针运算的一种表现形式，如a[10]等价于*(a+10)。此外，还讲解了指针如何引用字符串，包括字符数组存储字"符串和字符指针指向字符串的具体方法和区别。

Exercise 习　题

一、单选题

1. 假如有以下程序段：int *p,a=9,b=2; p=&a; a=*p+b;，执行该程序段后，a的值为（　　）。

   A. 20　　B. 11　　C. 9　　D. 编译出错

2. 若有以下定义和语句：double k=88 , *p = &k;*p = k;，则下列描述中正确的是（　　）。

   A. 以下两处的 *p 含义相同，都说明给指针变量 p 赋值

   B. 在 double k=88,*p=&k 中，把 k 的地址赋值给了 p 所指向的存储单元

   C. 语句 *p=k 取变量 k 的值放回 k 中

   D. 语句 *p=k 把变量 k 的值赋给指针变量 p

3. 以下程序运行后，输出结果是（　　）。

```
#include <stdio.h>
int main()
{
    char *p="hello";
    p+=4;
    printf("%d\n",*p);
    return 0;
}
```

   A. ello　　B. 字符 o 的 ASCLL 码值

   C. 字符 o 的地址　　D. 报错

4. 下列关于指针说法的选项中，正确的是（　　）。

   A. 指针是用来存储变量值的类型　　B. 指针类型只有 int 型

   C. 指针不可以指向字符串　　D. 指针变量可以与整数进行相加或相减

5. 下列关于指针变量的描述，不正确的是（　　）。

   A. 在没有对指针变量赋值时，指针变量的值是不确定的

   B. 同类指针类型可以进行相减操作

   C. 在使用没有赋值的指针变量时，不会出现任何问题

   D. 可以通过指针变量来取得它指向的变量值

6. 对于基类型相同的两个指针变量之间，不能进行的运算是（　　）。

   A. <　　B. =　　C. +　　D. −

7. 若有如下程序段：

```
int arry[3]={0,1,2}; int *p=arry;
```

则下面对数组元素 arry[1] 引用不正确的是（　　）。

A.*(&arry[1])　　B.p[p-arry+1]　　C.*(arry+1)　　D. *arry+1

8. 以下程序的输出结果为（　　）。

```
#include <stdio.h>
int main()
{
    char a[10]={'0','1','2','3','4','5','6','7','8', 9 }, *p;
    int k=8, p=a+k;
    printf("%s\n",p-3);
    return 0;
}
```

A. 5　　B. 5678　　C. ' 5'　　D. 56789

9. 有如下定义：

```
char *a[2]={"hello","HELLO"};
```

那么以下说法中正确的是（　　）。

A. a 是指针变量，它指向含有两个数组元素的字符型一维数组

B. a 数组成元素的值分别是 "hello" 和 "HELLO"

C. a 数组的两个元素分别存放的是含有 4 个字符的一维字符数组的首地址

D. a 数组的两个元素中各自存放了字符 'h' 和 'H' 的地址

10. 若有如下程序段：

```
int arry[2][2]={0,1,2,3,4};
int *p=arry;
```

则下面对数组元素 arry[1][1] 引用不正确的是（　　）。

A. *(*(arry+1)+1)　　B. p[1][1]　　C. (*(p+1))[1]　　D. *(p+1)[1]

二、程序分析题

1. 下面程序的运行结果是什么？

```
# include <stdio.h>
int main()
{
    int  a[5] = {0, 1, 2, 3, 4 };
    int  *p ;
    p=a;
    printf("%d", *(p+2 ) );
    return 0;
}
```

2. 下面程序的运行结果是什么？

```
#include <stdio.h>
int main()
{
    char x[]="123456";
    char *p;
    p=x;
    printf(" 字符 =%c\n",*(p+5));
```

```
    return 0;
}
```

3. 下面程序的运行结果是什么？

```
#include <stdio.h>
int main()
{
    int a[]={1, 2, 3, 4, 5, 6};
    int  x, y, *p;
    p = &a[0];
    x = *(p+2);
    y = *(p+4);
    printf("*p=%d, x=%d, y=%d\n", *p, x, y);
    return 0;
}
```

4. 下面程序的运行结果是什么？

```
#include <stdio.h>
char s[]="123";
int main()
{
    char *p;
    for(p=s;p<s+4;p++)
    printf("%c  %s ",*p,p);
    return 0;
}
```

5. 请修改下面程序中的错误。

```
#include <stdio.h>
int main()
{
    char a='c';
    char*p1;
    *p1=a;
    printf("*p1=%c",*p1);
    return 0;
}
```

三、编程题

1. 创建一个整型数组，并使用指针找出数组中的最大元素并打印出来。
2. 编写一个程序，使用指针将一个整型数组中的元素逆序排列。
3. 定义一个字符串，使用字符指针遍历该字符串并打印每个字符的ASCII值。
4. 定义一个二维整型数组，使用指针遍历该数组并打印所有元素。
5. 使用完全指针方法实现求字符串长度的功能，即实现strlen函数的功能。
6. 使用完全指针方法实现复制字符串的功能，即实现strcpy函数的功能。
7. 请使用完全指针方法实现连接两个字符串的功能，即实现strcat函数的功能。
8. 请使用完全指针方法实现比较两个字符串大小的功能，即实现strcmp函数的功能。

Chapter 7

Function

第 7 章

函　数

Introduction 主要内容

Function is a vivid representation of C language modular programming idea, they not only optimize the structure of the code, significantly improve the program maintainability, but also greatly enhance the reusability of the code. In this chapter, we will take an in-depth look at the definition, declaration, and call mechanisms of functions. Meanwhile, we will also address the unique nature of local and global variables and how they are stored. In particular, we will focus on the programming techniques and handling mechanisms of arrays and pointers as function parameters, so as to demonstrate their flexibility and efficiency in passing data through the function. In addition, we will introduce the method of recursive program design, which cleverly simplifies the solution of some specific problems, and demonstrates the powerful charm of function calls.

函数是C语言模块化编程思想的生动展现，它们不仅优化了代码的结构，显著提升了程序的可维护性，还极大地增强了代码的可重用性。在本章中，我们将深入剖析函数的定义、声明与调用机制。同时，我们还将解析局部变量与全局变量的独特性质及其存储方式。特别地，我们将重点阐述数组与指针作为函数参数的编程技巧及其处理机制，以展示它们在函数传递数据时的灵活性和高效性。此外，我们还将介绍递归程序设计的方法，它巧妙地简化了某些特定问题的解决方案，展示了函数调用的强大魅力。

Objectives 学习目标

Students should have a thorough understanding of the definition, declaration and calling mechanism of functions, master the scope and storage mode of local and global variables, focus on the use of arrays and pointers as function parameters, as well as the flexible use of pointers in functions, They should also understand the principle of recursive functions and be able to design simple recursive functions, and improve the modular programming ability and code reusability through practice programming.

学生应深入理解函数的定义、声明与调用机制，掌握局部变量与全局变量的作用域和存储方式；重点掌握数组和指针作为函数参数的使用方法，以及指针在函数中的灵活运用；理解递归函数的原理并能进行简单设计；通过实践编程，提高模块化编程能力和代码复用性。

7.1 Definition of Function 函数的定义

7.1节视频

For a simple program, it can be implemented in a single module, but in the face of complex large-scale software systems, modular design is essential. The idea of modular programming can make the program structure clear and hierarchical, and improve the efficiency of programming. As a structured programming language, C language realizes modular design through functions.

对于简单程序，可整体实现于单一模块，但面对复杂的大型软件系统，模块化设计至关重要。模块化编程思想可令程序结构清晰、层次分明，提升编程效率。C 语言作为结构化程序设计语言，通过函数实现模块化设计。

7.1.1 Why Use Functions（为什么要用函数）

Before formally learning functions, let's think about why we need functions to achieve modularity? How to achieve modularity?

According to the previous way of programming, if the program has many functions and all of them are written in the main function, what problems will arise? There will be at least the following problems:

(1) Confused the program structure. As functionality increases, the main function becomes long and complex, making it difficult to read and understand.

(2) Difficult maintenance. When a function needs to be modified or added, it is necessary to find the corresponding code segment in the huge main function, which is error-prone and inefficient.

(3) Write code repeatedly. If the same function needs to be implemented many times in the program, we have to write the same code repeatedly, which not only increases the workload, but also leads to the rapid expansion of the program length.

In order to solve these problems, we can use the idea of modular programming design to achieve the modularization of the program by creating functions. For a more intuitive understanding, we use a diagram to show the role of the function, as shown in Figure 7-1.

From Figure 7-1, we can see that the main function has a lot of repeated code. This not only makes the code longer but also makes it harder to read and maintain. To improve this, we used a modular approach. We took the repeated code segments and put them into separate modules, which we call functions. We gave each function a short and descriptive name, which is called a function name. With this method, we can now call these functions directly in the main function instead of using the original repetitive code blocks. This makes the program structure simpler and makes the code easier to reuse and maintain.

在正式学习函数之前，让我们先思考为什么需要函数来实现模块化？如何实现模块化？

按照以往的编程方式，如果程序的功能繁多且都集中写在主函数中，将会出现什么问题呢？将会至少存在如下问题：

（1）程序结构混乱。随着功能的增加，主函数将变得冗长复杂，难以阅读和理解。

（2）维护困难。当需要修改或添加功能时，需要在庞大的主函数中查找相应的代码段，容易出错且效率低下。

（3）重复编写代码。如果程序中需要多次实现同一功能，我们不得不反复编写相同的代码，这不仅增加了工作量，也导致程序长度迅速膨胀。

为了解决这些问题，我们可以采用模块化程序设计思想，通过创建函数来实现程序的模块化。为了更为直观地理解，我们通过图示法来展示函数的作用，如图7-1所示。

从图7-1中我们发现，主函数有许多重复的代码，这不仅增加了代码的冗长性，还降低了其可读性和可维护性，为了优化这种结构，我们采取了模块化的方法，将那些重复的代码段提取出来，封装成独立的模块，称为函数。并为其赋予一个简洁且描述性的名字，称为函数名。通过这种方法，我们可以在主函数中直接调用这些函数，替代原本冗余的代码块。这样的重构不仅使程序结构更加简洁，还提高了代码的重用性和可维护性。

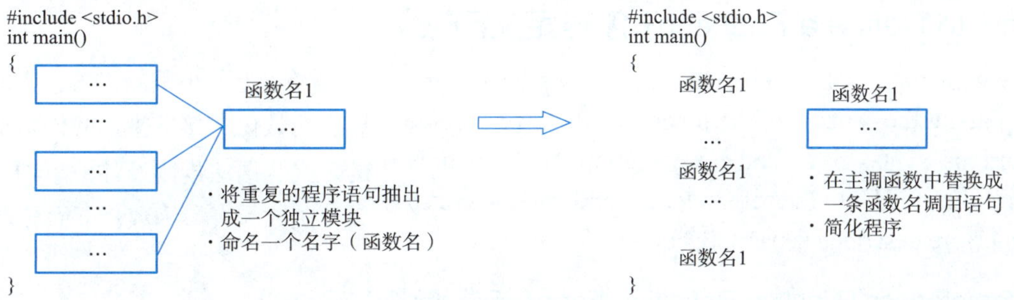

图7-1 函数示意图

The summary function has the following advantages:

(1) Code reuse. By defining a function, we can encapsulate the code segment that implements a specific function, and call it multiple times when needed, avoiding the situation of writing the same code repeatedly.

(2) Clear structure. The program is divided into a number of functions, each function is responsible for achieving a specific function, making the program structure more clear, easy to read and maintain.

Functions typically have two categories:

(1) Library functions are provided by the system. These functions are predefined in the C language standard library or other libraries, and can be used directly in the program. For example, the printf and scanf functions we learned earlier are library functions for input and output.

(2) User-defined function. According to the actual needs, we can define our own functions to achieve specific functions. User-defined functions require us to declare them first and then call them in the program.

Before learning how to define a function, we need to understand and strengthen the concept of a function. A function just likes $f(x)$ in math, it takes the input data(arguments), after a series of calculations or operations, it returns the output data(that return value). We can simplify the program structure and improve the programming efficiency by extracting the repeated code segments into a function and calling the function when needed.

To sum up, the function is an important concept in C language programming, which can help us achieve code reuse and improve the readability and maintainability of the program. In the following study, we will focus on how to define and call functions, and how to use functions to achieve modular design of the programs.

总结函数具有以下优点：

（1）代码复用。通过定义函数，我们可以将实现特定功能的代码段封装起来，并在需要时多次调用，避免了重复编写相同代码的情况。

（2）结构清晰。将程序划分为多个函数，每个函数负责实现一个特定的功能，使得程序结构更加清晰，易于阅读和维护。

函数通常有两个类别：

（1）系统提供的库函数。这些函数是C语言标准库或其他库中预先定义好的，可以直接在程序中使用。例如，我们之前学过的printf和scanf函数就是用于输入输出的库函数。

（2）用户自定义函数。根据实际需要，我们可以自己定义函数来实现特定的功能。自定义函数需要我们先进行声明，然后在程序中调用。

在学习如何定义函数之前，我们需要先了解和加强函数的概念。函数就像数学中$f(x)$，它接收输入数据（参数），经过一系列计算或操作后，返回输出数据（返回值）。通过将重复出现的代码段抽离成函数，并在需要时调用该函数，我们可以简化程序结构，提高编程效率。

综上所述，函数是C语言编程中的重要概念，它能够帮助我们实现代码的复用、提高程序的可读性和可维护性。在后续的学习中，我们将重点学习如何定义和调用函数，以及如何利用函数来实现程序的模块化设计。

7.1.2 How to Define a Function（怎样定义函数）

After the explanation of the previous section, we have some understanding about the concept of the function. Next, we'll delve into the definitions, syntax rules, and considerations of a function. Functions can be divided into the parameterless and parameterized functions, which are described separately below.

经过上一小节的讲解，大家对函数概念已有所认识。接下来，我们将深入探讨函数的定义、语法规则及注意事项。函数可分为无参函数和有参函数，下面分别介绍。

1. Parameterless Function Definition Form（无参函数定义形式）

```
type name function name()
{
    body of function
}
```

```
类型名  函数名 ()
{
    函数体
}
```

Syntax Description（语法说明）:

(1) A blank in() indicates that the function has no arguments. Note that the() cannot be omitted here, it is an integral part of the function definition syntax.

(2) The content in {} is the body of the function, which is responsible for implementing the specific functional logic.

(3) The function name is defined by the user, and it needs to follow the naming rules of the identifier.

(4) The type name refers to the type of the return value of the function. If the function has a return value, the specific type should be specified explicitly, and each function can only return one data after the execution ends. If the function does not return a value, the type name should be written as void, indicating that it is an empty type or a function that does not return a value.

（1）()里的空白代表函数没有参数。请注意，这里的()不能省略，它是函数定义语法中不可或缺的一部分。

（2）{}中的内容则是函数体部分，负责实现函数的具体功能逻辑。

（3）函数名是由用户自行定义的，它需要遵循标识符的命名规则。

（4）类型名指的是函数返回值的类型，如果函数有返回值，则应明确指定具体的类型，并且每个函数在执行结束后只能返回一个数据。如果函数没有返回值，那么类型名应写为void，表示这是一个空类型或无返回值的函数。

For example（举例）:

```
void print_star()
{
    printf("**************\n");
}
```

This is a concrete example of a no-parameter function definition. Enclosed with {} is the code part of the function, which is the body of the function. "print_star" is the function name, and "void print_star()" is the function header. This function has no return value, so the type name is void.

这是一个具体的无参函数定义示例，用{}括起来的是函数代码部分，是本函数的函数体。print_star是函数名，void print_star()是函数首部。此函数无返回值，故类型名用void。

2. Parametric Function Definition Form（有参函数定义形式）

```
type name function name(Formal parameter table
column)
{
    Body of function
}
```

```
类型名  函数名（形式参数表列）
{
    函数体
}
```

Syntax Description (语法说明):

(1) The function of the formal parameter list is to receive the external data. If there are multiple parameters, which needs to be separated by commas, the parameters of each formal parameter list are actually local variables, so they are consistent with the variable definition form.

(2) If there is a return value, use the return statement at the end of the function to return specific data. Note that the return value can only return one data.

(1) 形式参数表列的作用是接收外部数据，若有多个参数用逗号隔开，每个形参表列的参数实际上是局部变量，故与变量定义形式一致。

(2) 若有返回值则函数末尾使用return语句返回具体数据，注意返回值仅能返回一个数据。

For example (举例):

```
int max(int x,int y)
{
    int z;
    z=x>y?x:y;
    return z;
}
```

The int max(int x, int y) function is specifically used to find the maximum of two integers. When you call it, you only need to simply pass in any two integers for x and y, and it quickly evaluates and returns the larger of them.

In this function, we define a local variable called z to store the result of the comparison. By using the conditional operator ?:, you can succinctly decide whether x or y is larger and store the result in z. Finally, this maximum value is returned to the caller with a return statement.

Note that max is the name of the function, and(int x, int y) is the list of its formal parameters. These parameters are used to receive data from the outside for the inside computation of the function. Each formal parameter must be preceded by an explicit data type to ensure the correctness and safety of the function. These parameters are treated as local variables inside the function and can be used without the additional declaration.

int max(int x, int y)这个函数专门用于找出两个整数中的最大值。当调用它时，只需为x和y传入任意两个整数，它就会迅速计算并返回它们中的较大值。

在这个函数中，我们定义了一个名为z的局部变量，用于存储比较结果。通过使用条件运算符?:，可以简洁地判断x和y哪个更大，并将结果存储在z中。最后，通过return语句，将这个最大值返回给调用者。

值得注意的是，max是这个函数的名称，而(int x, int y)则是它的形参列表。这些形参用于从外部接收数据，以便函数内部进行计算。每个形参前都必须明确指定其数据类型，以确保函数的正确性和安全性。这些形参在函数内部被视为局部变量，可以直接使用而无须额外声明。

Case 7_1 (案例 7_1)

Define two functions without parameters to realize the printing a row of asterisks and the outputting of a string “I LOVE CHINA” respectively.

定义两个无参函数，分别实现打印一行星号、输出一个字符串“I LOVE CHINA”的功能。

Thinking (解题思路):

We can create two functions without passing in the parameters based on the form of the function definition. First,define a function to print a line of asterisks. We can intuitively

我们可以根据函数的定义形式，创建两个无须传入参数的函数。首先，定义一个函数用于打印一行星号，我们可以直观

name this function print_star. Since this function does not need to return any value, its type name should be void. Within the body of the function, we simply use a printf statement to print the asterisk.

Next, we define a function that outputs information such as “I LOVE CHINA”. Similarly, this function does not require the parameters or a return value, so the type name is also void. We can name this function print_message to express its function more intuitively. Inside the function body, we also use a printf statement to output the required information.

地将这个函数命名为print_star。由于这个函数不需要返回任何值，因此其类型名应为void。在函数体内，我们只需使用一条printf语句来实现打印星号的功能。

接着，我们再定义一个函数，用于输出“I LOVE CHINA”这样的信息。同样地，这个函数也不需要参数和返回值，因此类型名也为void。我们可以将这个函数命名为print_message，以更直观地表达其功能。在函数体内，我们同样使用一条printf语句来输出所需的信息。

Program Code（程序代码）:

Case7_1.c

```
1    #include <stdio.h>
2    int main()
3    {
4        void print_star();                  //函数声明
5        void print_massage();               //函数声明
6        print_star();                       //函数调用
7        print_massage();                    //函数调用
8        print_star();                       //函数调用
9        return 0;
10   }
11   void print_star()                       //函数定义
12   {
13       printf("************\n");
14   }
15   void print_massage()                    //函数定义
16   {
17       printf("I LOVE CHINA\n");
18   }
```

Program Output（运行结果）:

```
************
I LOVE CHINA
************
```

Case Analysis（案例解析）:

(1) Lines 4 and 5 declare the print_star function and the print_massage function. Functions should be declared before using, as described in a later section.

(2) Lines 6 and 8 call the print_star function, and line 7 calls the print_massage function. Only the function is called, the function body executes, which will be described in the later sections.

(3)Lines 11-14 define the print_star function, and lines 15-18 define the print_massage function. Pay attention to the correct syntax of the definition. Note that functions cannot be nested, that is, you cannot define one function inside another.

（1）第4、5行是声明print_star函数、print_massage函数；函数使用前要声明，后面章节会介绍。

（2）第6行和第8行是调用print_star函数，第7行是调用 print_massage函数，函数调用才能执行函数体部分，后面章节会介绍。

（3）第11~14行为print_star函数定义，第15~18行为print_massage函数的定义。注意定义的正确语法格式。注意函数不可以嵌套定义，即不可以将一个函数定义在另外一个函数内部。

Case Summary（案例总结）:

(1) Functions, as the code modules that implement specific functions, have a simple and direct use way, just calling their names. This kind of the modular design makes the whole program structure clear and logical.

(2) When the multiple function definitions are involved, we should note that these functions should be defined side by side, not nested. That is, the definition of a function cannot appear in the body of another function. For example, the print_star function and the print_message function cannot be defined within the main function. This is because the hierarchy and independence of function definitions require that each function should exist independently to facilitate its own invocation and management. Such specifications help to maintain the stability and readability of the program and ensure that the interaction among the functions is clear.

（1）函数作为实现特定功能的代码模块，其使用方式简单而直接，只需调用其名称即可。这种模块化的设计方式，使得整个程序结构清晰、逻辑明了。

（2）在涉及多个函数定义时，我们应当注意，这些函数应当并列定义，而非嵌套定义。也就是说，一个函数的定义不能出现在另一个函数的函数体内。例如，print_star函数和print_message函数就不能定义在主函数内。这是因为函数定义的层次性和独立性要求，每个函数都应独立存在，以便于各自的调用和管理。这样的规定有助于维护程序的稳定性和可读性，确保函数之间的交互清晰明了。

Case 7_2（案例 7_2）

Use the function to find the maximum value of two integers read from the keyboard.

用函数实现求从键盘读入的两个整数最大值。

Thinking（解题思路）:

In the above introduction, the definition of the max function has been introduced, and this case integrates it into a complete code implementation. If it is defined as a function, the following points need to be considered:

(1) How to name the function name? As far as we can tell, we name it max.

(2) Is it necessary to receive the external data? It needs to receive two integers passed externally, so it needs two formal parameters of int type.

(3) Is a return value required? The function can return the maximum of the two values, so the return value is set to int.

(4) How to implement inside the function body? By comparing the two, find the maximum value and return it using return.

在上述介绍中，已经介绍了max函数的定义，本案例将其整合成一个完成的代码实现。若定义成函数需要考虑如下几点：

（1）函数名如何命名？尽可能见名知意，我们命名为max。

（2）是否需要接收外部数据？需要接收外部传递的两个整数，故需要两个int型的形参。

（3）是否需要返回值？函数可以返回求得的两者最大值，故设定返回值，类型为int。

（4）函数体内部如何实现？通过比较两者，求得最大值并使用return返回。

Program Code（程序代码）:

Case7_2.c

```c
#include <stdio.h>
int main()
{
    int max(int x,int y);                          // 函数声明
    int a,b;                                       // 定义两个整型变量
    printf("Please enter a two-digit comparison size:\n");
    scanf("%d",&a);                                // 从键盘读入数据
    scanf("%d",&b);
```

```
9        printf("max=%d",max(a,b));              //输出函数的返回值
10       return 0;
11   }
12   int max(int x,int y)
13   {
14       int z;
15       z=x>y?x:y;
16       return(z);
17   }
```

Program Output(运行结果):

```
Please enter a two-digit comparison size:
-34 12
max=12
```

Case Analysis(案例解析):

(1) Line 4 is a function declaration, and lines 5-8 input two integer data from the keyboard.

(2) Lines 11-16 are the max function definitions. Two int formal parameters, and they are defined in the same form as variables. Variable names follow the rules for naming identifiers, multiple variables are separated by commas. Note that each variable is preceded by a data type.

(3) The function call in line 9 is written in the output list of printf function, the return value is output after the function is called(executed), and the output is the maximum value of the two.

（1）第4行为函数声明，第5~8行从键盘输入两个整数数据。

（2）第11~16行是max函数定义。两个int型形参，形参变量的定义形式跟变量一样，变量名遵循标识符命名规则，多个变量用逗号隔开，且要注意每个变量前面都要加数据类型。

（3）第9行函数调用写在了printf的输出表列中，则先调用函数（执行函数）后输出其返回值，输出的是两者中的最大值。

7.2 Function Declaration and Call 函数的声明和调用

7.2节视频

Now that we've covered the definition of a function, let's learn how to call it correctly. The C language follows the principle of defining before using. In fact, the function definition is often separated in different files, or the function definition is after the function call, so the function needs to be declared before the function call. Learn more about function declaration and call through this section.

我们已经介绍完了函数的定义，接下来学习如何正确地调用函数。C语言遵循着先定义后使用的原则，实际上函数定义常常分开在不同文件，或者函数定义在调用之后，故在函数调用之前需要先声明函数。通过本节的学习一起深入了解函数的声明和调用。

7.2.1 What is Function Declaration and Call(什么是函数的声明和调用)

1. Function Declaration General Form (函数声明一般形式)

```
type name function name(formal parameter list
column);
```

```
类型名 函数名(形参表列);
```

Function declaration effects:

(1) Inform the compiler of the existence, return type and parameter information of the function, so that the compiler can correctly handle the function call during compilation.

函数声明的作用：

（1）通知编译器函数的存在、返回类型及参数信息，使编译器在编译时能够正确处理函数调用。

(2) Perform type checking in the compilation phase to ensure that the parameter type is consistent with the declaration when the function is called.

（2）在编译阶段进行类型检查，确保函数调用时参数类型与声明一致。

For example（举例）:

```
int max(int x,int y);                    //案例 7_2 中的该语句即为函数声明语句
```

A function declaration is a way to specify a function in advance by simply adding a semicolon to the end of the function header. This declaration is mainly used to inform the compiler of the existence, type and number of parameters of the function in advance, without checking the specific names of the formal parameters. Therefore, when declaring a function, the name of the formal parameter variable can be omitted, for example: int max(int, int). However, when defining a function, the name of the formal parameter variable cannot be omitted, because the function definition needs to completely describe the function and implementation details of the function.

函数声明是对函数进行预先说明的一种方式，它仅需在函数首部末尾添加分号即可。这种声明方式主要起到提前告知编译器函数的存在、类型以及参数个数的作用，而不会对形参的具体名称进行检查。因此，在函数声明时，形参变量名是可以省略的，例如：int max(int, int);。但在函数定义时形参变量名不可省略，因为函数定义需要完整地描述函数的功能和实现细节。

2. Function Call General Form（函数调用一般形式）

```
Function name(argument table column);
```

```
函数名（实参表列）;
```

What a function call does: It passes a parameter value to a formal parameter and executes a specific function. For a function with a return value, it is necessary to receive its return value in time to ensure the correct execution of the program.

函数调用的作用：传递实参值给形参并执行特定函数。对于有返回值的函数，需及时接收其返回值信息以确保程序正确执行。

We use the example of finding the maximum of two integers to understand and master the declaration and call of functions.

我们通过求两个整数最大值的例子来理解和掌握函数的声明和调用。

```
#include <stdio.h>
int max(int x,int y);                          //全局函数声明
int main()
{
    //int max(int x,int y);                    //局部函数声明
    int a,b,c;
    printf("Please enter a two-digit comparison size:\n");
    scanf("%d",&a);
    scanf("%d",&b);
    c=max(a,b);                                //函数调用
    printf("max=%d",c);
    return 0;
}
int max(int x,int y)                           //函数定义
{
    int z;
    z=x>y?x:y;
    return(z);
}
```

Program Output（运行结果）:

```
Please enter a two-digit comparison size:
```

```
12 -34
max=12
```

This is the function code for finding the maximum of two integers.

这是求两个整数最大值的函数代码。

Code parsing:

代码解析：

(1) Line 2 is a declaration of the max function, which can be done by adding a semicolon after the function header. A function declaration is necessary if the function definition appears after the call or in another file. If the function definition precedes the call, the function declaration can be omitted because the compiler recognizes the function definition at compile time.

（1）第2行是对max函数的声明，只需在函数首部后添加分号即可完成。若函数定义出现在调用之后或者位于其他文件中，就必须进行函数声明。若函数定义位于调用之前，则函数声明可以省略，因为编译器在编译时会先识别到函数的定义。

(2)The location of the declaration is flexible and can be in global scope, as shown in line 2 of the code, or be in local scope, as shown in line 5 of the code. When choosing the declaration location, we only need to determine according to the actual needs. A global declaration allows all functions from the beginning of the declaration to call and use the function, while a local declaration is valid only within the function where it is located, and has a higher degree of locality.

（2）声明的位置具有灵活性，既可以位于全局范围，如代码的第2行所示，也可以置于局部范围，如代码的第5行所示。在选择声明位置时，我们只需根据实际需求来确定。全局声明允许从声明开始之后的所有函数都能调用和使用该函数，而局部声明则仅在其所在函数内部有效，具有更高的局部性。

(3) The 10th line is the function call statement. Execute max(a, b) to generate a function call, pass the values of the actual parameters a and b to the formal parameters x and y, and then return the maximum value after calculating. For example, input 12 and −34, bring back the 12 after calculating the maximum value, receive the return value 12 in time with the c variable, and finally output.

（3）第10行为函数调用语句。执行max(a,b)产生函数调用，将实参a和b的值传给形参x和y，然后求得最大值之后返回，如输入12和−34，求得最大值为12后带回，用c变量及时接收返回值12，最后输出。

3. Function's Formal and Actual Parameters（函数的形参和实参）

(1) When defining a function, the variable in parentheses after the function name is a formal parameter, which is a virtual parameter, and its value is obtained by passing the actual parameter. Note that the formal parameter should be written as: int max(int x, int y); not int max(int x, y);.

（1）在定义函数时，函数名后面括号中的变量为形参，是虚拟的参数，其值通过实参传递来获得。注意形参的写法应写为：int max(int x,int y);不要写成int max(int x,y);。

(2) The parameters in the function call are actual parameters. The parameter can be a constant, variable, or expression, and is used to pass the data to a function formal parameter. Note that the actual parameters should not be written as: max(int a, int b). The actual parameters are specific constants, variables, or expressions, and should be written as max(a, b), it represents to pass the values of a and b variables to the formal parameters x and y. It can be constants for the actual parameters , such as max(12, −34), 12 and −34 are passed to the formal parameters x and y in turn.

（2）函数调用中的参数为实参。实参可以是常量、变量或表达式，作用是为函数形参传递数据。注意实参的写法不要写成：max(int a,int b);。实参为具体的常量、变量或表达式，应写成max(a,b);。将a和b变量的值传递给形参x和y，也可以放常量，如max(12,−34);，12和−34依次传递给形参x和y。

(3) The value of the actual parameter is passed to the value of the formal parameter in one direction, and the type and number of the formal parameter should correspond to the actual parameter .

（3）实参值单向传递给形参值，形参与实参类型和个数应一一对应。

7.2.2 How Does the Function Work（函数如何工作）

Let's use Figure 7-2 to understand the entire execution of a function call. In the figure, we observe that the function2 is called inside the function1, and the next three steps are needed.

Step 1: Save the execution state of function1 and push the current parameters, variables and execution address of function1 into the stack.

Step 2: Transfer control to function2 and execute the body of function2.

Step 3: After the function2 function is finished, pop the parameters, variables, and return addresses saved in step 1 out the stack, and continue to execute the remaining statements of the calling function function1.

Based on the working principle of the above function call, let's straighten out the calling process of the max function again. When the program is executed to the max function call in sequence, the scene of the calling function(main function) will be saved first, and the control right will be given to the max function, that is, the function jump will be generated. In this process, the actual parameters a and b will be passed to the formal parameters x and y one by one, as shown in Figure 7-3. After the max function is executed, return to the calling function, assign the return value 12 to the variable c, and continue to execute the remaining statements in the main function.

我们结合图7-2来理解函数调用的整个执行过程。图中，我们观察到在function1函数内部产生了function2函数的调用，接下来需要做三个步骤的操作。

步骤1：保存好function1的执行状态、将function1当前的参数、变量和执行地址压入栈中。

步骤2：将控制权转移给function2，执行function2函数体部分。

步骤3：function2函数结束后，将步骤1中保存好的参数、变量和返回地址依次出栈，继续执行主调函数function1剩下的语句。

根据上述函数调用工作原理，让我们再次理顺max函数的调用过程。当程序按顺次执行到max函数调用时，会先保存主调函数（主函数）的现场，并将控制权给max函数，即产生函数跳转，在这个过程中实参a和b会一一传递给形参x和y，如图7-3所示。执行完max函数后返回主调函数将返回值12赋值给变量c，继续执行主函数中剩下的语句。

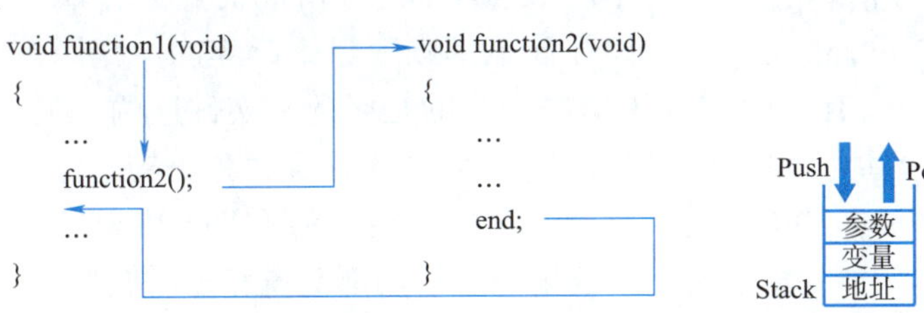

图7-2 函数的工作原理

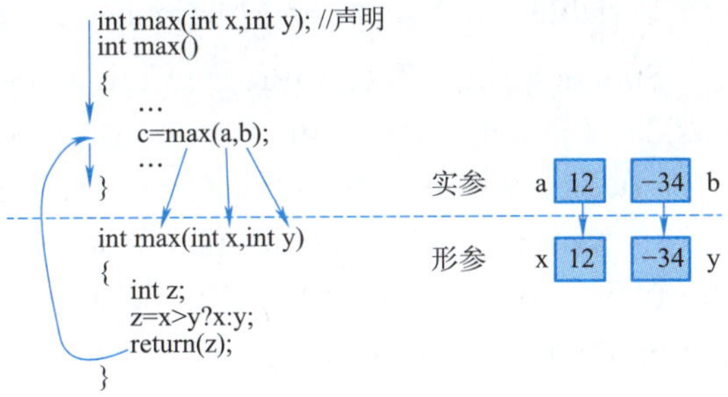

图7-3 max函数调用流程

Think about whether the change of the formal parameter affects the actual parameter. Specifically, in this example, x and y are formal parameters, and a and b are actual parameters. If the values of x and y change, do a and b also change?

The answer is no. Because the value transfer relationship is one-way, and we can observe from Figure 7-3 that x and y, a and b are independent, the change of x and y will not affect a and b, and x and y belong to local variables, which are created

请思考形参改变是否会影响实参。具体地，在这个例子中，x和y是形参，a和b是实参，如果x和y值变了，a和b是否也跟着变化?

答案是不会。因为值的传递关系是单向的，且从图7-3我们观察到x和y，a和b空间是独立的，x和y改变不会影响a和b，且x

only when the max function is called and released when the max function ends. Therefore, their relationship is one-way transmission and space independence, so the change of formal parameters in this case will not affect the change of actual parameters.

Other description:

(1) We know that function definitions cannot be nested, but function calls allow nested calls. For example, the main function calls funciton1, and the function1 function can also call function2, following the principle of function calls.

(2) Because the main function is a program entry, a program has only one main function, and it can only be called by the operating system, not by other functions.

和y属于局部变量，只有当max函数被调用时被创建，max函数结束时释放。故单向传递且空间独立，故本案例中的形参改变不会影响实参改变。

其他说明：

（1）我们已知函数定义不可以嵌套定义，但函数调用允许嵌套调用，如主函数调用funciton1，function1函数中还可以调用function2，遵循函数调用原理。

（2）因主函数（main函数）是程序入口，一个程序有且仅有一个main函数，且其只能被操作系统调用，而不能被其他函数调用。

7.3 Local and Global Variables 局部变量与全局变量

7.3节视频

Variables are often used in programs. In the process of reading and writing programs, we find that variables are sometimes defined inside the main function or outside the main function. Even if there are variables with the same name in different functions, what is the difference between them? Can a variable defined in one function be referenced in another function? To what extent are variables defined in different locations valid? These questions will be answered in this section.

变量在程序中是经常使用的，在阅读程序和编写程序过程中，我们发现变量有时候定义在主函数内部，也可以定义在主函数外部。甚至不同的函数内有同名的变量，它们之间有什么区别呢？在一个函数中定义的变量，在其他函数中能否被引用？在不同位置定义的变量，在什么范围内有效？这些疑问将在本节中给出解答。

7.3.1 Scope and Lifetime（作用域和生存期）

Variables have scope and lifetime.

Scope(space): Scope refers to the scope of the variable, outside of which the variable has no effect. Scope object refers to the two categories of functions and files in C programs. If we classify variables according to their scope, they can be divided into two categories: local variables and global variables. This section focuses on local and global variables.

Lifetime(time): It refers to the age of a variable. Variables are also classified according to their lifetime, which can be divided into two categories: dynamic storage and static storage. We will talk about this in the next section.

变量均有作用域和生存期。

作用域（空间）：指该变量的作用范围，离开该范围则变量不起作用。作用范围在C程序中指的是函数和文件这两大类。如果我们按照作用域的角度对变量进行分类，可分成两大类：局部变量和全局变量。本节重点讲解局部变量和全局变量。

生存期（时间）：指变量的存在时间。同样按照生存期的角度对变量进行分类，可分成两大类：动态存储方式和静态存储方式。这个我们下小节会讲到。

7.3.2 Local variables（局部变量）

Variables defined inside a function or within a compound statement are called local variables.

Local variable scope: Variables defined within a function are only valid within the function. Similarly, variables defined

在函数内部或者在复合语句内定义的变量称为局部变量。

局部变量作用域：在函数内部定义的变量只在本函数内有效，同样地在复合语

within a compound statement are only valid within the scope of the compound statement. If the scope is exceeded, they are invalid, that is, they are locally valid.

句内定义的变量也只在复合语句的范围内有效，超过该范围则无效，即局部有效。

Local variable lifetime: During the execution of a function or coincidence statement. If the function or compound statement ends, the local variable is released. A compound statement is a statement that is enclosed by {}.

局部变量生存期：函数或复合语句执行期间。若函数或复合语句结束，则局部变量会被释放。复合语句即为{}括起来的语句。

For example（举例）:

```
int main()
{
    int m,n;
    …
}
```

The m and n variables defined inside the main function are local variables, and m and n are valid only inside the main function, and are invalid in other functions. The memory space is created when the definition of the variable is executed, and the m and n memory spaces are released after the main function is finished.

在主函数内部定义的m和n变量为局部变量，m和n仅在主函数内有效，在其他函数无效。执行到变量定义时开辟内存空间，主函数结束后m和n内存空间会被释放。

```
float f1(int a)
{
    int b,c;
    …
}
char f2(int x,int y)
{
    int i,j;
    …
}
```

Where int a, int x, and int y are formal parameters of the function, which are also local variables. The variables b, c, i, j defined in the body of the function are also local variables, which are valid in the function they define. When the function is called, the space of the local variables are opened up, and when the function ends, the space of local variables are released. Specifically, the a, b, c local variables are defined within f1 and are valid only within the function f1, and the x, y, i, j local variables are defined within f2 and are valid only within the function f2.

其中int a、int x和int y是所在函数的形参，同样属于局部变量。函数体内定义的变量b、c、i、j也是局部变量，在其所定义的函数内有效，函数被调用时局部变量开辟空间，函数结束时局部变量释放空间。具体地，a、b、c局部变量定义在f1内，仅在函数f1内部有效，x、y、i、j局部变量在f2内定义，仅在函数f2内部有效。

```
int main()
{
    int a,b;
    …
    {   //此为复合语句
        int c;
        c=a+b;
        …
```

```
    }
    …
}
```

Wherein int a and b are local variables defined inside the main function, and the scope is the scope of the main function. c is defined inside a compound statement and is valid only inside this compound statement. It is invalid when it is used outside a compound statement.

其中int a,b是定义在主函数内部，是属于主函数内部的局部变量，作用域是主函数范围。c定义在复合语句内部，仅在这个复合语句内有效，在复合语句外面使用将无效。

Through the above example, we can understand the definition, scope and lifetime of local variables. Can different scopes have the same name? The answer is yes. Look at a simple example.

通过上述例子，理解了局部变量的定义、作用域和生存期，那不同的作用域是否可以同名呢？答案是可以。看一个简单的例子。

```
int main()
{
    int a,b;
    …
}
float f1( int a)
{
    int b,c;
    …
}
```

The local variables a and b are defined inside the function f1 and are valid only within the scope of the function f1. a and b with the same name are defined inside the main function, and a and b are valid only inside the main function. Local variables with the same name are defined inside different functions(scopes). In fact, they do not interfere with each other and each have its own storage space, just like two people with the same name in real life. Although they have the same name, they are actually two independent and unrelated people.

在函数f1内部定义了局部变量a和b，该a和b只在函数f1范围内有效。在主函数内部定义了同名的a和b，此a和b仅在主函数内部有效，在不同的函数（作用域）内部定义了同名的局部变量，实际上两者互不干扰，各自有独立的存储空间，就像现实生活中同名的两个人，虽然同名，但实际为两个独立的互不相干的人。

7.3.3 Global Variables（全局变量）

A variable defined outside a function is a global variable.

在函数外定义的变量即为全局变量。

Global variable scope: Overrides all regions from the beginning of the definition to the end of the current file, which means that it can be accessed and used by any subsequently defined function in the same file.

全局变量作用域：覆盖从定义开始直至当前文件结束的所有区域，这意味着它可以被同一文件中的任何后续定义的函数访问和使用。

Global variable lifetime: Throughout the execution of the program, from the beginning of the program to the end of the program.

全局变量生存期：贯穿于整个程序的执行过程，从程序开始运行直至程序结束。

For example（举例）:

```
#include <stdio.h>
int x=1,y=2;                    //定义在函数外部的变量，即为全局变量
int f1(int a)
{
```

```
5        int b;                                     //定义函数内部，为局部变量
6        …
7    }
8    float c1=3,c2=4;                               //全局变量
9    float(int x1,y1)                               //局部变量
10   {
11       …
12   }
13   int main()
14   {
15       int c3,c4;                                 //局部变量
16       …
17       return 0;
18   }
```

From the definition and scope of global variables above, we can see that int x and int y in line 2 are global variables, and the definition is valid from the position where the variable is defined to the end of the source file. The float c1 and float c2 in line 8 are also global variables. Similarly, the definition is valid from the position where the variable is defined to the end of the source file. Because it is defined after the f1 function, it is invalid in f1.

Thinking: If a global variable and a local variable have the same name, how does the system tell?

When a global variable has a name conflict with a local variable, that is, they have the same name, the rule of the C compiler is “local first”. This means that within the scope of a local variable, the global variable is “masked” or “hidden”, that is, the compiler will recognize and use the local variable first.

由我们上面根据全局变量的定义和作用范围可知：第2行的int x和int y为全局变量，从定义变量的位置到源文件结束该定义都有效。而第8行的float c1和float c2也为全局变量，同样地从定义变量的位置到源文件结束该定义都有效，因其定义在f1函数之后故在f1中是无效的。

请思考：若全局变量和局部变量同名，系统如何去判别呢？

当全局变量与局部变量发生名称冲突时，即它们拥有相同的名称，C编译器的规则是“局部优先”。这意味着在局部变量的作用域内，全局变量会被“屏蔽”或“隐藏”，即编译器将优先识别和使用局部变量。

For example（举例）：

```
#include <stdio.h>
int a,b=5;                                  //全局变量
int main()
{
    int min(int a,int b);
    int a=9;                                //局部变量
    printf("min =%d\n", min(a,b));          //a 为局部变量，b 为全局变量
    valuereturn 0;
}
Int min(int a, int b)                       //这里的 a 和 b 为局部变量，按收实参值 9 和 5
{
    int c;
    c=a<b?a:b;
    return c;
}
```

Program output（运行结果）：

```
min=5
```

Important notes:

(1) Assignment problem: If the global variable is not assigned, the value assigned to the numeric system is 0. If it is a character, the value assigned to the system is '\0'. This is different from a local variable, which is not automatically assigned to an unassigned local variable, so its value is uncertain. For example, if the global variable a in line 2 is not initialized, the system automatically assigns a value of 0. The local variable c in line 12 is initialized and has a random value.

(2) In line 6, we define a local variable a that has the same name as the global variable a defined in line 2. The C language allows local variables to exist with the same name as global variables, but when used, local variables "mask" global variables. Therefore, when the min(a, b) function is called at line 7, the first argument a, is actually the value 9 of the local variable, and the second argument b, is the value 5 of the global variable.

(3) Local variables a and b with the same name as the global variable are also defined inside the min function. These local variables also mask global variables. Their values are derived from the actual arguments passed to the min function, that is, the value of a is 9 and the value of b is 5.

In summary, when a local variable has the same name as a global variable, the local variable takes precedence in its scope, while the global variable is temporarily "hidden".

重要说明：

（1）赋值问题：如果全局变量未赋值，对于数值型系统赋值为0，如果是字符型，系统赋值为'\0'。这和局部变量不同，系统不会为未赋值的局部变量自动赋值，所以其值不确定。如第2行的全局变量a未初始化，系统自动赋值0。第12行的局部变量c定义时为初始化，为随机值。

（2）在第6行中，我们定义了一个局部变量a，它与第2行定义的全局变量a同名。C语言允许局部变量与全局变量同名存在，但在使用时，局部变量会“屏蔽”全局变量。因此，在第7行调用min(a,b)函数时，作为第一个实参的a实际上是局部变量的值9，而第二个实参b则是全局变量的值5。

（3）在min函数内部也定义了与全局变量同名的局部变量a和b。这些局部变量同样会屏蔽全局变量。它们的值来源于传递给min函数的实参，即a的值为9，b的值为5。

总结来说，当局部变量与全局变量同名时，局部变量在其作用域内会优先被使用，而全局变量则会被暂时“隐藏”。

7.4 Storage of Variables 变量的存储方式

7.4节视频

The previous section describes the scope and lifetime of variables, as well as local and global variables divided by scope. This section looks at variables from a lifetime perspective, depending on how they are stored and the type of storage. The main contents of this section include the storage method of variables, the storage category of local variables, and the storage category of global variables.

上节介绍了变量的作用域和生存期，以及按照作用域划分的局部变量和全局变量。本节将从生存期的角度来观察变量，这跟变量的存储方式和存储类别有关。本节主要内容有变量的存储方式、局部变量的存储类别、全局变量的存储类别。

7.4.1 What Is the Storage of Variables（什么是变量的存储方式）

Variables are divided into local variables and global variables according to the perspective of scope, and divided into static storage mode and dynamic storage mode according to the perspective of lifetime.

按照作用域区分变量的角度，将变量分为局部变量和全局变量，而按照生存期的角度可以分为静态存储方式和动态存储方式。

Static storage mode: The system allocates fixed storage space from the beginning of the program until the end of the program, and the storage mode has a long lifetime. For example,

静态存储方式：由系统从程序开始时就分配固定的存储空间直到该程序结束便释放，该存储方式的生存期长。如全局变

global variables and static local variables are stored in this way, and the area opened up is a static storage area.

Dynamic storage mode: The storage space is dynamically allocated during the running of the program. The storage space is dynamically allocated for the variable from the time the function is called until the storage space of the variable is released at the end of the function. Local variables are stored in this way, and the area opened up is a dynamic storage area.

Table 7-1 shows the division of the storage area, and Figure 7-4 shows the schematic diagram. Among them, static local variables and global variables belong to static storage mode, which are allocated fixed space by the system, have a long lifetime, and do not change much during the running of the program. Local varia bles belong to the dynamic storage mode, which will change with the call and end of the function during the running of the program, and have a short life span.

表 7-1 变量存储区域划分

用户存储区			
程序存储区	数据存储区		
程序代码	静态存储区		动态存储区
	static 局部变量	全局变量	局部变量

量和static局部变量采用就是这样的存储方式，所开辟的区域为静态存储区。

动态存储方式：在程序运行期间动态分配存储空间，从调用函数开始动态为变量分配存储空间，直到函数结束该变量的存储空间便释放。局部变量采用的就是这样的存储方式，所开辟的区域为动态存储区。

表7-1给出了存储区域的划分，图7-4给出了示意图。其中static局部变量和全局变量属于静态存储方式，由系统分配固定的空间、生存期长，程序运行期间变化不大。局部变量属于动态存储方式，程序运行期间会随着函数的调用和结束而变化，生存期短。

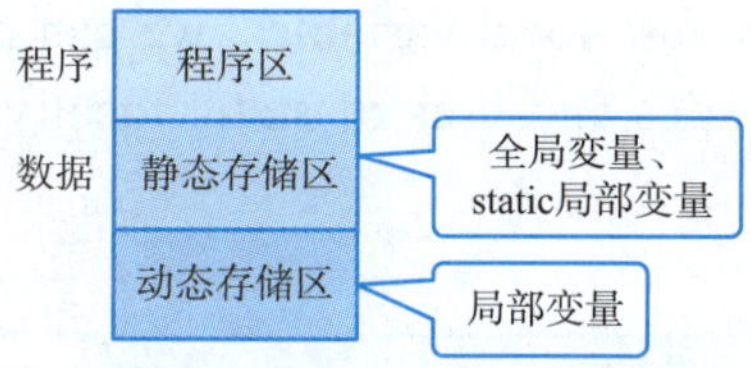

图7-4 内存空间存储示意图

When we dig deeper, we find that the variables not only have the property of data type(which is used to determine the amount of needed storage space), but also have another important property-storage category. The storage category actually determines how the variable is stored, that is, it determines whether the variable is placed in static storage or dynamic storage. This property is critical to understanding the lifecycle of a variable and its behavior in a program, as shown in Table 7-2, it illustrates all of the properties of a variable.

深入理解后，我们发现变量不仅具备数据类型这一属性（用于确定所需存储空间的大小），还具备另一个重要属性——存储类别。存储类别实际上决定了变量的存储方式，也就是说，它决定了变量是存放在静态存储区还是动态存储区。这一属性对于理解变量的生命周期以及其在程序中的行为至关重要，见表7-2，展示了变量的所有属性。

表 7-2 变量的属性

存储类型	存 储 类 别					
int、float、double、char 等	auto		static		extern	
	auto 局部变量	函数形参	static 外部变量	static 局部变量	将作用域扩展到本文件内	将作用域扩展到文件外

7.4.2 Storage Categories of Local Variables（局部变量的存储类别）

Storage categories of local variables mainly include dynamic(auto) storage category and static(static) storage category.

Where the local variable defaults to the auto storage category. When a function is called, it temporarily creates storage

局部变量的存储类别主要有动态（auto）存储类别和静态（static）存储类别。

其中，局部变量默认是auto存储类别。函数被调用时在动态存储区临时开辟存储

space in the dynamic storage area, and when the function call ends, it releases the storage space. This process is dynamic.

空间，函数调用结束则释放存储空间，这个过程是动态的。

For example（举例）:

```
int f()
{
    auto int a, b=5;
    …
}
```

Inside the function f, the local variables a and b belong to the auto storage category by default. Since auto is the default storage category, it can be omitted when declared. In fact, the local variables we defined earlier are of type auto unless they explicitly specify a storage type. This means that these variables allocate memory when the function is called and automatically release memory when the function returns.

在函数f的内部，局部变量a和b默认属于auto存储类别。由于auto是默认的存储类别，因此在声明时可以省略。实际上，我们之前所定义的局部变量，若未明确指定存储类别，它们都是auto类别的。这意味着这些变量在函数被调用时分配内存，并在函数返回时自动释放内存。

Next, we will focus on static local variables. By comparing them with auto local variables, we can deeply understand the characteristics and uses of static local variables, as shown in Table 7-3.

下面我们将重点聚焦于静态（static）局部变量，通过将其与auto局部变量进行对比，来深入理解和认识static局部变量的特性与用途，见表7-3。

表 7-3　auto 和 static 局部变量的区别

对比项	auto 局部变量	static 局部变量
空间	在动态存储区开辟空间	在静态存储区开辟空间
生存期	生存期短：函数执行期间	生存期长：整个程序执行期间
作用域	作用域：函数范围内	作用域：函数范围内
再次调用	再次调用函数时，上次函数调用时已释放，重新开辟空间，重新赋值	再次调用函数时，上次函数结束未释放（空间仍在），使用上次函数调用后的结果

According to the comparison of Table 7-3, we can conclude that when a local variable is added to static, it will be stored in the static storage area, and its lifetime is long. Specifically, when the function is called again, the auto local variables will release all the memory space they occupy after the last function call, so each call needs to open up space for these variables and get new values. In contrast, static local variables do not release their memory space at the end of the function call, so when the next function call occurs, these variables remain the value at the end of the last function call, and there is no need to re-open the space. In the following, we will explain this feature in detail through specific cases.

根据表7-3的对比，我们可以得出结论当局部变量加上static后其将存储到静态存储区，且生存期边长。具体地，当函数被再次调用时，auto局部变量在上次函数调用结束后会全部释放其占用的内存空间，因此每次调用都需要重新为这些变量开辟空间并获取新的值。与之不同的是，static局部变量在函数调用结束后不会释放其内存空间，因此在下一次函数调用时，这些变量仍然保持上次函数调用结束时的值，无须重新开辟空间。下面，我们将通过具体的案例来详细解释这一特性。

Case 7_3（案例 7_3）

Factorial using static local variables.

利用静态局部变量求数的阶乘。

Program Code（程序代码）:

Case7_3.c

```
#include <stdio.h>
```

```
2     int main()
3     {
4         int fac(int);                         // 函数声明，声明时可以省略形参变量名
5         int i;
6         for(i=1;i<=3;i++)
7             printf("%d\n",fac(i));            // 调用了 3 次 fac 函数
8         return 0;
9     }
10    int fac(auto int n)
11    {
12        static int f=1;                       //static 局部变量
13        f=f*n;
14        return f;
15    }
```

Program Output（运行结果）:

```
1
2
6
```

Case Analysis（案例解析）:

(1) Lines 6 and 7 are a for loop that repeats the call to the fac function, and to test the result after each function call. The loop is executed three times to compute the factorial of 3(3!).

(2) The fac function is called for the first time when i is equal to 1. Inside the fac function, the formal parameter n is an auto local variable for which a new memory space is opened each time the function is called. On this call, n gets an argument value 1. And f is a static local variable that is allocated memory at compile time and released at the end of the program. Therefore, after f is assigned a value 1, the value of f is retained even the function ends. Then, after the execution of line 13, the value of f becomes 1, the function returns this value through the return statement, and the memory space of n is released. After returning to the main function, the output return value is 1.

(3) When i is equal to 2, the fac function is called again. Parameter n gets the new memory space again and receives the argument value 2. Since f is a static local variable, it retains its value 1 from the last function call. After executing line 13, the value of f is updated to 2, and then the function ends and releases the memory space of n. At this point, the fac function returns a value 2, and the main function outputs this value.

(4) When i equals 3, the fac function is called for the third time. The parameter n again gets a new memory space, while the value of f remains the result of the last call, and is 2. After the function is executed, the value of f is updated to 6, and the memory space of n is released. Finally, the fac function returns a value of 6, and the main function outputs this value, marking the end of the entire program.

（1）第6、7行是一个for循环，用于重复执行fac函数的调用，以测试每次函数调用后的结果。循环共执行三次，目的是计算3的阶乘（3!）。

（2）当i等于1时，fac函数首次被调用。在fac函数内部，形参n是一个auto局部变量，每次函数调用时都会为其开辟新的内存空间。此次调用时，n获得的实参值为1。而f是一个静态局部变量，它在编译时就被分配了内存空间，并且会在整个程序结束时才被释放。因此，当f被赋值为1后，即使函数结束，f的值也会被保留。随后，第13行代码执行后，f的值变为1，函数通过return语句返回这个值，同时n的内存空间被释放。返回主函数后，输出返回值为1。

（3）当i等于2时，fac函数再次被调用。形参n再次获得新的内存空间，并接收实参值2。由于f是静态局部变量，它保留了上次函数调用后的值1。执行第13行代码后，f的值更新为2，然后函数结束并释放n的内存空间。此时，fac函数返回值为2，主函数输出这个值。

（4）当i等于3时，fac函数进行第三次调用。形参n再次获得新的内存空间，而f的值保持为上次调用的结果2。函数执行完毕后，f的值更新为6，n的内存空间被释放。最后，fac函数返回值为6，主函数输出这个值，标志着整个程序的结束。

Case Summary（案例总结）:

Throughout the process, because f is a static local variable, its value is retained among the function calls, allowing us to accumulate the result of each call and end up with a factorial of 3.

在整个过程中，由于f是静态局部变量，它的值在函数调用之间得以保留，使得我们能够累积每次调用的结果，最终得到3的阶乘。

Considerations When Using Static Local Variables 使用静态局部变量时的注意事项

(1) Although the space occupied by a static local variable will not be released, it is still a local variable confined to the interior of a function and can only be accessed and used within the function in which it is defined.

(2) The initial value of a static local variable is determined at compile time, and if the programmer does not explicitly assign a value to it, the system automatically initializes it to 0. This is consistent with the processing mechanism of global variables when initial values are assigned.

(3) Static local variables should be used with caution and avoid unnecessary use. This is because static local variables are not released at the end of a function call, have a long lifetime, and consume memory resources. In addition, when a function is called frequently, the values of static local variables can become difficult to trace, which can reduce the readability and maintainability of the program. Therefore, it is not recommended to use static local variables too much unless there is a specific need or reason to do so.

（1）尽管静态局部变量所占用的空间不会被释放，但它依然是局限于函数内部的局部变量，只能在定义它的函数内部被访问和使用。

（2）静态局部变量的初始值是在编译时确定的，如果程序员没有显式地为其赋值，系统会自动将其初始化为0。这与全局变量在赋初值时的处理机制是一致的。

（3）应谨慎使用静态局部变量，并避免非必要的使用。这是因为静态局部变量不会在函数调用结束时被释放，其生存期较长且会占用一定的内存资源。此外，当函数被频繁调用时，静态局部变量的值可能会变得难以追踪，这会降低程序的可读性和可维护性。因此，除非有特定的需求或理由，否则不建议过多地使用静态局部变量。

7.4.3 Storage Categories of Global Variables（全局变量的存储类别）

Global variables are stored in the static storage area by default, and their scope extends to the end of the file where they are located. However, we can use the extern keyword to extend the scope of a global variable to be accessible not only within the file in which it is defined, but also across files. In contrast, the static keyword is used to limit the scope of global variables. When we declare a global variable as static, its scope is limited to the inside of the file in which it is defined, and it cannot be accessed in other files. This helps to avoid potential conflicts of global variables among different files and improves the modularity and maintainability of the code.

Therefore, by properly using the storage category keywords extern and static, we can more flexibly control the scope of global variables to meet the needs of different programming scenarios. Table 7-4 gives a description of the storage categories for global variables.

全局变量默认存储在静态存储区，其作用域自定义处起延伸至所在文件的末尾。然而，我们可以使用extern关键字可以扩展全局变量的作用域，不仅可以在定义它的文件内部访问，还可以跨文件访问。相反，static关键字则用于限制全局变量的作用域。当我们将一个全局变量声明为static时，它的作用域仅限于定义它的文件内部，无法在其他文件中访问。这有助于避免全局变量在不同文件之间的潜在冲突，提高代码的模块化和可维护性。

因此，通过合理使用extern和static这些存储类别关键字，我们可以更灵活地控制全局变量的作用域，满足不同编程场景的需求。表7-4给出了全局变量的存储类别说明。

表 7-4 全局变量的存储类别

存储区域	作用域	可扩展作用域（extern）		可限制作用域（static）
静态存储区	从定义到本文件末尾	在一个文件内扩展	扩展到其他文件	仅在所定义的文件内可见

Let's further illustrate the usage scenarios of extern and static with examples.

我们通过示例，进一步展示extern和static的使用场景。

1. Extend Scope Within Same File（同一文件内扩展作用域）

Extend a scope within a file. If a global variable is not defined at the beginning of the file and a previous function wants to reference it, you can use the extern keyword to extend it. If A and B are global variables defined in the middle, if the main function wants to refer to them, it can use extern statement extension. The program code is as follows:

在一个文件内部扩展作用域，若全局变量不在文件的开头定义，之前的函数又想引用，可使用extern关键字扩展。如A、B是在中间定义的全局变量，若主函数想引用，则可使用extern语句扩展。程序代码如下：

```
#include <stdio.h>
int main()
{
    int max();
    extern int A,B;                     //使用 extern 语句扩展变量 A 和 B 的作用域
    …
}
int A, B=2;
int max()
{…}
```

2. Extend Scope Between Different Files（不同文件间扩展作用域）

The scope can also extend to other files. In practice, programs often consist of more than one source file. If one file refers to the global variables of another file, you can use the extern keyword extension. If a global variable A is defined in the file1.c file, and you want to refer to it in the file2.c file, we can add extern A statement in the file2.c file to extend the scope of A. The program code is as follows:

作用域还可扩展到其他文件。实际上程序常常由多个源文件组成的，若一个文件引用另一个文件的全局变量，则可使用extern关键字扩展。如file1.c文件内定义了一个A全局变量，若file2.c文件内想引用，则在file2.c文件中加上extern A;语句即可将A的作用域扩展到该文件内。程序代码如下：

```
// 文件 file1.c
#include <stdio.h>
int A;
int main()
{
    …
}
// 文件 file2.c
extern A;                    // 将变量 A 作用域扩展到本文件内
int max()
{
    …
}
```

3. Local Variables are Limited in This Document（局部变量限定在本文件内）

If you want to restrict a global variable to this file and use it only in this file, you can add the static keyword when defining the variable. For example, A in the file1 is preceded by static. Then A is not extendable to other files. If the file2 file uses A, an error is reported. The program code is as follows:

若想把某个全局变量限制在本文件内，希望只在本文件中使用，可以在定义变量时加static关键字。如file1文件内的A前面加了static。则A不可扩展到其他文件。file2文件使用A则报错。程序代码如下：

```
//文件 file1.c
#include <stdio.h>
static int A;                          //将变量 A 的作用域限定在本文件内，不允许扩展到其他文件使用
int main()
{
    …
}
//文件 file2.c
extern A;                              //报错
void fun(int n)
{
    …
    A=A*n;
    …
}
```

Summary: This section describes the storage categories of global variables, distinguishing them from those of local variables. The static global variable is only used in this file, and extern statement can extend the global variable to other files. It can be used flexibly according to the actual programming requirements.

总结：本小节介绍了全局变量的存储类别，注意与局部变量的存储类别相区分。static全局变量仅限定在本文件内使用，extern可以令全局变量扩展到其他文件内使用。根据实际编程需求灵活使用。

7.5 Arrays as Function Parameters 数组作为函数参数

7.5节视频

Arrays can be combined with the loops to efficiently process the same type of data, which greatly improves the convenience of programming. So can an array be passed as an argument to a function? The answer is yes. By passing the array to the function, we can manipulate and process the array elements inside the function to achieve more complex logical functions.

数组能够结合循环高效处理同类型的数据，极大地提高了编程的便捷性。那么，数组能否作为函数的参数进行传递呢？答案是肯定的。通过传递数组到函数中，我们可以在函数内部对数组元素进行操作和处理，实现更为复杂的逻辑功能。

7.5.1 Array Elements as Function Parameters（数组元素作为函数实参）

The array element is essentially a variable, and its characteristics are the same as ordinary variables. As a function argument, its value is passed to the corresponding formal parameter, which is a one-way passing of values. Array elements are unique compared to ordinary variables in being able to access and manipulate different variables with different subscripts.

数组元素的实质即是变量，其特性与普通变量无异。作为函数实参时，其值被传递给对应的形参，这是值的单向传递过程。与普通变量相比，数组元素的独特之处在于能够通过不同的下标来访问和操作不同的变量。

Case 7_4（案例 7_4）

Input 10 numbers, how to determine the largest element and its position?

输入10个数，如何确定其中的最大元素及其位置？

Thinking（解题思路）：

The purpose of this case is to find the maximum value of 10 numbers through the passing of array elements, and determine the position of the value in the array. First, define an array of 10 elements to store the input values. Next, design a helper function, such as the max function, to compare two numbers and return the larger of the two. Finally, we use the algorithm to traverse each element in the array, and obtain the value of the largest element and its position in the array by calling the max function many times.

本案例旨在通过数组元素的传递来找出10个数中的最大值，并确定该值在数组中的位置。首先，定义一个包含10个元素的数组，用于存放输入的数值。接着，设计一个辅助函数（如max函数），用于比较两个数的大小，并返回其中的较大者。最后，采用“打擂台算法”遍历数组中的每个元素，通过多次调用max函数，求得最大元素的值及其在数组中的位置。

Program Code（程序代码）：

Case7_4.c

```
1   #include <stdio.h>
2   int max(int num1, int num2);                          // 函数声明
3   int main()
4   {
5       int numbers[10], maxValue, maxIndex, i;
6       printf("Please enter 10 integers:\n");
7       for(i=0; i<10; i++)
8       {
9           scanf("%d", &numbers[i]);
10      }
11      printf("\n");
12      for(i=1, maxValue=numbers[0],maxIndex=0; i<10; i++)
13      {
14          if(max(maxValue, numbers[i])>maxValue)        // 使用打擂台算法
15          {
16              maxValue=numbers[i];                      //maxValue 存储最大值
17              maxIndex=i;                               //maxIndex 存储最大值下标
18          }
19      }
20      printf("Maximum number:%d\nIt is located at:%d\n", maxValue, maxIndex+1);
21      return 0;
22  }
23  int max(int num1, int num2)
24  {
25      return(num1>num2?num1:num2);
26  }
```

Program Output（运行结果）：

```
Please enter 10 integers:
1 3 5 7 6 9 10 8 2 4
Maximum number:10
It is located at:7
```

Case Analysis（案例解析）：

(1) In line 5, we define a one-dimensional array numbers of

（1）在第5行代码中，我们定义了一个

integers of length 10 to store the 10 values of the input. At the same time, we also define the variable maxValue to store the maximum value currently found, and the variable maxIndex to record the subscript of the maximum value in the array. In addition, we define a loop variable i that controls the traversal of the array elements.

长度为10的整型一维数组numbers，用于存储输入的10个数值。同时，我们还定义了变量maxValue用于存放当前找到的最大值，变量maxIndex用于记录最大值在数组中的下标。此外，我们还定义了一个循环变量i，用于控制数组元素的遍历。

(2) In lines 7 to 10, we assign values to the 10 elements of array numbers in turn by keyboard input.

（2）在第7~10行代码中，我们通过键盘输入的方式，依次为数组numbers的10个元素赋值。

(3) The for loop in lines 12 to 19 implements the logic of the challenge algorithm. At the beginning of the loop, we assign initial values to i, maxValue, and maxIndex via the for statement on line 12. The loop condition is i<10, ensuring that we can traverse to every element in the array. Inside the loop body, we call the max function each time to compare the value of maxValue and numbers[i] and return the larger of the two. If the returned value is greater than maxValue, which means that the current value of numbers[i] is greater, we update the values of maxValue and maxIndex.

（3）第12~19行的for循环实现了打擂台算法的逻辑。循环开始时，我们通过第12行的for语句为i、maxValue和maxIndex赋予初值。循环条件是i<10，确保我们能够遍历到数组中的每一个元素。在循环体内部，我们每次调用max函数比较maxValue和numbers[i]的大小，并返回两者中的较大者。如果返回的值大于maxValue，即表示当前numbers[i]的值更大，我们则更新maxValue和maxIndex的值。

(4) It should be pointed out that this case is mainly to demonstrate the use of array elements as function arguments, so the code may not be the most efficient way to implement it. In practical applications, we may need to optimize the algorithm according to specific needs to improve the efficiency of code execution.

（4）需要指出的是，本案例主要是为了演示数组元素作为函数实参的用法，因此代码可能并不是最高效的实现方式。在实际应用中，我们可能需要根据具体需求对算法进行优化，以提高代码的执行效率。

7.5.2 Arrays as Function Parameters（数组作为函数参数）

When an array is used as a function parameter(formal parameter), the C compiler system treats it as a pointer to the array element to receive the first address of the passed array for the sake of simplifying the processing. The array name is actually the first address of the array space. Therefore, during the process of function call, we often use the name of the array as the actual parameter, and pass the address of the first element of the array to the formal parameter. Inside the function, the essence is that the pointer manipulates the array. In programming practice, this usage is very common, and it gives us the flexibility to pass array data between functions. Next, we will use an example to show the concrete application of array as a function parameter.

当数组作为函数参数（形参）时，C编译系统出于简化处理的考虑，将其视为指向数组元素的指针，用以接收传递的数组首地址。而数组名实际为数组空间的首地址。故在函数调用过程中，我们常以数组名作为实参，将数组的首元素地址传递给形参。而在函数内部本质为指针操作数组。在编程实践中，这样的用法非常普遍，它使得我们能够灵活地在函数间传递数组数据。接下来，我们将通过一个实例来详细展示数组作为函数参数的具体应用。

Case 7_5（案例 7_5）

This case requires using an array as a parameter to write a function to calculate the average score of 10 students.

本案例要求使用数组作为参数，编写一个函数来计算10个学生的平均成绩。

Thinking（解题思路）:

(1) First, define a function called average, which will be responsible for calculating and returning the average of all the

（1）首先，定义一个名为average的函数，该函数将负责计算并返回数组中所有

students' grades in the array.

(2) When calling the average function, we use the array name as an argument to pass the array that stores the student's score to the function.

(3) In the definition of the average function, we use an array as a formal parameter so that each element in the array can be accessed and manipulated inside the function.

(4) Inside the average function, we accumulate the scores of all students by traversing the array elements, and finally calculate the average.

学生成绩的平均值。

（2）在调用average函数时，我们使用数组名作为实参，将存储学生成绩的数组传递给该函数。

（3）在average函数的定义中，我们将数组作为形参，以便在函数内部能够访问并操作数组中的各个元素。

（4）在average函数内部，我们通过遍历数组元素，累加所有学生的成绩，并最终计算平均值。

Program Code（程序代码）:

Case7_5.c

```
1   #include <stdio.h>
2   int main()
3   {
4       float average(float array[10]);          //函数声明
5       float score[10],aver;
6       int i;                                   //下标应为整数
7       printf("Enter scores for 10 students:\n");
8       for(i=0;i<10;i++)
9           scanf("%f",&score[i]);
10      aver=average(score);
11      printf("Average score of 10 students:%5.2f\n",aver);
12      return 0;
13  }
14  float average(float array[10])
15  {
16      int i;
17      float aver,sum=array[0];
18      for(i=1;i<10;i++)
19          sum=sum+array[i];
20      aver=sum/10;
21      return(aver);
22  }
```

Program Output（运行结果）:

```
Enter scores for 10 students:
85 83 76 75 66.5 78 90 86 79.5 80.5
Average score of 10 students: 79.95
```

Case Analysis（案例解析）:

(1) Line 4 is a function declaration statement. Since the definition of the average function is after its call, it needs to be declared in advance to ensure that the compiler knows about the existence of the function before it is called.

(2) Line 5 defines a one-dimensional array for storing 10 float-type data. Also, an aver variable is defined to store the calculated average.

(3) An integer variable i is defined to control the subscript

（1）第4行是函数声明语句，由于average函数的定义位于其调用之后，因此需要进行提前声明，以确保编译器在调用前知晓该函数的存在。

（2）第5行定义了一个一维数组，用于存放10个float型数据。同时，定义了一个aver变量，用于存储计算出的平均值。

（3）定义了一个整型变量i，用于在后

change of array elements in subsequent loops. Note that the array subscript should be an integer.

(4) The for loop on lines 8 to 9 is used to enter the values of 10 elements from the keyboard and store them in the array in turn.

(5) Line 10 is the function call statement. Executing average(score) calls the average function and passes the actual parameter score(the name of the array, that is, the first address of the array). The aver variable is used to receive the average value returned by the function.

(6) Lines 14 to 22 define the average function. The declaration float average(float array[10]) in the header of the function indicates that the return type of the function is float, and the array float array[10] is used as a formal parameter. During compilation, the system treats the array formal parameter array as a pointer variable, which is actually equivalent to float *array. When average(score) is called, the first address of the score array is passed to array, which points to the score array defined in the main function. Therefore, array[i] is equivalent to *(array+i) and *(score+i) or score[i], which means that what is accessed through array[i] is actually the i-th element of the score array in the main function, as shown in Figure 7-5. Because the scope of the score array is limited to the main function, we access the score array indirectly through array inside the average function. Inside the body of the function, the values of the array elements are accumulated through a for loop, and the average is calculated, which is finally used as the return value of the function.

(7) The main function outputs the calculated average value, and then the program ends. By viewing the running results, we can verify the correctness of the calculation results.

续循环中控制数组元素的下标变化。请注意，数组下标应为整型。

（4）第8、9行的for循环用于从键盘输入10个元素的值，并将它们依次存储在数组中。

（5）第10行是函数调用语句。执行average(score)会调用average函数，并传递实参score（数组名，即数组首地址）。aver变量用于接收函数返回的平均值。

（6）第14~22行定义了average函数。该函数首部的声明float average(float array[10])指明了函数的返回类型为float，数组float array[10]作为形参。在编译过程中，系统会将数组形参array视为指针变量处理，实际上等同于float *array。当调用average(score)时，score数组首地址被传递给array，此时array指向主函数中定义的score数组。因此，array[i]等同于*(array+i)，也等同于*(score+i)或score[i]，这意味着通过array[i]访问的实际上是主函数中score数组的第i个元素，如图7-5所示。由于score数组的作用域仅限于主函数内，所以在average函数内部我们通过array来间接访问score数组。在函数体内部，通过for循环累加数组元素的值，并计算平均值，最终将其作为函数的返回值。

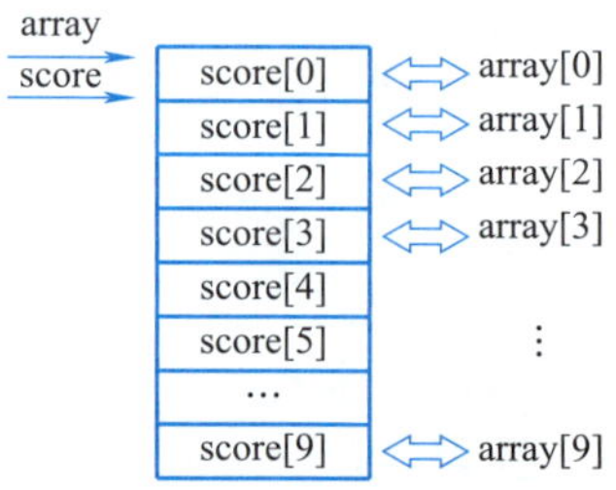

图7-5　数组作为函数参数示意图

（7）主函数输出计算得到的平均值，随后程序结束。通过查看运行结果，可以验证计算结果的正确性。

Case Summary（案例总结）:

When an array is used as a function formal parameter, the function itself does not actually create a copy of the array, but instead treats it as a pointer variable to an array element. During the function call, the actual argument is not passed the entire array, but the first address of the array. This first address is assigned to the pointer variable in the formal parameter, allowing indirect access to the array defined in the calling function. In this way, the operation inside the function is actually

数组作为函数形参时，函数内部实际上并不创建数组的副本，而是将其处理为一个指向数组元素的指针变量。在函数调用过程中，实参并没有传递整个数组，而是传递了数组的首地址。这个首地址被赋值给形参中的指针变量，使得函数内部能够间接访问主调函数中定义的数组。通过

carried out on the array in the calling function, realizing the sharing and passing of data.

这种方式，函数内部的操作实际上是对主调函数中的数组进行的，实现了数据的共享和传递。

Case Development（案例拓展）:

If we change the value of the array element in the average function, will it affect the value of the score array element? If array[0]=70 is executed inside the function, return to the main function and test the value of score[0] after the function ends. The code is as follows:

若我们在average函数中改变array数组元素的值，是否会影响到score数组元素的值呢？假如在函数内部执行array[0]=70，函数结束后，返回到主函数测试score[0]的值。代码如下：

```
int main
{
    …
    aver=average(score);
    printf("%d",score[0]);
    …
}
float average(float array[])
{   …
    array[0]=70;
    …
}
```

After the function is executed, the output score[0] value is 70, which fully proves that the change of the formal parameter array does affect the actual parameter array. This is because an array operated by the formal parameters is essentially an array of actual parameters. Inside the function, we indirectly access and modify the score array through the array pointer variable. Since the array and score point to the same memory space, any operation on array is reflected directly on the score array.

函数执行后，输出的score[0]值为70，这充分证明了形参数组的改变确实影响了实参数组。这是因为形参操作的数组本质上就是实参数组。在函数内部，我们通过array这个指针变量间接访问和修改了score数组。由于array和score指向的是同一块内存区域，所以对array的任何操作都会直接反映到score数组上。

In short, we are always working with the array in the main call function. The formal parameter array just gets the address(first element pointer) of the actual parameter array score, and does not create a copy. Therefore, whether you use array or score, you are operating on an array in the same block of memory. Changes to the formal parameter array affect the actual parameter array because they all point to the same memory space, that is the array in the calling function.

简而言之，我们始终在操作的是主调函数中的数组。形参数组array只是获取了实参数组score的地址（首元素指针），并没有创建副本。因此，无论使用array还是score，都是在操作同一块内存中的数组。形参数组的改变会影响实参数组，因为它们都指向同一块内存空间，即主调函数中的数组。

7.6 Pointers as Function Parameters 指针作为函数参数

7.6节视频

When an array is used as a formal parameter, the C compiler treats it as a pointer variable and passes it as an argument. Now that we understand that pointer variables can be used as function parameters, this section will delve into the specific use of pointer variables as function parameters. The core role of a pointer type parameter is to pass the address of a variable to a function so that the function can access and modify the variable indirectly.

数组作为形参时，C编译器会将其处理为指针变量作为参数传递。既然我们已经了解指针变量可以作为函数参数，那么本节将深入探讨指针变量作为函数参数的具体用法。指针类型参数的核心作用是将变量的地址传递给函数，以便函数能够间接访问和修改该变量。

The pointer variable is used as a function parameter, and its form is shown as void swap(int *p1, int *p2), which is declared as a pointer type in the formal parameter list. A pointer variable is still essentially a variable, but it is unique in that it stores an address rather than a value like a normal variable. Therefore, when a pointer variable is used as a formal parameter, the actual parameter should pass the address of the variable, not the actual value of the variable. Next, we will discuss it in detail through specific examples.

关于指针变量作为函数参数，其形式如void swap(int *p1, int *p2);所示，在形参列表中声明为指针类型。指针变量本质上仍是一个变量，但其独特之处在于它所存储的是地址，而非像普通变量那样存储值。因此，当指针变量作为形参时，实参应传递变量的地址，而非变量的实际值。接下来通过具体的实例进行详细探讨。

Case 7_6（案例 7_6）

It is required to write a function to realize the function of exchanging two integer data by using pointer variable to pass parameters.

要求采用指针变量传参的方式，编写一个函数来实现两个整型数据交换的功能。

Thinking（解题思路）:

(1) First, we need to give this function a proper name, here we call it swap.

(2) Next, in order to implement data exchange, we need to define two pointer variables of type int*as formal parameters of swap function. These two pointer variables will be used to receive the addresses of the actual parameter variables.

(3) Inside the swap function, we will indirectly access the pointed integer variables through these two pointer variables, and use temporary variables to realize the data exchange between them. In this way, when the function is executed, the values of the original two integer variables will be successfully exchanged.

（1）首先，我们需要给这个函数起一个合适的名字，这里将其命名为swap。

（2）接着，为了实现数据交换，我们需要定义两个int*类型的指针变量作为swap函数的形参。这两个指针变量将用于接收实参变量的地址。

（3）在swap函数内部，我们将通过这两个指针变量间接访问所指向的整型变量，并利用临时变量来实现它们之间的数据交换。这样，当函数执行完毕后，原先的两个整型变量的值就会被成功交换。

Program Code（程序代码）:

Case7_6.c

```
#include <stdio.h>
int main()
{
    void swap(int *p1,int *p2);                    //函数声明
    int a,b;
    int *p_1=NULL,*p_2=NULL;                       //定义指针变量
    printf("Please enter a and b:\n");
    scanf("%d%d",&a,&b);
    p_1=&a; p_2=&b;                                //存储a和b的地址
    swap(p_1, p_2);                                //函数调用
    printf("a=%d,b=%d\n"a,b);                      //查看是否交换
    return 0;
}
void swap(int *p1,int *p2)                         //指针变量作为形参
{
    int temp;                                      //辅助交换的变量
    temp=*p1;                                      //实现交换
    *p1=*p2;
```

```
19        *p2=temp;
20    }
```

Program Output（运行结果）:

```
Please enter a and b:
5 9
a=9,b=5
```

Case Analysis（案例解析）:

(1) The fourth line is the function declaration. The function definition should be declared first after the call.

（1）第4行为函数声明，函数定义在调用之后，应先进行声明。

(2) Line 5 defines two integer variables, a and b, which are used to store the data to be exchanged. Next, in line 6, two integer pointer variables, p_1 and p_2, are defined and initialized to NULL to ensure that the pointer does not point to any valid memory address in the initial state, thereby improving the safety of the program.

（2）第5行定义了两个整型变量a和b，分别用于存储需要交换的数据。接着在第6行，定义了两个整型指针变量p_1和p_2，并将它们初始化为NULL，以确保指针在初始状态下不指向任何有效内存地址，从而提高程序的安全性。

(3) Line 8 reads the values of a and b from the keyboard through standard input, which represent the data to be exchanged.

（3）第8行通过标准输入从键盘读取a和b的值，这两个值代表了将要进行交换的数据。

(4) In line 9, we make the pointer variable p_1 point to the address of variable a and p_2 point to the address of variable b.

（4）第9行中，我们使指针变量p_1指向变量a的地址，使p_2指向变量b的地址。

(5) Line 10 calls the swap function and passes p_1 and p_2 as the actual parameters, actually passing the addresses of the variables a and b. Since the swap function has no return value, we don't need to receive any return value and just call it.

（5）第10行调用了swap函数，并将p_1和p_2作为实参传递，实际上传递的是变量a和b的地址。由于swap函数无返回值，因此无须接收任何返回值，直接调用即可。

(6) Line 11 is an output statement that verifies that the values of a and b have been successfully swapped after the swap function call. Through the running results, it is observed that the data exchange is completed.

（6）第11行是一个输出语句，用于验证swap函数调用后a和b的值是否已成功交换。通过运行结果观察到，完成了数据交换。

(7) Line 14 begins the definition of the swap function, which returns no value and receives two pointers to integer variables as the formal parameters. These pointers receive the addresses of the variables a and b from the main function.

（7）第14行是swap函数的定义开始，该函数无返回值，并且接收两个指向整型变量的指针作为形参。这些指针从主函数中接收了变量a和b的地址。

(8) Inside the swap function, since we receive the addresses of a and b, we can indirectly access and modify a and b variables in the main function through the dereference operations(p1 and p2). In this way, we exchange the values of a and b inside the body of the function.

（8）在swap函数内部，由于我们接收了a和b的地址，因此可以通过解引用操作（p1和p2）来间接访问和修改主函数中的a和b变量。通过这种方式，我们在函数体内部实现了a和b值的交换。

Case Summary（案例总结）:

When the swap function is called, the program makes a function jump and passes the real parameter value to the corresponding formal parameter value accurately. At this point, the distribution of memory space is shown in Figure 7-6. From

当swap函数被调用时，程序发生函数跳转，将实参值准确无误地传递给对应的形参值。此时，内存空间的分布情况如图7-6所示，从图中可以看到p1和p2指向了

the figure, we can see that p1 and p2 point to a and b. During the execution of the swap function, the data of a and b are exchanged through indirectly accessing. Figure 7-7 shows the detailed state of the function execution. When the swap function ends, the memory space occupied by the two formal parameters p1 and p2 is released, as shown in Figure 7-8. At this point, although p1 and p2 no longer exist, the data of the variables a and b they point to have been exchanged and the expected function has been achieved.

a和b。而在swap函数执行过程中通过间接访问交换了a和b的数据，图7-7展现出函数执行时的详细状态。当swap函数结束，p1和p2这两个形参所占据的内存空间得到释放，如图7-8所示。此刻，尽管p1和p2已不复存在，但它们所指向的变量a和b的数据却已完成交换，实现了预期的功能。

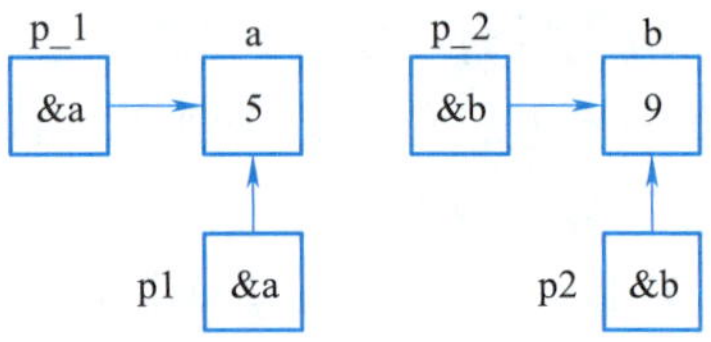

图7-6　swap函数被调用时空间分布情况

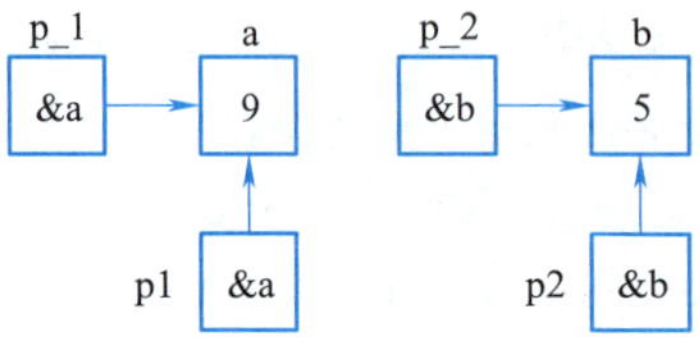

图7-7 swap函数调用期间空间分布情况

图7-8 swap函数结束后空间的分布情况

Case Development（案例拓展）:

Think about why you can't use a and b to swap directly inside the swap function instead of accessing them indirectly through a pointer variable?

请思考：为什么不能在swap函数内部直接使用a和b进行交换，而要通过指针变量间接访问？

This is because a and b are local variables of the main function, and their visibility and lifetime are limited to the inside of the main function. And in the swap function, a and b become not directly accessible. However, the swap function can indirectly modify the values of these variables by accepting the pointers to them as the formal parameters. This is the subtlety of the pointers as function arguments, which allows you to manipulate variables without directly accessing them.

这是因为a和b作为main函数的局部变量，其可见性和生命周期仅限于main函数内部。一旦进入swap函数，a和b就变得不可直接访问。然而，swap函数通过接收指向这些变量的指针作为形参，能够间接地修改它们的值。这正是指针作为函数参数的精妙之处，它使得在不直接访问变量本身的情况下也能对它们进行操作。

It is worth noting that although the swap function does not directly return a value, it indirectly modifies the value of the variables in the calling function through a pointer parameter, implementing an implicit return value mechanism. This flexible approach is not only practical, but also helpful to understand the important role of the pointers in function parameter passing.

值得注意的是，尽管swap函数没有直接返回值，但它通过指针参数间接地修改了主调函数中变量的值，实现了一种隐式的返回值机制。这种灵活的方式不仅实用，且有助于深入理解指针在函数参数传递中的重要作用。

Comparison and Analysis of Several Rewritten Forms　几种改写形式对比与分析

(1) If ordinary variables are used as the function parameters, can the data be exchanged? The code is as follows.

（1）若普通变量作为函数参数，是否能交换数据？代码如下：

```
#include <stdio.h>
int main()
{
    void swap(int p1,int p2);              //函数声明
    ...
    swap(a,b);                             //函数调用
    printf("a=%d,b=%d\n"a,b);              //查看是否交换
```

```
    return 0;
}
void swap(int p1,int p2)                                  //普通变量作为形参
{
    int temp;                                             //辅助交换的变量
    temp=p1;                                              //实现交换
    p1=p2;
    p2=temp;
}
```

Figure 7-9 details the space allocation during function execution, where p1 and p2, as well as a and b, occupy separate spaces. During a function call, data is passed in one direction, that is, the value of the actual parameter is passed to the formal parameter, and there is no reverse pass. Although the values of p1 and p2 are exchanged inside the function, this does not change the actual data of a and b in the calling function. After the function finishes executing, the space occupied by p1 and p2 is released, while the values of a and b remain unchanged. Therefore, the implementation of the swap function fails to achieve the expected data exchange effect.

图7-9详细展示了函数执行时的空间分配情况，其中p1和p2，以及a和b各自占据独立空间。在函数调用过程中，数据以单向方式传递，即实参的值传递给形参，不存在反向传递。尽管函数内部完成了p1和p2的值交换，但这并未改变主调函数中a和b的实际数据。函数执行完毕后，p1和p2所占用的空间将被释放，而a和b的值保持不变。因此，这种swap函数的实现方式未能达到预期的数据交换效果。

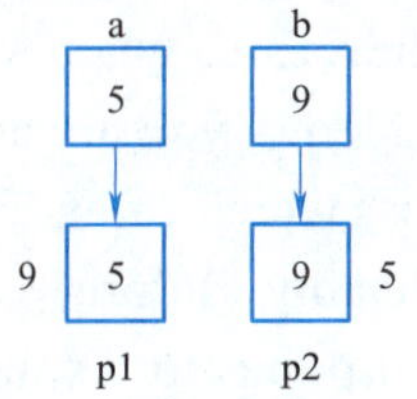

图7-9　普通变量作为参数空间分布情况

When an ordinary variable is used as a function formal parameter, it actually receives a copy of the actual parameter, which means that the formal parameter and the actual parameter occupy the separate spaces. Therefore, any modification to the formal parameter inside the function does not affect the value of the actual parameter. However, when a pointer variable is used as a formal parameter, it does not directly receive the value of the actual parameter, but receives the address of the actual parameter, so that it can indirectly access and modify the corresponding variable in the calling function. This mechanism makes the pointer variable as a formal parameter in the function call to achieve data sharing and modification, which is obviously different from the principle of ordinary variables as a formal parameter.

普通变量作为函数形参时，实际上接收的是实参的副本，这意味着形参与实参各自占用独立的空间。因此，在函数内部对形参所做的任何修改都不会影响到实参的值。而指针变量作为形参时，它并不直接接收实参的值，而是接收实参的地址，从而能够间接访问和修改主调函数中对应的变量。这种机制使得指针变量作为形参在函数调用中能够实现数据的共享和修改，与普通变量作为形参的原理有着明显的区别。

(2) In order to grasp the usage of pointer variables as function parameters, we need to carefully analyze the correctness of different rewriting forms.

（2）为了深入掌握指针变量作为函数参数的用法，我们需仔细分析不同改写形式的正确性。

First, consider the following rewriting of the swap function:

首先，考虑如下swap函数的改写：

```
void swap(int *p1, int *p2)
{
    int *p;
    p=p1;
    p1=p2;
```

```
    p2=p;
}
```

This rewriting is wrong. Because this code only exchanges the addresses pointed to by the formal parameters p1 and p2, this exchange has no effect after the function returns, and has no effect on the variables a and b in the calling function. The exchange of formal parameters does not affect the actual parameters, because they are simply copies of the actual parameter values.

这种改写是错误的。因为这段代码仅仅交换了形参p1和p2所指向的地址，这种交换在函数返回后即失去作用，对主调函数中的变量a和b没有任何影响。形参的交换并不会影响实参，因为它们只是实参值的副本。

Next, we consider another form of rewriting:

接下来，我们考虑另一种改写形式：

```
void swap(int *p1, int *p2)
{
    int *temp;
    *temp=*p1;
    *p1=*p2;
    *p2=*temp;
}
```

The temp here is an int pointer variable, but this code is also wrong. The problem is that the temp pointer is not initialized to point to a valid memory address prior to assignment. Therefore, when you try to access and modify values through *temp, you are actually accessing an undefined memory address, which often causes the program to crash or behave unpredictably. Before you can use a pointer, you must make sure that it points to a valid memory address. This again emphasizes that the pointer variables must be given explicit pointers before they can be used.

这里的temp是一个int指针变量，但这段代码同样是错误的。问题在于，temp指针在赋值之前没有被初始化为指向有效的内存地址。因此，尝试通过*temp来访问和修改值时，实际上是在访问一个未定义的内存地址，这通常会导致程序崩溃或产生不可预测的行为。在使用指针之前，必须确保它指向了一个有效的内存地址。这再次强调了在使用指针变量之前必须给予它们明确的指向。

In summary, when a pointer is used as a function parameter, we must have a clear understanding of what the pointer points to and how to manipulate the data through the pointer. Before you use a pointer at any time, you must make sure that it points to a valid memory address.

总结来说，指针作为函数参数时，我们必须清晰地理解指针的指向以及如何通过指针来操作数据。任何时候使用指针之前，都必须确保它指向了有效的内存地址。

7.7 Application of Pointer in Function 指针在函数中的应用

7.7节视频

Pointers are the key feature in C that allows us to directly access and manipulate memory addresses. Pointer variables store the memory addresses of other variables, as well as the entry addresses of functions. We can call the function through the pointer by assigning the entry address of the function to the pointer variable. A pointer can also pass a memory address as the return value type of a function. This section focuses on the advanced uses of pointers in functions, including the pointers to functions and the functions that return pointer values.

指针是C语言中的关键特性，它允许我们直接访问和操作内存地址。指针变量存储其他变量的内存地址，也可以存储函数的入口地址。我们可以将函数的入口地址赋给指针变量，从而通过指针调用函数。指针还可以作为函数的返回值类型，传递内存地址。本节将聚焦于指针在函数中的高级应用，包括指向函数的指针和返回指针值的函数。

7.7.1 Pointer to Function（指向函数的指针）

A callback function is an advanced application of a function pointer, it gives the ability to dynamically call a predefined function by another function when a specific event or condition is triggered. This mechanism shows great practical values in asynchronous operation, event-driven programming and other scenarios. In C, we can easily implement the mechanism of a callback function by defining a function pointer variable and passing it as a parameter.

回调函数是函数指针的进阶应用，它赋予了在特定事件或条件触发时，由其他函数动态调用预定义函数的能力。这种机制在异步操作、事件驱动编程等场景中展现出巨大的实用价值。在C语言中，通过定义函数指针变量并将其作为参数传递，我们可以轻松地实现回调函数的机制。

The close relationship between function pointers and callback functions constitutes the essence of C language programming, showing its flexibility and power. Function pointers, as a unique form of pointer type, point to functions rather than ordinary data. In C, the function name is essentially the starting address of the function in memory, so we can define a pointer variable to store this address, and this special pointer variable is called the function pointer. Through function pointers, we can dynamically select and call different functions at runtime, and even pass them to other functions as the parameters, which greatly improves the flexibility and reusability of the code.

函数指针和回调函数之间的紧密关系共同构成了C语言编程的精髓，展现出其灵活性和强大性。函数指针作为指针类型的一种独特形式，其指向的是函数而非普通数据。在C语言中，函数名本质上就是该函数在内存中的起始地址，因此我们可以定义一个指针变量来存储这个地址，这个特殊的指针变量称为函数指针。通过函数指针，我们能够在运行时动态地选择并调用不同的函数，甚至将其作为参数传递给其他函数，极大地提升代码的灵活性和可重用性。

In the C language, the definition of a function pointer usually follows the following format:

```
data type(* pointer variable name)(function
argument list);
```

在C语言中，函数指针的定义通常遵循以下格式：

```
数据类型(*指针变量名)(函数参数列表);
```

For example, defining a pointer variable p to a function that takes two integer formal parameters and returns an integer result, we can write: int(* p)(int, int); Here, p is a pointer variable to a function, which can only point to functions whose return type is an integer and which accept two integer parameters.

例如，要定义一个指向接受两个整型参数并返回整型结果的函数的指针变量p，我们可以这样写：int(*p)(int, int);。这里的p就是一个指向函数的指针变量，它只能指向那些返回类型为整型且接受两个整型参数的函数。

Case 7_7（案例 7_7）

Writing a function to find the greater of two integers that requires using a function pointer to call the function indirectly.

编写函数求两个整数中的较大者，要求使用函数指针间接调用函数。

Thinking（解题思路）:

First, define the max function, which has two int parameters and receives any two integer data from the outside. In the function body, use the conditional operator to find the maximum value of the two, and bring the result back as the return value, so the return value is also int. Define a pointer variable pointing to max inside the main function, call the

首先定义max函数，该函数有两个int参数，接收外部传入的任意两个整型数据，函数体内使用条件运算符求得两者的最大值，并将结果作为返回值带回，故返回值也为int型。在主函数内部定义一个指向max的指针变量，通过指针变量间接调

function indirectly through the pointer variable, and finally output the viewing result. Through this case, understand and master the specific usage of pointer variable pointing function.

用函数，最后输出查看结果。通过本案例理解并掌握指针变量指向函数的具体用法。

Program Code（程序代码）:

Case7_7.c

```
1    #include <stdio.h>
2    int max(int x,int y)
3    {
4        return(x>y)?x:y;
5    }
6    int main()
7    {
8        int a,b,c;
9        int(*p)(int,int);                  //定义指向函数的指针变量
10       p=max;                             //函数名即为函数首地址
11       printf("Please enter a and b:");
12       scanf("%d%d", &a, &b);
13       c=(*p)(a,b);                       //间接调用函数
14       printf("max=%d\n",c);
15       return 0;
16   }
```

Program Output（运行结果）:

```
Please enter a and b:5 9
max=9
```

Case Analysis（案例解析）:

(1) In lines 2-5 of the code, we define the max function, and since it precedes the code that calls it, the explicit function declaration for it can be omitted.

(2) In line 8, we define the variables a and b, which are used to store two integers read from the keyboard. At the same time, the variable c is defined to receive the return value of the max function.

(3) In line 9, we declare a pointer variable named p that points to a function type that takes two int parameters and returns an int value. The pointer variable p can point to any function of this type.

(4) In line 10, we assign the address of the max function(or function pointer) to p. Since max is the name of the function, it is resolved to the entry address of the function(or the first address of the function) at compile time, so the pointer variable p points to the max function.

(5) Line 12 is responsible for reading the values of two integer variables a and b from the keyboard.

(6) "c=(*p)(a,b)" in line 13 is the way to call the max function indirectly. Here, we dereference the function pointer p,

（1）在代码的第2~5行，我们定义了max函数，由于它位于调用它的代码之前，因此可以省略对其的显式函数声明。

（2）在第8行，我们定义了变量a和b，它们用于存储从键盘读入的两个整型数据。同时，定义了变量c，用于接收max函数的返回值。

（3）第9行中，我们声明了一个名为p的指针变量，它指向的函数类型接受两个int型参数并返回一个int型值。指针变量p可以指向任何符合这种类型的函数。

（4）在第10行，我们将max函数的地址（或称为函数指针）赋值给p。由于max是函数名，它在编译时会被解析为函数的入口地址（或称为函数首地址），因此此时指针变量p就指向了max函数。

（5）第12行负责从键盘读入两个整型变量a和b的值。

（6）第13行中的"c=(*p)(a,b);"是间接调用max函数的方式。这里，我们解引用函数指针p，并传入参数a和b，然后将

pass in the parameters a and b, and store the returned result in c.

返回的结果存储在c中。

(7) Finally, on line 14, we output the return value of the max function(i.e., the value stored in c) to the screen.

（7）最后，在第14行，我们将max函数的返回值（即存储在c中的值）输出到屏幕上。

Case Summary（案例总结）:

Case7_7 provides a concise overview of the core concepts and applications of function pointers. First, we have learned to define pointers to functions of a particular type, which is fundamental to understanding how they work. Secondly, we have mastered the method of assigning a value to a function pointer and realized the association between a function and a pointer. Finally, the flexibility and maintainability of the code are improved by calling the function indirectly through the function pointer.

案例7_7精练地概述了函数指针的核心概念与应用。首先，我们学会了定义指向特定类型函数的指针，这是理解其运作的基础。其次，我们掌握了为函数指针赋值的方法，实现了函数与指针的关联。最后，通过函数指针间接调用函数，提升了代码的灵活性和可维护性。

Case 7_8（案例 7_8）

The user needs to input two integers and decide which function to call to compare the two numbers based on the choice. If the user selects 1, the max function is called to find and output the larger of the two integers, and if the user selects 2, the min function is called to find and output the smaller of the two integers. In this way, users can flexibly obtain the maximum or minimum value between two numbers according to their needs.

用户需要输入两个整数，并根据选择来决定调用哪个函数来比较这两个数。如果用户选择1，则调用max函数来找出并输出这两个整数中的较大值；如果用户选择2，则调用min函数来找出并输出这两个整数中的较小值。通过这种方式，用户可以灵活地根据需求获取两个数之间的最大值或最小值。

Thinking（解题思路）:

First, we need to define two functions, max and min, which are used to find the maximum and minimum values of two integers respectively. Next, in the main function, we define a function pointer variable. When the user selects to input 1, we point the pointer variable to the max function. When the user selects to input 2, we point it to the min function. In this way, we can indirectly call the corresponding function through the function pointer variable. Finally, we execute the function called by the pointer variable and output the result. Through this case, we can better understand the advantages of function pointers in dynamically calling functions, and how it can improve the flexibility and maintainability of the code.

首先，我们需要定义两个函数：max和min，分别用于求取两个整数的最大值和最小值。接着，在主函数中，我们定义一个函数指针变量。当用户选择输入1时，我们将该指针变量指向max函数；而当用户选择输入2时，则将其指向min函数。这样，我们便能通过这个函数指针变量来间接地调用相应的函数。最终，我们执行这个通过指针变量调用的函数，并输出其结果。通过本案例，我们能够更深入地理解函数指针在动态调用函数方面的优势，以及它如何提升代码的灵活性和可维护性。

Program Code（程序代码）:

Case7_8.c

```
#include <stdio.h>
int max(int x,int y)
{
    return(x>y)?x:y;
}
```

```
6    int min(int x,int y)
7    {
8        return x<y?x:y;
9    }
10   int main()
11   {
12       int(*p)(int,int);                          //定义指针变量
13       int a,b,n,c;
14       printf("Please enter data:");
15       scanf("%d%d%d",&a,&b,&n);
16       if(n==1)
17           p=max;
18       else if(n==2)
19           p=min;
20       c=(*p)(a,b);                               //调用函数
21       printf("a=%d,b=%d\n",a,b);
22       if(n==1)
23           printf("max=%d\n",c);
24       else
25           printf("min=%d\n",c);
26       return 0;
27   }
```

Program Output（运行结果）:

When 1 is entered（当输入1时）:

```
Please enter data:10 20 1
a=10,b=20
max=20
```

When 2 is entered（当输入2时）:

```
Please enter data:10 20 2
a=10,b=20
min=10
```

Case Analysis（案例解析）:

(1) In lines 2 to 9 of the code, we define the max and min functions to calculate the maximum and minimum values of two integers, respectively.

(2) On line 12, we declare a function pointer variable p that points to a function type that accepts two integer parameters and returns an integer value.

(3) Line 13 defines four integer local variables a, b, n, and c. Where the values of a, b, and n are read from the keyboard, where n is used to specify the action the user wishes to perform(1 for maximum and 2 for minimum).

(4) Based on the value of n input by the user, we dynamically set the pointing of the function pointer p. If n is 1, the address of the max function is assigned to p on line 17 so that p points to the max function, and if n is 2, the address of the min function is assigned to p on line 18 so that p points to the min function.

（1）在代码的第2~9行，我们定义了max和min两个函数，分别用于计算两个整数的最大值和最小值。

（2）在第12行，我们声明了一个函数指针变量p，该指针指向的函数类型接收两个整型参数并返回一个整型值。

（3）第13行定义了四个整型局部变量a、b、n和c。其中，a、b和n的值从键盘读入，其中n用于指定用户希望执行的操作（1表示求最大值，2表示求最小值）。

（4）根据用户输入的n值，我们动态地设置函数指针p的指向。如果n为1，则在第17行将max函数的地址赋给p，使得p指向max函数；如果n为2，则在第18行将min函数的地址赋给p，使得p指向min函数。

(5) Line 20 calls the function(max or min) indirectly through the function pointer p and stores the return value of the function in the variable c.

（5）第20行通过函数指针p间接地调用函数（max或min），并将函数的返回值存储在变量c中。

(6) On lines 22 to 25, we first output the values of a and b, and then output c according to the value of n. If n is 1, c is the maximum value obtained by calling the max function. If n is 2, c is the minimum value obtained by calling the min function. In this way, we show users how to flexibly select and invoke functions based on their needs.

（6）在第22~25行，我们首先输出a和b的值，然后根据n的值输出c。如果n为1，则c是调用max函数得到的最大值；如果n为2，则c是调用min函数得到的最小值。通过这种方式，我们向用户展示了如何灵活地根据需求选择并调用函数。

Function Pointer as Function Argument 函数指针作为函数参数

Variables that point to functions play a vital role in programming, especially as function parameters, which greatly improve the flexibility and versatility of the code. In a procedure-oriented high-level language like C, we are constantly exploring different ways to write more flexible and reusable functions. Among them, it is a powerful means to achieve this goal that the pointer variable of the function is used as a formal parameter.

指向函数的指针变量在编程中扮演着至关重要的角色，尤其是作为函数参数时，它们极大地提升了代码的灵活性和通用性。在C语言这种面向过程的高级语言中，我们不断探索各种方法以编写出更加灵活和可重用的函数。其中，将指向函数的指针变量作为形参，正是实现这一目标的有力手段。

Here is an example that further illustrates how to use a pointer variable to a function as a function parameter.

下面是一个示例，进一步说明如何使用指向函数的指针变量作为函数参数。

```
void fun(int(*x1)(int), int(*x2)(int, int))          //指向函数的指针变量作为函数参数
{
    (*x1)(…);                                         //函数内部调用函数
    (*x2)(…);
}
int main()
{
    fun(f1, f2);                                      //传递函数名作为实参
    fun(f3, f4);
    …
}
```

The fun function takes two parameters, which are pointer variables to functions, wherein the first parameter is to a function of type integer with one integer parameter, and the second pointer variable is to a function of type integer with two integer parameters. When the fun function is passed the f1 and f2 function names as parameters, the f1 and f2 functions are called in turn inside the function. Similarly, if you pass in the names of the f3 and f4 functions, the fun function calls the f3 and f4 functions instead. This flexibility not only makes the code more versatile, but also improves its maintainability and reusability. By using a function pointer as a parameter, we can easily modify the behavior of the fun function without making any modifications to itself.

fun函数有两个参数，均是指向函数的指针变量，其中第一个参数是指向类型为整型且具有一个整型参数的函数，第二个指针变量是指向类型为整型且具有两个整型参数的函数。当向fun函数传递f1和f2函数名作为参数时，函数内部会依次调用f1和f2函数。同样地，如果传入的是f3和f4函数的函数名，则fun函数会转而调用f3和f4函数。这种灵活性不仅使得代码更加通用，还提高了代码的可维护性和可重用性。通过使用函数指针作为参数，我们可以轻松地修改fun函数的行为，而无须对其本身进行任何修改。

Case 7_9（案例 7_9）

The user needs to input two integers and then select one of the options 1 or 2. If the user selects 1, the program calls the max function to find and output the larger of the two numbers; if the user selects 2, the program calls the min function to find and output the smaller of the two numbers.

用户需要输入两个整数，并随后从选项1或2中选择一个。如果用户选择1，程序将调用max函数来找出并输出这两个数中的较大值；而如果用户选择2，程序则会调用min函数来找出并输出这两个数中的较小值。

Thinking（解题思路）:

By cleverly using function pointers as parameters, we can flexibly solve the above problems. First, we need to define two auxiliary functions: the max function is used to find the larger of two integers, and the min function is used to find the smaller of two integers. Next, we define a function called fun that accepts two integers and a pointer variable to the function as parameters. This pointer variable can point to either the max or min function, allowing the fun function to perform different comparisons depending on the actual need.

In the main function, the user will input two integers and select the actual parameter passed when calling the fun function(that is, a pointer to the max or min function). The fun function will call the appropriate function based on the user's selection and output the appropriate result. This design not only improves the flexibility and reusability of the code, but also makes the program easier to understand and maintain.

通过巧妙地利用函数指针作为参数，我们可以灵活地解决上述问题。首先，我们需要定义两个辅助函数：max函数用于求取两个整数的较大值，min函数用于求取两个整数的较小值。接下来，我们定义一个名为fun的函数，它接受两个整数和一个指向函数的指针变量作为参数。这个指针变量可以指向max或min函数，从而允许fun函数根据实际需求执行不同的比较操作。

在主函数中，用户将输入两个整数，并选择调用fun函数时传递的实参（即指向max或min函数的指针）。fun函数将根据用户的选择调用相应的函数，并输出相应的结果。这种设计不仅提高了代码的灵活性和可重用性，还使得程序更加易于理解和维护。

Program Code（程序代码）:

Case7_9.c

```
#include <stdio.h>
int max(int x,int y)
{
    return x>y?x:y;
}
int min(int x,int y)
{
    return x<y?x:y;
}
void fun(int x,int y,int(*p)(int,int))
{
    printf("%d\n",(*p)(x,y));
}
int main()
{
    int a,b,n,c;
    printf("Please enter data:\n");
    scanf("%d%d%d",&a,&b,&n);
    if(n==1)
        fun(a,b,max);
    else if(n==2)
```

```
22            fun(a,b,min);
23        return 0;
24    }
```

Program Output（运行结果）：

```
Please enter data:
34 99
1
99
```

Case Analysis（案例解析）：

(1) In lines 2 to 9, we define two functions, max and min, to calculate and return the maximum and minimum of two integers, respectively.

（1）在第2~9行中，我们定义了max和min两个函数，分别用于计算并返回两个整数中的最大值和最小值。

(2) In lines 10 to 13, we define the fun function, which takes two integer parameters and a function pointer as parameters. The fun function uses function pointers internally to make indirect function calls, which allows the behavior of the function to change dynamically according to the function pointers passed in, thus improving the flexibility and reusability of the code.

（2）在第10~13行，我们定义了fun函数，该函数接收两个整型参数和一个函数指针作为参数。fun函数内部使用函数指针进行间接函数调用，这种方式使得函数的行为可以根据传入的函数指针动态改变，从而提高了代码的灵活性和可复用性。

(3) The main function section is similar to the previous case, but in the if...else statements on lines 19 to 22, we call the fun function based on the user's choice (the value of variable n). When n is 1, we pass the address of the max function as a parameter to the formal parameter p of the fun function, so that the call to the max function is executed inside the fun function. Similarly, when n is 2, we pass the address of the min function to p, and the fun function internally executes the call to the min function. This design allows us to dynamically choose different functions to execute according to our needs, which improves the flexibility and extensibility of the code.

（3）主函数部分与前一个案例类似，但在第19~22行的if...else语句中，我们根据用户的选择（通过变量n表示）来调用fun函数。当n为1时，我们将max函数的地址作为实参传递给fun函数的形参p，这样fun函数内部就会执行max函数的调用。同样地，当n为2时，我们将min函数的地址传递给p，fun函数内部则会执行min函数的调用。这种设计允许我们根据需求动态地选择执行不同的函数，提高了代码的灵活性和可扩展性。

7.7.2 Functions That Return Pointer Values（返回指针值的函数）

In C, a function can not only perform an operation and return the result of a calculation, but it can also return a pointer, that is a reference to a memory address. Functions that return pointers are often used to create dynamically allocated memory blocks, reference static or global variables, or return addresses of complex data types, such as arrays and structures.

在C语言中，函数不仅可以执行操作并返回计算结果，还可以返回指针，即某个内存地址的引用。这种能够返回指针的函数，通常用于创建动态分配的内存块、引用静态或全局变量，或是返回数组、结构体等复杂数据类型的地址。

The general form of a function that returns a pointer value:

```
Type name*function name(parameter table column);
```

返回指针值的函数的一般形式：

```
类型名 *函数名(参数表列);
```

Important Principles（重要原则）:

When designing a function that returns a pointer value, an important principle to follow is not to return a pointer to a local variable. This is because local variables are allocated on the stack when the function is called, and when the function finishes executing and returns, the local variables on the stack are automatically released. If a pointer points to these freed local variables at this time, the pointer will become a dangling pointer, and the memory area it points to is uncertain, and accessing such a memory area may cause the program to crash or other unpredictable behavior.

当设计返回指针值的函数时，必须遵循一个重要的原则，即不要返回指向局部变量的指针。这是因为局部变量是在函数被调用时分配在栈上的，当函数执行完毕并返回时，栈上的局部变量会被自动释放。如果此时有指针指向了这些已经释放的局部变量，那么这个指针将成为悬垂指针，它指向的内存区域是不确定的，访问这样的内存区域可能会导致程序崩溃或其他不可预测的行为。

Therefore, when you write a function that returns a pointer value, you should make sure that the returned pointer points to a stable area of memory, such as a dynamically allocated block of memory through a function such as malloc, calloc, or realloc, or the address of a static or global variable. This ensures that the returned pointer is still valid after the function returns.

因此，在编写返回指针值的函数时，应当确保返回的指针指向的是稳定的内存区域，如通过malloc、calloc或realloc等函数动态分配的内存块，或者是静态或全局变量的地址。这样可以确保返回的指针在函数返回后仍然有效。

Case 7_10（案例 7_10）

Find the maximum value and its address in a 3×3 two-dimensional array.

求3×3二维数组中的最大值及其地址。

Thinking（解题思路）:

To find the maximum value in a 3×3 two-dimensional array and return its address, we need to write a function max whose return value type should be int *. This function iterates through the entire two-dimensional array, compares the elements to find the maximum value, and returns the address of the maximum value in the array. This method not only returns the maximum value itself, but also provides the exact position of the value in the array, increasing the flexibility and understandability of the program.

为了找到3×3二维数组中的最大值并返回其地址，我们需要编写一个函数max，该函数的返回值类型应为int *。该函数将遍历整个二维数组，通过比较各个元素来找出最大值，并返回该最大值在数组中的地址。这种方法不仅返回了最大值本身，还提供了该值在数组中的确切位置，增加了程序的灵活性和可理解性。

Program Code（程序代码）:

Case7_10.c

```
#include <stdio.h>
int *max(int b[3][3])
{
    static int m=b[0][0],int i,j;
    for(i=0;i<3;i++)
    {
        for(j=0;j<3;j++)
        {
            if(m<b[i][j]) m=b[i][j];
        }
    }
    return &m;
```

```
13    }
14    int main()
15    {
16        int a[3][3];
17        int i,j;
18        for(i=0;i<3;i++)
19        {
20            for(j=0;j<3;j++) scanf("%d",&a[i][j]);
21        }
22        printf("Max:%d",*max(a));
23        return 0;
24    }
```

Program Output（运行结果）:

```
1 2 3
4 5 6
7 8 9
Max: 9
```

Case Analysis（案例解析）:

(1) In lines 2 to 13, we define a function called max whose return value is of type int *. Internally, this function implements a "challenge" algorithm that iterates through a two-dimensional array in a double loop to find the maximum value m. Once the maximum is found, the program returns the address of the maximum in the array, which is &m.

(2) In the main function(lines 14-24), we define a 3×3 two-dimensional array a and input its elements one by one using a two-dimensional loop. Then the max function is called. When the array is used as the formal parameter of the function, the actual parameter passes the name of the array. At this time, the array operated by the formal parameter is the real parameter group, that is, the array a is accessed and obtained in the max function. The address of its maximum value is brought back to the main function as the return value, and then the dereference operator*is used to obtain the maximum value and output it.

（1）在第2~13行中，我们定义了一个名为max的函数，其返回值类型为int *。这个函数内部实现了“打擂台”算法，通过双重循环遍历二维数组以寻找最大值m。一旦找到最大值，程序将返回该最大值在数组中的地址，即&m。

（2）在主函数（第14~24行），我们定义了3×3二维数组a并使用二维循环逐一输入其元素。随后调用max函数，当数组作为函数形参时，实参传递数组名，此时形参操作的数组即为实参数组，即在max函数中访问并求得的是数组a，将其最大值的地址作为返回值带回主函数，随后利用解引用操作符*获取最大值并输出。

Case Development（案例拓展）:

What happens if you change line 4 “static int m=b[0][0]” in the max function to just using an ordinary local variable int m = b[0][0];

When we try to return & m(the address of the local variable m), we are actually returning a pointer to an area of memory that is about to be destroyed. Because the local variable m is defined inside the function, its lifetime is limited to the duration of the execution of the max function. Once the max function finishes executing and returns, the memory space occupied by the local variable m is released, and the contents of the memory pointed to by the return address are uncertain,

若将max函数中第4行从“static int m=b[0][0];”更改为直接使用一个普通局部变量int m = b[0][0];，则会出现什么问题呢？

当我们尝试返回&m（即局部变量m的地址）时，实际上返回的是一个指向即将被销毁的内存区域的指针。因为局部变量m是在函数内部定义的，它的生命周期仅限于max函数的执行期间。一旦max函数执行完毕并返回，局部变量m所占用的内存空间将被释放，此时返回的地址指向的内

which may cause the program to crash or produce unpredictable behavior.

The workaround is to extend the lifetime of the variable, using static local variables or global variables. Because these two types of variables are not released at the end of the function.

存内容是不确定的，这可能导致程序崩溃或产生不可预测的行为。

解决方法是延长变量的生命周期，可以使用静态局部变量，也可以使用全局变量。因为这两类变量不会随着函数的结束而释放。

7.8 Recursive Programming 递归程序设计

7.8节视频

We have learned the core knowledge of C language, including three structures, functions, arrays and pointers. Now, we will focus on a powerful programming technique called recursion. By mastering recursion, we will further improve our programming thinking and problem-solving skills.

我们已经深入学习了C语言的核心知识，包括三大结构、函数、数组和指针等。现在，我们将聚焦于一种强大的编程技术——递归。通过掌握递归，我们将进一步提升编程思维和解决问题的能力。

7.8.1 What Is Recursion（什么是递归）

First, let's look at what recursion is.

The essence of recursion is an efficient solution strategy that fits human logical thinking, especially for those "big problems" that can be decomposed into several similar "sub-problems". In a nutshell, the core idea of recursion is to "divide and conquer": to reduce a large complex problem to a number of similar and more tractable sub-problems. This way of thinking is not only close to the natural logic of human beings, but also shows excellent efficiency in solving a specific class of problems.

The core feature of recursion is the ability of a function to call itself directly or indirectly. Specifically, when a function calls itself directly within its body of definition, we call it direct recursion. If the function calls itself indirectly through the intermediary of another function, it is called indirect recursion. These two forms together constitute different expressions of recursive calls, showing the flexibility and depth of recursion.

In order to better understand recursion, we can use the graphic method to visualize it. As shown in Figure 7-10.

首先让我们看一下什么是递归。

递归，其本质在于一种契合人类逻辑思维的高效求解策略，特别适用于那些能够被拆解为若干个相似"子问题"的"大问题"。简而言之，递归的核心思想在于"分而治之"——将复杂的大问题简化为若干相似且更易于处理的子问题。这种思维方式不仅贴近人类的自然逻辑，而且在解决一类特定问题时展现出卓越的效率。

递归的核心特性在于函数能够直接或间接地调用自身。具体而言，当函数在其定义体内直接调用自身时，我们称之为直接递归；若该函数通过其他函数作为中介间接地调用自身，则称为间接递归。这两种形式共同构成了递归调用的不同表达，展现了递归的灵活性和深度。

为了更好地理解递归，我们可以采用图示法来形象化地认识它，如图7-10所示。

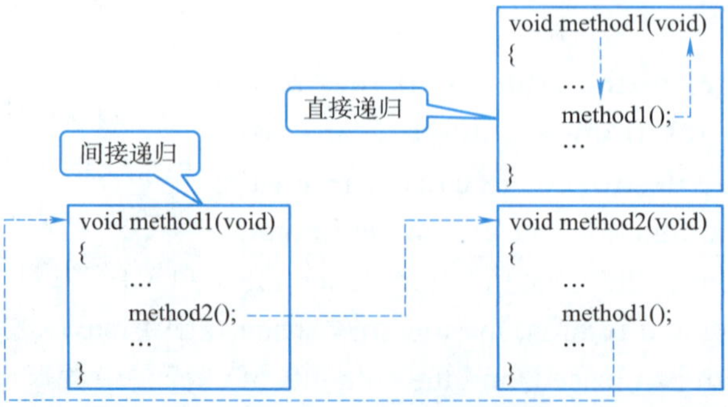

图7-10 什么是递归

When the method1 function calls itself directly in its definition, this self-referencing phenomenon is called direct recursion. On the other hand, if the method1 function calls method2, which in turn calls method1, this indirect self-reference through another function is called indirect recursion. In programming practice, direct recursion is more commonly used because of its intuitive and succinct nature.

In practice, recursion is widely used in various scenarios, such as sorting algorithms(such as merge sort, quick sort), traversing data structures(such as tree traversal), solving mathematical problems(such as factorial calculation, Fibonacci sequence) and so on. By mastering recursion technology, we can solve complex problems more flexibly and improve our programming ability and thinking level.

Finally, in order to deepen the understanding of the concept of recursion, we will show the application of recursion through a concrete program example, which is to solve the mathematical problem of factorial of n(n!). This example is not only easy to understand, but also can clearly reflect the core idea of recursion, and help you to grasp the powerful programming technology of recursion more deeply.

当method1函数在其定义中直接调用自身时，这种自引用现象被称为直接递归。另一方面，如果method1函数调用method2，而method2又转而调用method1，则这种通过另一个函数实现的间接自引用称为间接递归。在编程实践中，直接递归因其直观性和简洁性而更为常用。

在实际应用中，递归被广泛应用于各种场景，如排序算法（如归并排序、快速排序）、遍历数据结构（如树的遍历）、求解数学问题（如阶乘计算、斐波那契数列）等。通过掌握递归技术，我们能够更加灵活地解决复杂问题，提升编程能力和思维水平。

最后，为了深化对递归概念的理解，我们将通过一个具体的程序实例——求解n的阶乘（n!）的数学问题，来直观展示递归的应用。这一实例不仅易于理解，而且能够清晰地体现递归的核心思想，帮助大家更加深入地掌握递归这一强大的编程技术。

```
int fun(int n)
{
    if(1==n || 0==n)
        return 1;
    else
        return fun(n-1)*n;                    //自己调用自己
}
```

This fun function is a function that specializes in computing the factorial of n(n!). Its internal uniqueness is that it implements a self-calling process, which is the core feature of recursive programs. In this way, the fun function can simplify the problem step by step until it reaches an easy-to-solve benchmark case(1! Or 0!), which can efficiently and accurately compute the factorial of n.

So how exactly does the recursion work? How to find n! through this code? Next, let's explore how the recursion works.

这个fun函数是一个专门用于计算n的阶乘（n!）的函数。其内部的独特之处在于它实现了一种自我调用的过程，这即是递归程序的核心特点。通过这种方式，fun函数能够逐步简化问题，直至达到一个易于解决的基准情况（1!或者0!），从而高效且准确地计算出n的阶乘。

那么递归到底是如何工作的呢？如何通过这段代码实现求n!呢？接下来，让我们共同探讨递归的工作原理。

7.8.2 How Does Recursion Work（递归如何工作）

Although recursive programs call themselves, they still follow the calling process of ordinary functions. We still take the function n! as an example to explore the working principle of recursion in depth. In the fun function body statement, when n is 1 or 0, it returns 1 directly. Otherwise, it executes the fun

递归程序虽然是自己调用自己，但仍旧遵循着普通函数的调用过程。我们仍以求n!函数为例，来深入探索递归的工作原理，在fun函数体语句中，当n为1或者为0时则直接返回1，否则执行fun(n-1)*n语

(n−1)*n statement, which generates a recursive call.

As shown in Figure 7-11, by calling the fun(5) function, the main function can find that the factorial of 5 is 120. How to find the result?

Let's take a closer look at the calling process between functions, as shown in Figure 7-12.

句，此时产生递归调用。

如图7-11所示，主函数通过调用fun(5)函数，可以求得5的阶乘为120，如何求得该结果的呢？

我们来详细审视一下函数之间的调用过程，具体过程如图7-12所示。

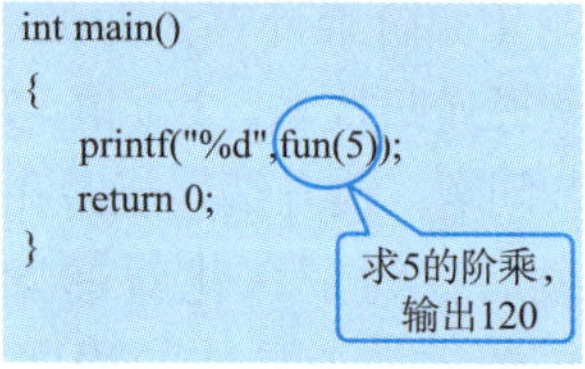

图7-11　主函数调用fun函数

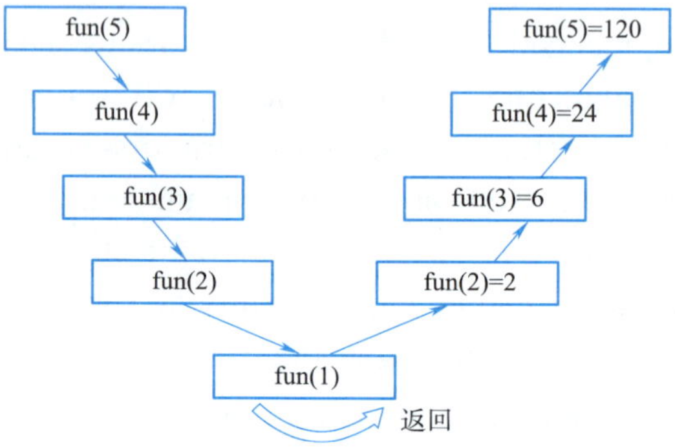

图7-12　求5!递归过程

First, the main function initiates a call to fun(5). Depending on the mechanism of the function call, the program jumps to the fun(5) function to executing. Since n is not equal to 1 or 0 at this step, the statement fun(4)*5 is executed inside the function, which again triggers the call to fun(4). The program then jumps to the fun(4) function. Since n is also not equal to 1 or 0, fun(3)*4 is evaluated inside the function. This results in yet another fun function call, fun(3). In the fun(3) function, the program executes fun(2)*3. Next, enter the fun(2) function again, this time executes fun(1)*2. At this time, n is 1, which satisfies the termination condition of the recursion, so fun(1) returns 1 directly.

Following the executing flow of the function call, the program returns to the previous call, fun(2). In fun(2), 1*2 is computed using the returned 1, resulting in 2, and ends fun(2) execution. Subsequently, the return value 2 of fun(2) is passed to its upper level call fun(3), which evaluates 2*3 to the result 6, and ends. Next, the program returns to fun(4), calculates 6*4 with the 6 returned by fun(3), obtains the result 24, and ends fun(4). Finally, the return value 24 of fun(4) is passed to fun(5), which computes 24*5 to get the final result 120, which is returned to the main function. The main function receives this value, outputs it to the console, and the program ends.

Through the example just now, let's analyze the working principle of recursion in depth. The core of recursion is embodied in two intertwined processes:

首先，主函数发起了对fun(5)的调用。根据函数调用的机制，程序跳转到fun(5)函数开始执行。由于此时n不等于1或0，函数内部会执行fun(4)*5的语句，这再次触发了对fun(4)的调用。随后，程序跳转到fun(4)函数。由于n同样不等于1或0，函数内部会计算fun(3)*4。这导致了又一次的fun函数调用，即fun(3)。在fun(3)函数中，程序执行fun(2)*3。接着，再次进入fun(2)函数，这次会执行fun(1)*2。此时，n为1，满足了递归的终止条件，因此fun(1)直接返回1。

按照函数调用的执行流程，程序将返回至上一层调用，即fun(2)。在fun(2)中，使用返回的1计算1*2，得到结果2，并结束fun(2)的执行。随后，fun(2)的返回值2被传递给其上一层调用fun(3)，fun(3)计算2*3得到结果6，并结束。接着，程序返回到fun(4)，用fun(3)返回的6计算6*4，得到结果24，并结束fun(4)。最后，fun(4)的返回值24被传递给fun(5)，fun(5)计算24*5得到最终结果120，并返回给主函数。主函数接收到这个值后，将其输出到控制台，随后程序结束。

通过刚才的例子，我们一同来深入剖析递归的工作原理。递归的核心体现在两个相互交织的过程上：

(1) Decomposition process. That is to say, a complex problem is gradually transformed into a number of similar but smaller sub-problems. Taking fun(5) as an example, we first decompose it into fun(4), and then fun(4) is decomposed into fun(3) and fun(2), and so on, until we reach a minimum problem that can be solved directly or easily, that is, fun(1). This process continues to simplify the problem until the recursive baseline case is reached.

（1）分解过程。即把复杂的大问题逐步转化为若干与之相似但规模更小的子问题。以求解fun(5)为例，我们首先将其分解为fun(4)的求解，随后fun(4)又被分解为fun(3)和fun(2)，如此层层递进，直至达到一个可以直接或易于求解的最小问题，即fun(1)。这个过程不断将问题简化，直至到达递归的基准情况。

(2) Evaluation process. This is a bottom-up process. When the smallest sub-problem fun(1) is solved, the result is passed up level by level to compute the sub-problem of the previous level. For example, the result of fun(1) is returned to fun(2), the result of fun(2) is computed, and it is passed up until the result of fun(5) is finally computed. In this process, the final solution of the whole problem is constructed step by step by using the results of the solved sub-problems.

（2）求值过程。这是一个自下而上的过程。当最小的子问题fun(1)解决后，其结果会逐层向上传递，用于计算上一层的子问题。例如，fun(1)的结果返回给fun(2)，进而计算出fun(2)的结果，并继续向上传递，直至最终计算出fun(5)的结果。这个过程通过不断利用已求解的子问题结果，逐步构建出整个问题的最终解。

Now, let's review again the recursive procedure for solving 5!. Starting from the decomposition process, we gradually refine the problem until fun(1) can be solved directly; then, in the evaluation process, we start from the smallest sub-problem and calculate upward step by step, and finally get the result of 5!. It is worth noting that the whole recursion process is automatically executed by the computer without human intervention, which reflects the efficiency and convenience of recursion.

现在，我们再次回顾一下求解5!的递归过程。从分解过程开始，我们逐步将问题细化，直至可以直接求解的fun(1)；随后，在求值过程中，我们从最小的子问题开始，逐层向上计算，最终得出5!的结果。值得注意的是，整个递归过程是由计算机自动执行的，无须人为干预，这体现了递归的高效性和便捷性。

7.8.3 How Is Recursion Designed（递归如何设计）

So far, we have gained insight into the concept and workings of recursion through the example of n factorial(n!). Not only does recursion make code easy to understand and clear, it is also a common technique in programming. So how can we get a better handle on the design of recursive programs? Fortunately, recursive programming does follow a fixed structure.

至此，我们已经通过n的阶乘（n!）的例子深入了解了递归的概念和工作原理。递归不仅使得代码易于理解、清晰简洁，还是编程中常用的一种技术。那么，我们如何更好地掌握递归程序的设计呢？幸运的是，递归程序设计确实遵循着一种固定的结构。

The key of recursion is to make two important points clear:

递归的关键在于明确两个要点：

Recursive exit: It is the termination condition for recursive decomposition, and avoids stack overflow caused by infinite recursion. In general, the recursive exit corresponds to the smallest problem that can be solved directly or conveniently. Such as 0! or 1!, that is, the smallest problem, just returns 1.

递归出口：这是递归分解的终止条件，避免无限递归导致的栈溢出。通常，递归出口对应着可以直接或方便求解的最小问题。如0！或1！即为最小的问题，直接返回1即可。

Recursive relation: This describes how the original problem(the big problem) is decomposed into sub-problems(the small problems) and the relationship between the big and small problems, e.g., n! =(n−1)!*n.

递归关系：这描述了原始问题（大问题）如何分解为次一级问题（小问题），以及大问题和小问题之间的关系。如n！=(n−1)!*n。

Next, let's write a program using a recursive structure, using the example of finding the minimum of a sequence of integers.

First of all, we determine whether we can use recursion to solve the problem. The size of the sequence is different, but the method of solving the problem should be the same, so we can try to use the recursive structure.

We define the function f(a, i) to find the minimum of the i numbers from a[1] to a[i] in the array a. To find out the recursive structure, we need to first determine the recursive exit. In this problem, when i is 1, there is only one number, the number itself is the minimum, which is our recursive exit.

Next, we need to determine the recursion relation. Generally speaking, we need to locate the original problem(big problem), that is, f(a, i), and then find the next problem(small problem), that is, f(a, i−1). Then, we need to find out the relationship between these two problems. In this example, to solve the minimum of i numbers, we only need to solve the minimum of the first i−1 numbers, and then compare this minimum with the i-th number a[i], the smaller one is the minimum of i numbers. This is our recursion relation.

In this way, we can clearly construct the structure of the recursive program, as shown in Figure 7-13. Recursive functions typically use if...else constructs to connect recursive exits or recursive relations. When a condition is met, the recursive decomposition terminates and the evaluation process begins. Otherwise, the recursion continues to decompose. The complete code implementation looks like this:

接下来，让我们以求解一个整数序列的最小值为例，使用递归结构来编写程序。

首先我们判定是否能够使用递归求解，序列的规模不同，但求解问题的方法应该是一致的，故可以尝试使用递归结构。

我们定义函数f(a,i)，用于求解数组a中从a[1]到a[i]这i个数的最小值。要找出递归结构，我们需要首先确定递归出口。在这个问题中，当i为1时，即只有一个数时，这个数本身就是最小值，这就是我们的递归出口。

接下来，我们需要确定递归关系。一般来说，我们需要定位到原始问题（大问题），即f(a, i)，然后找出次一级的问题（小问题），即f(a, i−1)。然后，我们需要找出这两个问题之间的关系。在这个例子中，要求解i个数中的最小值，我们只需要先求解前i−1个数中的最小值，然后将这个最小值与第i个数a[i]进行比较，较小者即为i个数中的最小值。这就是我们的递归关系。

通过这种方式，我们可以清晰地构建出递归程序的结构，如图7-13所示。递归函数一般使用if...else结构来连接递归出口或递归关系。当满足某个条件时，递归分解将终止，并开始求值过程；否则，递归将继续分解。

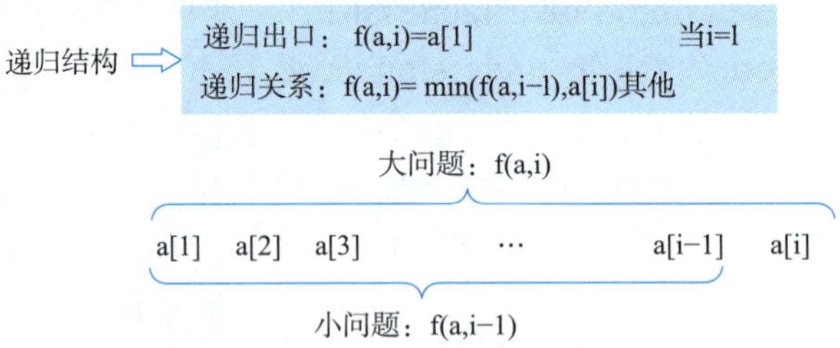

图7-13 求序列最小值问题的递归结构

Program Code（程序代码）:

```
#include <stdio.h>
//假设数组a从索引1开始，i表示要查找最小值的元素数量
int f(int a[], int i)
{
    //递归出口：当只有一个元素时，直接返回该元素作为最小值
    if(i==1)
    {
        return a[1];
    }
```

```
        else
        {
            int m=f(a,i-1);                         //自己调用自己，分解为更小的规模
            //再将m与第i个数a[i]比较，取较小值作为i个数的最小值
            return(m<a[i])?m:a[i];
        }//递归关系：先求解前i-1个数的最小值
    }
    int main()
    {
        int a[]={5, 2, 9, 1, 5, 6};                 //示例数组
        int n=sizeof(a)/sizeof(a[0]);
        /*n为数组元素个数，sizeof(a)求整个数组所在字节大小，sizeof(a[0])单个元素所占自己大小*/

        int minVal=f(a, n);                         //调用递归函数查找最小值
        printf("The minimum value in the array is: %d\n", minVal);
        return 0;
    }
```

Program Output(运行结果):

```
The minimum value in the array is:1
```

In this example, the f function is recursive and takes an integer array a and an integer i as parameters. If i is equal to 1(that is, there is only one element to compare), the function returns that element directly as the minimum(recursive exit). Otherwise, the function calls itself recursively, finds the minimum of the first i-1 elements, and stores it in the variable m. The function then compares m to the current element a[i], returning the smaller of the two as the minimum of the i elements(recursive relation).

In the main function, we define an illustrated array a and calculate the number of its elements, and that is n. We then call the f function to find the minimum value in the array and print the result to the console.

在这个示例中，f函数是递归函数，它接受一个整数数组a和一个整数i作为参数。如果i等于1（即只有一个元素需要比较），则函数直接返回该元素作为最小值（递归出口）。否则，函数会递归调用自身，查找前i-1个元素的最小值，并将其存储在变量m中。然后，函数将m与当前元素a[i]进行比较，返回两者中的较小值作为i个元素的最小值（递归关系）。

在main函数中，我们定义了一个示例数组a，并计算了其元素个数n。然后，我们调用f函数来查找数组中的最小值，并将结果打印到控制台上。

7.9 Comprehensive Cases 综合案例

下面将通过几个精选案例深化对递归原理的理解，并熟练掌握递归设计结构的应用，以便编写出高效且正确的递归程序。这些案例有助于揭开递归的神秘面纱，以期望达到灵活运用递归解决复杂问题的目的。

综合案例 7_9_1 递归求数列和。

案例描述： 使用递归方法求一个数列的和。

解题思路： 在解决此类问题时，首先识别出它适合采用递归方法，因为问题可以被分解为更小的相似问题。接下来，我们确定递归结构，包括递归出口和递归关系。递归出口对应最小问题，即当没有数据项时，和为0；而递归关系则是大问题与其次小问题之间的关系，通过求解次小问题（即n-1个数据项的和）并加上最后一个数据项，来求解大问题（n个数据项的和）。最后使用if...else结构来构建递归程序，当达到

递归出口时返回0，否则继续递归调用并累加数据项。这种递归方法能够清晰、有效地解决此类问题。

程序代码：

Case7_9_1.c

```
1    #include <stdio.h>
2    int sum(int a[],int n)                              // 大问题
3    {
4        if(0==n) return 0;
5        else return sum(a,n-1)+a[n-1];                  // 次小问题加上 a[n-1] 解决了大问题
}
6    int main()
7    {
8        int a[5]={1,2,3,4,5};
9        printf(" 和为:%d",sum(a,5));
10       return 0;
11   }
```

运行结果：

```
和为:15
```

案例解析：

（1）第2~6行是一个递归实现的sum求和函数，接受一个整数数组a[]和元素个数n作为参数，返回整数类型的和。当n为0时，递归分解结束并返回0；否则，递归调用sum(a, n−1)求得前n−1个元素的和，再与当前最后一个元素a[n−1]相加。

（2）第9行主函数内定义并初始化了一个用于测试的数组。第10行调用了sum函数，求这5个数据之和，最终结算结果为15。

（3）下面给出计算整数数组a[] = {1, 2, 3, 4, 5}的和详细调用过程。

递归分解过程如下：

①初始调用：sum(a, 5)。

递归开始，此时n为5，不等于0，所以不会直接返回0。进入递归逻辑，需要计算sum(a, 4)+a[4]（a[4]是5）。

②第一次递归调用：sum(a, 4)。

递归继续，此时n为4。进入递归逻辑，需要计算sum(a, 3)+a[3]（a[3]是4）。

③第二次递归调用：sum(a, 3)。

递归继续，此时n为3。进入递归逻辑，需要计算sum(a, 2)+a[2]（a[2]是3）。

④第三次递归调用：sum(a, 2)。

递归继续，此时n为2。进入递归逻辑，需要计算sum(a, 1)+a[1]（a[1]是2）。

⑤第四次递归调用：sum(a, 1)。

递归继续，此时n为1。进入递归逻辑，需要计算sum(a, 0)+a[0]（a[0]是1）。

⑥第五次递归调用（基准情况）：sum(a, 0)。

递归到达基准条件，因为n为0，直接返回0。即将进入求值过程。

递归求值过程如下：

①回溯第四次递归调用：

由于sum(a, 0)返回了0，可以计算0 + a[0]，即0 + 1 = 1。因此，sum(a, 1)返回1。

②回溯第三次递归调用：

使用sum(a, 1)返回的1，可以计算1 + a[1]，即1 + 2 = 3。因此，sum(a, 2)返回3。

③回溯第二次递归调用：

使用sum(a, 2)返回的3，可以计算3 + a[2]，即3 + 3 = 6。因此，sum(a, 3)返回6。

④回溯第一次递归调用：

使用sum(a, 3)返回的6，可以计算6 + a[3]，即6 + 4 = 10。因此，sum(a, 4)返回10。

⑤回溯初始调用：

使用sum(a, 4)返回的10，可以计算10 + a[4]，即10 + 5 = 15。因此，sum(a, 5)返回15，这是整个数组的和。

（4）若第4行的递归出口改为只有1个数的情况，代码改为if(1==n)return a[0];，答案也是对的，只要符合逻辑即可。

综合案例 7_9_2 递归求偶数个数。

案例描述： 使用递归方法统计一个数组中偶数的个数。

解题思路： 该案例可以通过递归方法将问题逐步分解为更小的子问题来求解。定义一个递归函数EvenNum(int a[], int n)来计算数组a中前n个元素的偶数个数。递归过程包括两个关键部分：

（1）递归出口（最小问题）：当n为0时，没有元素需要检查，偶数个数为0。

（2）递归关系（大问题和次小问题的关系）：考虑当前元素a[n−1]，如果它是偶数，则在EvenNum(a, n−1)（次小问题）的结果上加1；否则保持不变，则可解决大问题EvenNum(a, n)。

通过递归调用，我们逐步缩小问题的规模，直到满足递归出口条件，最终得到整个数组中偶数的个数。

程序代码：

Case7_9_2.c

```
1    #include <stdio.h>
2    //递归函数，用于计算数组中偶数的个数
3    int EvenNum(int a[], int n)
4    {
5        if(n==0)                              //递归出口
6            return 0;
7        else                                  //递归关系
8            return(a[n-1]%2==0)?1+EvenNum(a,n-1):EvenNum(a,n-1);
9    }
10   int main()
11   {
12       int array[]={1, 2, 3, 4, 5, 6, 7, 8, 9, 10};
13       int size=sizeof(array)/sizeof(array[0]);            //计算数组大小
14       int evenCount=EvenNum(array, size);                 //调用递归函数计算偶数个数
15       printf("The number of even elements in the array is: %d\n", evenCount);
16       return 0;
17   }
```

运行结果：

```
The number of even elements in the array is: 5
```

案例解析：

（1）第3行开始定义递归函数EvenNum：函数接收一个整数数组a和一个整数n作为参数。n表示当前正在检查的元素个数。函数首先检查递归出口条件：如果n为0（即没有元素需要检查），则返回0。如果n不为0，函数会检查当前元素a[n−1]是否为偶数。如果是偶数，函数会递归调用自身EvenNum(a, n−1)来计算剩余元素的偶数个数，并在此结果上加1。如果不是偶数，函数会简单地递归调用自身EvenNum(a, n−1)来计算剩余元素的偶数个数。

（2）第12行定义了一个整数数组array，并计算了其大小size（第13行）。

（3）第14行调用 EvenNum函数来计算数组中的偶数个数，并将结果存储在evenCount变量中。

（4）第15行使用printf函数输出结果。

（5）下面给出执行过程：

①当EvenNum(array, size)被调用时，由于size为10（数组有10个元素），所以不会立即返回。

②函数首先检查第10个元素（a[9]），即10，它是偶数，所以递归调用EvenNum(array, 9) 并加1。

③然后EvenNum(array, 9)会检查第9个元素（a[8]），即8，也是偶数，所以递归调用 EvenNum(array, 8)并加1。

④上述过程会继续进行，直到n变为0，递归出口被触发，然后逐层返回并累加偶数的个数。最终，EvenNum函数返回整个数组中偶数的个数，这个值被main函数中的evenCount变量接收并打印出来。

综合案例 7_9_3 递归分解素因数。

案例描述：递归法分解素因数，即把一个合数写成几个素因数相乘的形式。例如，输入90，将打印出90=2*3*3*5；45=3*3*5；18=2*3*3。

解题思路：求素因子，首先要定义判断是否为素数的函数（IsPrime），假设数为num，则应遍历2~sqrt(num)的所有整数，检查它们是否是num的因数。如果找到任何因数，则返回0（表示num不是质数）。如果遍历完所有数字都没有找到因数，则返回1（表示num是素数）。

其次，定义素因数分解函数（prime_factorize）：对于输入的数num，从2开始遍历其可能的因数。如果找到因数i，则打印该因数，并递归调用prime_factorize(num/i)以处理剩下的部分。递归调用的终止条件是num为素数，因此时不需要继续分解。

最后在主函数（main）中测试：如果是素数，则提示用户重新输入。如果不是素数，则调用prime_factorize函数进行素因数分解，并打印结果。

程序代码：

Case7_9_3.c

```
#include <stdio.h>
#include <math.h>                                   //引入math.h库来使用sqrt函数
int IsPrime(int num)                                //若为素数返回1，否则返回0
{
    int i;
    if(num<=1) return 0;                            //0和1不是素数
    for(i=2; i<=sqrt(num); i++)                     //只检查到sqrt(num)
        if(num%i==0)return 0;                       //若有因子，则不是素数
    return 1;                       //循环结束都没有遇到return 0，则说明没有因子，为素数
}
void prime_factorize(int num)
{
     int i;
     if(IsPrime(num))                               //判断是否为素数
     {
         printf("%d",num);                          //若为素数直接输出
         return;                                    //终止递归
    }
    else
    {
        for(i=2;i<=sqrt(num);i++)                   //注意检查因子
        {
            if(num%i==0)
            {
                printf("%d*",i);                    //若是因子则输出
                prime_factorize(num/i);             //输出剩下的因子
                return;
            }
        }
    }
}
int main()
```

```
33 {
34     int number;
35     printf("请输入要分解的合数:");
36     scanf("%d", &number);
37     while(IsPrime(number))
38     {
39         printf("您输入的不是合数，请重新输入:");
40         scanf("%d", &number);
41     }
42     printf("分解后的结果为:\n%d=", number);
43     prime_factorize(number);                    //递归分解素因数函数调用
44     return 0;
45 }
```

运行结果：

```
请输入要分解的合数:5
您输入的不是合数，请重新输入:18
分解后的结果为:
18=2*3*3
```

案例解析：

（1）当程序输入5时，第37行，IsPrime(5)返回值为1，函数条件为真，则提示“您输入的不是合数，请重新输入：”，当输入18时，IsPrime(18)返回值为0，所以不会要求用户重新输入。

（2）程序按照顺序执行43行递归函数，进入函数内执行。

（3）第14~18行为递归终止条件，当合数不断分解，直到分解为素数时，则不需要继续分解。

（4）第19行开始为递归关系的代码，第12行的for循环，逐一检查num的因子，若找到第一个因子，即第23行的num%i==0条件成立，第25行输出该因子和“*”，同时继续递归prime_factorize(num/i);。

（5）不断执行递归逻辑，当分解到素数时结束。

（6）IsPrime函数中，使用了sqrt函数（需要包含math.h头文件）减少循环次数，而不用到num-1。该函数若有因子则返回0，代表不是素数；相反则为素数。

（7）以18为例，展示prime_factorize函数的详细执行过程：

①调用prime_factorize(number)，即prime_factorize(18)。在prime_factorize函数中，由于18不是素数（IsPrime(18)返回0），进入else分支。

②进入循环，从i=2开始检查18的因子。当i=2时，18不能被2整除，所以继续循环。当i=3时，18能被3整除。打印3*。

③调用prime_factorize(num/i)，即prime_factorize(18/3)，也就是prime_factorize(6)，继续分解。

④对于prime_factorize(6)，因为6不是素数（IsPrime(6)返回0），再次进入else分支。进入循环，从i=2开始检查6的因子。当i=2时，6能被2整除。打印2*。

⑤调用prime_factorize(num/i)，即prime_factorize(6/2)，也就是prime_factorize(3)，继续分解。

⑥对于prime_factorize(3)，因为3是素数（IsPrime(3)返回1），所以进入if分支。打印3。

⑦由于3是素数，函数返回，不再继续调用自身。

⑧回到上一次的prime_factorize(6)调用，由于已经打印了所有因子并返回，所以不再继续执行。同样地，回到最初的prime_factorize(18)调用，也不再执行后续代码。

⑨main函数结束，程序退出。最终输出：

```
请输入要分解的合数:18
分解后的结果为:
18=3*2*3
```

案例总结：在采用递归程序设计时，首要步骤是判断大型问题是否能被合理地分解为若干小型问题，这些小型问题能够使用与原始问题相同的算法进行求解。其次，关键在于精确地确定递归的终止条件（即递归出口）以及递归调用之间的逻辑关系（递归关系）。递归程序设计的核心在于顶层设计，只要逻辑设计正确，程序正确执行的可能性就会大大提高。

Summary 小　结

This chapter explores the core concepts of functional modular programming in depth. Firstly, it expounds the basic methods of function definition, declaration and call. Then, the practical applications of arrays and pointers as function parameters are analyzed in detail to enhance the flexibility and functionality of the function. In addition, this chapter also introduces the advanced application of functions, callback functions, which allow one function to be passed to another as a parameter and called when needed, greatly improving the dynamic and scalability of the program. Finally, the recursive program design principle and programming method are emphatically explained, and then a more efficient program can be written. Overall, this chapter is designed to help readers understand the essence of functional modular programming, master the advanced use of functions, and how to write more efficient and elegant code through the recursion and other techniques.

本章深入探讨了函数模块化编程的核心概念，首先阐述了函数的定义、声明和调用的基本方法。接着，详细解析了数组和指针作为函数参数的实际应用，增强函数的灵活性和功能性。此外，本章还介绍了函数的高级应用——回调函数，这种编程技巧允许一个函数作为参数传递给另一个函数，并在需要时被调用，极大地提升了程序的动态性和可扩展性。最后，着重讲解了递归程序设计原理及编程方法，进而编写更为高效的程序。整体而言，本章内容旨在帮助读者深入理解函数模块化编程的精髓，掌握函数的高级用法，以及如何通过递归等技巧来编写更加高效、优雅的代码。

Exercise 习　题

一、单选题

1. C语言中，凡未指定存储类别的局部变量的隐含存储类别是（　　）。
   A. 自动（auto）　B. 静态（static）　C. 外部（extern）　D. 寄存器（register）
2. 关于函数声明int fun(int arr[]);说法正确的是（　　）。
   A. fun 函数要传入一整个数组　B. fun 函数要传入一个指针
   C. fun 函数只能传入一个数组名　D. fun 函数要传入一个指针数组
3. 以下关于程序中函数的定义、调用和声明的叙述中，正确的是（　　）。
   A. 函数的定义必须放在该函数的调用之前
   B. 函数的声明必须放在该函数的调用之前
   C. 函数的定义必须放在该函数的声明之前
   D. 函数的声明必须放在该函数的定义之前
4. 以下关于函数定义时的形参说法正确的是（　　）。
   A. 形参只能一个　B. 形参不能为空
   C. 形参可以是任意类型　D. 形参不可以是指针

5. 以下函数的返回类型为（ ）。

```
int sum(double a, double b)
{
    return(a+b);
}
```

A. void B. int C. float D. double

6. 若有以下调用语句，则正确的fun函数首部是（ ）。

```
#include <stdio.h>
int main()
{
    …
    int a;
    float x;
    …
    fun(x,a);
    …
}
```

A. void fun(int m,float x) B. void fun(float a,int x)

C. void fun(int m,float x[]) D. void fun(int x,float a)

7. 以下程序的运行结果是（ ）。

```
#include <stdio.h>
int main()
{
    int k=4,m=1,p;
    p=func(k,m);
    printf("%d\n",p);
    p=func(k,m);
    printf("%d\n",p);
    return 0;
}
int func(int a,int b)
{
    static int m=0,i=2;
    i+=m+1;
    m=i+a+b;
    return m;
}
```

A. 8, 17 B. 8, 17 C. 8, 8 D. 4, 1

8. 以下程序的运行结果是（ ）。

```
#include <stdio.h>
void fun(int(*a)[2])
{
    printf("%d\n", a[0][0]);
}
int main()
{
    int a[][2]={1, 2, 3, 4};
    fun(a+1);
    return 0;
}
```

A. 1 B. 2 C. 3 D. =4

9. 以下程序的运行结果是（　　）。

```
#include <stdio.h>
int *fun(int(*a)[4])
{
    return(*(a+1)+2);
}
int main()
{
    int a[2][4]={5, 6, 2, 1, 3, 7, 8, 9};
    printf("%d", *(fun(a)+1));
    return 0;
}
```

A. 1　　B. 2　　C. 8　　D. 9

10. 以下程序的运行结果是（　　）。

```
#include <stdio.h>
void fun(int(*a)[2])
{
    printf("%d\n", a[0][0]);
}
int main()
{
    int a[][2]={1, 2, 3, 4};
    fun(a+1);
    return 0;
}
```

A. 1　　B. 2　　C. 3　　D. 4

二、程序分析题

1. 请改正程序中的错误，使程序能得出正确的结果。下列给定程序中，函数fun的功能是：分别统计字符串中大写字母和小写字母的个数。例如，给字符串s输入AAaaBBb123CCccccd，则应该输出结果：upper=6,lower=8。

```
#include <stdio.h>
void fun(char *s,int a,int b)
{
    while(*s)
    {
        if(*s>='a' && *s<='Z')  a++;
        if(*s>='a' && *s<='z')  b++;
        s++;
    }
}
int main()
{
    char s[100]; int upper=0,lower=0;
    printf("\nPlease enter a string:");gets(s);         //读入字符串的函数
    fun(s,&upper,&lower);
    printf("\nupper=%d lower=%d\n",upper,lower);
    return 0;
}
```

2. 请修正程序中的错误，使程序能得出正确的结果。下列程序中，fun函数的功能是：根据形参m，计算t=1+1/2+1/3+1/4+…+1/m的值。例如，若输入5，则应输出2.283333。

```
#include <stdio.h>
```

```
double fun(int m)
{
    double t=1.0;
    int i;
    for(i=2;i<=m;i++)
    t+=1.0/m;
}
int main()
{
    int m;
    printf("nPlease enter 1 integer number:");
    scanf("%d",&m);
    printf("\nThe result is %lf\n",fun(m));
    return 0;
}
```

3. 阅读下面的程序，分析程序是否能编译通过并正确运行，如果不能，说明原因；如果能，请写出运行结果。

```
#include <stdio.h>
int x=10;
void show()
{
    int x=20;
    printf("show() 函数中 x 的值为:%d\n",x);
}
int main()
{
    int x=30;
    printf("main() 函数中 x 的值为:%d\n",x);
    show();
    return 0;
}
```

4. 阅读下面的程序，分析程序是否能编译通过并正确运行，如果不能，说明原因；如果能，请写出运行结果。

```
#include <stdio.h>
int func2(int a, int b)
{
    int c;
    c = a*b%3;
    return c;
}
int  func1(int a, int b)
{
    int c;
    a+=a;
    b+=b;
    c=func2(a,b);
    return c*c;
}
int main()
{
    int x=11, y=19;
    printf("The final result is:%d\n", func1(x,y));
    return 0;
}
```

5. 阅读下面的程序，分析程序是否能编译通过并正确运行，如果不能，请说明原因；如果能，写出运行结果。

```
#include <stdio.h>
int globalVar=50;
int calculate(int localVar)
{
    int result=globalVar+localVar;
    globalVar=100;
    return result;
}
void printGlobalVar()
{
    printf("Global variable value: %d\n", globalVar);
}
int main()
{
    int localVar=25;
    int sum=calculate(localVar);
    printf("Local and global variable sum: %d\n", sum);
    printGlobalVar();
    return 0;
}
```

三、编程题

1. 定义一个函数，打印出1~100之间3与4的所有公倍数，并在主函数中调用。

2. 使用函数f(int x)，求分段函数的值。要求在主函数main中进行输入和输出。

$$y=\begin{cases}5x+1 & (x<10)\\ x & (x=10)\\ 5x-1 & (x>10)\end{cases}$$

3. 编写名为fun的函数，函数功能为：计算正整数num各位上数字之积，例如，若输入252，则输出应该是20，若输入202，则输出是0。

4. 编写一个函数，用户从键盘输入英文字母，如果是大写，将其转换成小写输出；如果是小写，将其转换成大写输出。

5. 编写一个函数，接收一个字符串作为参数，并统计其中字母'a'出现的次数。

6. 定义一个函数，接收两个整数作为参数，返回它们的最大公约数。

7. 编写一个函数，用于检查一个给定的年份是否是闰年，并在主函数中调用该函数进行测试。

8. 实现一个函数，用于将一个十进制数转换为二进制数，并在主函数中打印转换结果。

9. 编写一个函数，接收一个字符串作为参数，将该字符串反转并返回反转后的结果。

10. 编写一个函数，用于生成指定范围内的随机整数，并在主函数中调用该函数进行测试。

11. 五角数被定义为n(3n−1)/2，n=1,2,⋯，依此类推。因此，最初的几个五角数为1,5,12,22,⋯。编写函数返回五角数，并编写测试程序，使用该函数展示前100个五角数，每行打印10个。

12. 孪生素数是指两个素数之间的差值为2的一对素数。例如，3和5、5和7以及11和13都是孪生素数对。请编写一个函数，找到所有小于1 000的孪生素数。以如下形式显示：

```
(3,5)
(5,7)
...
```

Chapter 8

Structures

第 8 章

结 构 体

Introduction 主要内容

In C language, in addition to the basic data types, users can also customize types, such as the structure. A structure is a self-defined data type that combines different types of data. This chapter will focus on the definition of structure and the creation of variables, and explore the basic concepts and operations of the linked list. By learning the structure and linked list, we can manage data more flexibly and improve program efficiency.

在C语言中，除了基本数据类型，用户还可以自定义类型，如结构体。结构体是组合不同类型数据的自定义数据类型。本章将重点学习结构体的定义和变量创建，同时探讨链表这一常用数据结构的基本概念及其操作。通过学习结构体和链表，我们能更灵活地管理数据，提升程序效率。

Objectives 学习目标

Deeply understand the concept, definition and variable declaration of structure, master the access method of structure members, and effectively use structure array and pointer to operate. Understand the definition and working principle of the linked list, and master the basic operations commonly used in linked list, such as creation, output, insertion, deletion, etc. Theoretical learning needs to be combined with practice, through case analysis and programming exercises, to deepen the understanding of relevant knowledge, improve logical thinking and programming ability, so as to be able to flexibly apply the knowledge learned to solve the practical problems.

深入理解结构体的概念、定义及其变量声明方式，熟练掌握结构体成员的访问方法，并能有效应用结构体数组和指针进行操作。理解链表定义、工作原理，掌握链表常用的基本操作，如创建、输出、插入、删除等。理论学习需与实践相结合，通过案例分析和编程练习，不断加深对相关知识的理解，提高逻辑思维和编程能力，从而能够灵活应用所学知识解决实际问题。

8.1 Structure Type and Variable Definition 结构体的类型及变量定义

8.1节视频

8.1.1 Why Do We Need a Structure（为什么需要结构体）

In real life, data often appear in the form of groups, and there are often inherent logical links among the groups of data. For example, a student's data includes many aspects, such as student number, name, gender, age, grade and home address. In program, we usually use structure to represent such data objects with multiple attributes.

在现实生活中，数据常常以组的形式出现，各组数据之间往往存在着内在的逻辑联系。例如，一个学生的信息数据，就包含了学号、姓名、性别、年龄、成绩以及家庭地址等多个方面。在程序中，我们通常使用结构体来表示这种具有多个属性的数据对象。

The structure can combine multiple data items of different types into a whole, which is convenient for us to manage and operate in a unified way. Through the structure, we can more clearly represent the structure and relationship of data, and then achieve more efficiently data processing and analysis.

结构体能够将多个不同类型的数据项组合成一个整体，方便我们进行统一的管理和操作。通过结构体，我们可以更加清晰地表示数据的结构和关系，进而实现更高效的数据处理和分析。

In practical applications, we can define different structure types according to specific requirements, and create structure variables to store specific data. At the same time, we can also use the advanced features of structure array and pointer to realize batch data processing and dynamic management.

在实际应用中，我们可以根据具体需求定义不同的结构体类型，并创建结构体变量来存储具体的数据。同时，我们还可以利用结构体数组和指针等高级特性，实现对数据的批量处理和动态管理。

For example, the data of two students contains the following information: student number, name, gender, age, grade and home address. Assuming that students are represented by student1 and student2, the corresponding types of information in the program are shown in Table 8-1.

例如，有两名学生的数据，均包含如下信息：学号、姓名、性别、年龄、成绩、家庭地址。假设学生用student1和student2来表示，对应的信息在程序中的类型见表8-1。

表 8-1 两个学生数据和类型

变量	学号	姓名	性别	年龄	成绩	家庭住址
student1	int	char[]	char	int	double	char[]
student2	int	char[]	char	int	double	char[]

The characteristic of this set of student data is that it is a set of intrinsically related data, and the data types of these data items are different. In the program, we need a way to organize this data so that it can be easily accessed and processed. Using past knowledge, we might try to use separate variables to represent each of the student's attributes, for example:

这组学生数据的特点在于它是一组具有内在联系的数据，并且这些数据项的数据类型各不相同。在程序中，我们需要一种方式来组织这些数据，以便能够方便地访问和处理它们。若使用以往的知识，我们可能会尝试使用多个单独的变量来分别表示学生的各个属性，例如：

```
int studentID;
char name[50];
char gender;
int age;
```

```
    float score;
    char address[100];
```

However, this method has obvious disadvantages:

Data dispersion: Data is dispersed in multiple separate variables, resulting in inconvenient data management and difficult maintenance.

Ambiguous relationship: From the declaration of variables, we can not directly see that these variables belong to the same student, and the correlation of data is not strong.

Poor extensibility: If you need to add or remove an attribute, you need to modify the code and adjust the declaration of multiple variables.

Transmission difficulty: When data pass between functions, these variables need to be passed one by one, which increases the complexity of the code.

Arrays reflect the internal relationship and integrity of data, but arrays represent the same type, which can not meet the needs of solving actual problems.

To solve the above problem, we can use the structure to represent student data. The structure allows us to combine multiple variables of different types into a whole, forming a self-defined data type. Thus, each structure variable represents a student object with multiple attributes.

The benefits of using the structure to represent student data include:

Data concentration: All the data about students are concentrated in a structure variable, which is convenient for management and access.

Clear relationship: The definition of the structure clearly represents the relationship between the data, which improves the readability of the code.

然而，这种方法存在明显的弊端：

数据分散：数据被分散在多个单独的变量中，导致数据管理不便，难以维护。

关系不明确：从变量的声明中，我们无法直接看出这些变量是属于同一个学生的，数据的关联性不强。

扩展性差：如果需要添加或删除某个属性，就需要修改代码并调整多个变量的声明。

传递困难：在函数间传递数据时，需要逐个传递这些变量，增加了代码的复杂性。

数组体现了数据的内在联系具有整体性，但数组表示的是同一类型，并不能满足实际解决问题的需求。

为了解决上述问题，我们可以使用结构体（struct）来表示学生数据。结构体允许将多个不同类型的变量组合成一个整体，形成一个自定义的数据类型。这样，每个结构体变量就代表了一个具有多个属性的学生对象。

使用结构体表示学生数据的好处包括：

数据集中：所有关于学生的数据都集中在一个结构体变量中，方便管理和访问。

关系明确：结构体的定义清晰地表示了数据之间的关系，提高了代码的可读性。

8.1.2 What Is a Structure（什么是结构体）

A structure is a composite data type in C(and many other programming languages) that allows you to combine data of multiple different types or of the same type into a single unit of data. A structure can contain multiple members(or fields), and each member can be a basic data type(such as integer, floating point, character, and so on) or other user-defined data types(such as other structures). When you define a structure, you need to specify the name of the structure and the member variables it contains and their types.

结构体是C语言（以及许多其他编程语言）中的一种复合数据类型，它允许将多个不同类型或相同类型的数据组合成一个单独的数据单元。结构体可以包含多个成员（或称字段），每个成员可以是一个基本数据类型（如整型、浮点型、字符型等）或其他用户自定义的数据类型（如其他结构体）。在定义结构体时，需要指定结构体的名称以及它包含的成员变量及其类型。

Struct Definition Form（结构体定义形式）

```
struct struct name
{
    member table columns;
}
```

```
struct 结构体名
{
    成员表列；
};
```

Syntax Description（语法说明）

(1) Struct is a keyword used to define the type of structure in C language.

(2) The structure name is a user-defined identifier used to identify the specific structure type. According to the naming rules of identifiers, structure names are usually capitalized to distinguish them from other types of identifiers.

(3) The member list is the list of data members contained in the structure. Each data member has its own type and name, which is called the member or field of the structure.

(4) The semicolon at the end “;” is the end mark of the structure definition statement, indicating the end of the structure definition, and it is an essential part of the syntax.

(5) The structure name is regarded as a whole, which defines a new data type, that is, the structure type. Once defined, we can create a variable of the structure type and access its members through the “.” dot operator.

The structure provides a way to combine different types of data into a single data structure, which makes the organization and management of data more efficient and flexible. For example, a student has a student number, name, gender, age, grade, home address, etc., and can declare a student structure type, which is defined as follows:

（1）struct 是 C 语言中用于定义结构体类型的关键字。

（2）结构体名是用户自定义的标识符，用来标识这个特定的结构体类型。按照标识符的命名规则，结构体名通常采用首字母大写的方式，以区别于其他类型的标识符。

（3）成员表列是结构体中包含的数据成员列表，每个数据成员都有自己的类型和名称，称为结构体的成员或字段。

（4）结尾的分号“;”是结构体定义语句的结束标志，表明结构体定义的结束，是必不可少的语法部分。

（5）structure结构体名看作一个整体，它定义了一个新的数据类型，即结构体类型。在定义之后，我们可以创建该结构体类型的变量，并通过点操作符“.”来访问其成员。

结构体提供了一种将多个不同类型的数据组合成一个单独的数据结构的方式，使得数据的组织和管理更为高效和灵活。例如，一个学生有学号、姓名、性别、年龄、成绩、家庭地址等项，可声明一个学生结构体类型，定义如下：

```
struct Student
{
    int num;
    char name[20];
    char sex;
    int age;
    double score;
    char addr[30];
};
```

The compilation system only defines the structure type and does not allocate storage space for it. In fact, when you use this type to create specific variables, the system allocates the appropriate

编译系统仅定义结构体类型而不为其分配存储空间。实际上，是在使用这种类型创建具体的变量时，系统才会为这些变

storage space for those variables. Only after these variables have been allocated storage space, they can be used to store actual data. Therefore, we should not attempt to assign values to members during the definition phase of the structure. The member assignment should occur after the structure variables are created, storing data by assigning concrete values to the members of these variables.

量分配相应的存储空间。只有当这些变量被分配了存储空间后，它们才能用于存储实际的数据。因此，在结构体的定义阶段，我们不应该尝试给成员赋值。成员赋值应发生在结构体变量创建之后，通过为这些变量的成员分配具体的值来存储数据。

8.1.3 Structure Variable（结构体变量）

The structure type is a valid self-defined data type, and its status and role are equivalent to the basic data types(e.g., int, char, double, float), and its usage is basically the same. When using structure types, we can define the corresponding variables, which is mainly achieved through three forms.

结构体类型是一种有效的自定义数据类型，其地位和作用与基本数据类型（如int、char、double、float）相当，用法也大体相同。使用结构体类型时，我们可以定义相应的变量，这主要通过三种形式来实现。

(1) First declare the structure type, and then use that type to define the variable. This definition is consistent with the previous form of basic type definition variables. For example, define a structure variable named student1 as follows.

（1）首先声明结构体类型，随后使用该类型定义变量。这种定义方式与以往的基本类型定义变量的形式保持一致。例如，定义一个名为student1的结构体变量，其步骤如下：

```
1   struct Student                          //结构体类型声明
2   {
3       int num;
4       char name[20];
5       char sex;
6       int age;
7       float score;
8       char addr[30];
9   };                                      //分号不能省略
10  struct Student student1;
```

Lines 1-9 define the structure first, complete the declaration of the self-defined data type. In line 10, the structure variable student1 is created by using the defined struct Student.

第1~9行先定义了结构体，即完成了自定义数据类型的声明。第10行使用已经定义好的结构体类型struct Student创建了结构体变量student1。

(2)Define variables directly while declaring the structure type. This approach works well when writing small programs because it completes both the type declaration and variable definition steps:

（2）在声明结构体类型的同时直接定义变量。这种方式在编写小型程序时较为适用，因为它同时完成了类型声明和变量定义两个步骤：

```
struct Student
{
    int num;
    char name[20];
    char sex;
    int age;
    float score;
    char addr[30];
}student1;                  //在声明结构体类型的同时定义变量student1
```

(3)Define the nameless structure type and define the variable at the same time. In this case, because no name is specified for the structure, the structure type can only be used once and cannot be used again to define other variables:

（3）定义无名的结构体类型并同时定义变量。在这种情况下，由于没有为结构体指定名称，因此这个结构体类型只能被使用一次，无法再次用于定义其他变量。

```
struct
{
    int num;
    char name[20];
    char sex;
    int age;
    float score;
    char addr[30];
}student1;                    //定义无名的结构体类型并同时定义变量 student1
```

Among the above three definition methods, the first one is the most commonly used, because it separates the declaration of the structure type from the definition of the variable, makes the code structure clearer, and also facilitates the subsequent reuse of the structure type. The second method is suitable for the applets in which the structure type only needs to be used once. The third method is one-time, and the nameless structure type can not be used to define other variables later.

在上述三种定义方式中，第一种最为常用，因为它使得结构体类型的声明和变量的定义分离开来，使得代码结构更为清晰，也便于后续对结构体类型的重用。第二种方式适合在结构体类型只需要使用一次的小程序中采用。第三种方式则具有一次性，后续不可再用这个无名的结构体类型来定义其他变量。

8.1.4 Storage Space Allocation of Structure Variable（结构体变量的存储空间分配）

Supposing that you define two struct variables, student1 and student2.

假设定义了两个struct Student结构体变量student1、student2。

There is a statement: struct Student student1, student2; Then the space allocation of the two variables is as follows, and the storage space is opened up successively according to the order of member definition.

即有语句struct Student student1,student2;，则两个变量的空间分配情况如下所示，按照成员定义的顺序连续开辟存储空间。

student1	10001	Zhan san	M	19	91.5	Shang hai
student2	10002	Li si	M	19	91.0	Shang hai

The storage allocation for a structure variable depends on the type and number of its members. When you declare a structure variable, the compiler allocates storage for each member of the structure. This storage space is contiguous, that is, the members of the structure are arranged sequentially in memory.

结构体变量的存储空间分配取决于其成员的类型和数量。当声明一个结构体变量时，编译器会为结构体的每个成员分配存储空间。这个存储空间是连续的，也就是说，结构体的成员在内存中是依次排列的。

Here is another simple example to illustrate the storage space allocation of the structure variable, the following code is:

下面再举一个简单的例子来说明结构体变量的存储空间分配，有如下代码：

```
#include <stdio.h>
//定义一个表示教师信息的结构体
struct Teacher
{
```

```
    char name[50];                              //教师姓名，假设最大长度为 49 个字符
    int age;                                    //教师年龄
    char gender;                                //教师性别，'M' 表示男性，'F' 表示女性
    int teachingYears;                          //教龄
    double monthlySalary;                       //月薪
};
int main()
{
    struct Teacher t;                           //声明一个 Teacher 类型的变量 t
    //使用 sizeof 运算符获取 Teacher 结构体的大小
    printf("Size of struct Teacher: %lu bytes\n", sizeof(struct Teacher));
    return 0;
}
```

In this example, we define a structure called Teacher with five members: name, age, gender, teachingYears, and monthlySalary. When we declare a variable t of type Teacher, the compiler allocates storage for each member of t. Finally, we use the sizeof operator to get the size of the entire structure.

It is important to note that the size of the structure may be affected by padding. To improve the efficiency of memory access, the compiler may insert some extra bytes among the members of the structure. Therefore, the actual storage occupied may be greater than the sum of the member sizes. We can use compiler-specific options or tools to check the memory layout and alignment of the structure.

在这个例子中，我们定义了一个名为Teacher的结构体，它有5个成员：name、age、gender、eachingYears和monthlySalary。当我们声明一个Teacher类型的变量t时，编译器会为t的每个成员分配存储空间。最后，我们使用sizeof运算符来获取整个结构体的大小。

需要注意的是，结构体的大小可能会受到内存对齐（padding）的影响。为了提高内存访问的效率，编译器可能会在结构体的成员之间插入一些额外的字节。因此，实际占用的存储空间可能会大于成员大小的总和。可以使用编译器特定的选项或工具来查看结构体的内存布局和对齐情况。

8.1.5 Use of Structure Variables（结构体变量的使用）

A structure variable is essentially a reference to a set of member variables of a particular type. In practice, we focus on the individual member variables in the structure and access them through the member operator ".". The specific forms of access are:

结构体变量本质上是对一组具有特定类型的成员变量的集合的引用。在实际使用中，我们关注的是结构体中的各个成员变量，并通过成员运算符“.”来访问它们。具体的访问形式为：

```
struct variable name.member name
```

```
结构体变量名.成员名
```

For example, if a structure variable is named student1 and contains a member variable num, the member can be accessed through student1.num to read or assign it, thereby managing the internal data of the structure.

例如，若结构体变量名为student1，其包含成员变量num，则可通过student1.num访问该成员，实现对其的读取或赋值操作，从而管理结构体内部数据。

Notes（说明）:

(1)The structure does not support globally to input and output operations, so do not attempt to directly input and output its members through the structure name. Each member needs to be dealt with separately.

(2)Structure variables of the same type can be assigned to each other, for example, student1 = student2, which will make

（1）结构体不支持整体输入输出操作，因此请勿尝试通过结构体名直接输入输出其成员。每个成员需要单独处理。

（2）相同类型的结构体变量之间可以相互赋值，如student1 = student2，这将使

the value of each member of student1 the same as that of the corresponding member of student2.

student1的各成员值与student2的相应成员值相同。

Case 8_1（案例 8_1）

Read the following code and give the running results.

阅读下列代码，给出运行结果。

Case8_1.c

```
1   #include <stdio.h>
2   struct Student
3   {
4       int num;
5       char name[20];
6       char sex;
7   };
8   int main()
9   {
10      struct Student a={10101,"Li Lin",'M'};
11      Printf("No.:%d\nname:%s\nsex:%c\n",a.num,a.name,a.sex);
12      return 0;
13  }
```

Program Output（运行结果）:

```
No.:10101
name:Li Lin
sex:M
```

Case Analysis（案例解析）:

(1)Lines 2-7 show a simplified declaration of student structure types, which help students grasp the basic definition of structures and how members are accessed. Therefore, the structure type declaration part is relatively terse. Here, structure Student is the defined struct type.

（1）第2~7行展示了学生结构体类型的简化声明，旨在帮助学生掌握结构体的基本定义及成员的访问方式。因此，结构体类型声明部分相对简洁。这里struct Student即为定义的结构体类型。

(2)On line 10, we define and initialize a variable named a using the struct Student. The initialization form is similar to an array, using {} to assign data for each member of a in the order in which the members are declared.

（2）第10行，我们使用结构体类型struct Student定义了一个名为a的变量，并进行了初始化。初始化形式类似于数组，使用{}，将数据按照成员声明的顺序依次赋值给a的各个成员。

(3)Line 11 demonstrates how to output the members of the structure variable a. Member access requires the use of the variable name and the member operator ".", and follows the correct writing format. At output, the format control character needs to match the data type of the member in the same way that you would access an ordinary variable.

（3）第11行演示了如何输出结构体变量a的成员。成员的访问需要使用变量名和成员运算符"."，并遵循正确的书写格式。在输出时，格式控制符需要与成员的数据类型相匹配，这与访问普通变量的方式相同。

Difference and Connection Between Array and Structure 数组与结构体的区别与联系

(1) Definition and type:

（1）定义与类型：

①Structures are user-defined data types that allow different types of data to be combined into a single data structure.

①结构体是用户自定义的数据类型，允许将不同类型的数据组合成一个单一的数据结构。

②An array is a constructed type consisting of data elements

of the same type arranged in a certain order.

②数组是由相同类型的数据元素按照一定顺序排列而成的构造类型。

(2) Data combination and internal relationship:

（2）数据组合与内在联系：

①A structure embodies the grouping and inter-relatedness of data, and allows us to organize related data items together for easier management and manipulation.

①结构体体现了数据的组成和内在联系特性，它允许我们将相关的数据项组织在一起，以更方便地管理和操作。

②Arrays also embody the characteristics of data grouping, but they focus more on the collection of the same type of data and emphasize the order of data.

②数组也体现了数据成组的特性，但它更侧重于相同类型数据的集合，强调数据的顺序性。

(3) Member/element type:

（3）成员/元素类型：

①The data types of the members in the structure can be different from each other, and each member can have its own unique data type.

①结构体中的成员数据类型可以互不相同，每个成员可以有自己独特的数据类型。

②The elements in the array must be of the same type. All elements are of the same data type.

②数组中的元素类型必须相同，所有元素都属于同一数据类型。

(4) Storage space：

（4）存储空间：

①Whether it is a structure or an array, the storage space allocated in memory is contiguous.

①无论是结构体还是数组，它们在内存中分配的存储空间都是连续的。

②The storage size of a structure is the sum of the space occupied by all its members, and if there is a space alignment mechanism, it will be greater than the sum.

②结构体的存储空间大小是其所有成员占用的空间总和，若存在空间对齐机制，会大于总和。

③The storage size of an array is the number of its elements multiplied by the size of a single element.

③数组的存储空间大小是其元素个数乘以单个元素的大小。

(5) Input/output operation：

（5）输入输出操作：

①Neither a structure nor an array can be directly used for overall input and output operations.

①无论是结构体还是数组，都不能直接进行整体输入输出操作。

②For a structure, we need to access its members one by one and use the appropriate format controls to input and output.

②对于结构体，我们需要逐个访问其成员，并使用适当的格式控制符进行输入输出。

③For an array, we need to access its elements one by one, and input and output with format control characters that match the element type.

③对于数组，我们需要逐个访问其元素，并使用与元素类型相匹配的格式控制符进行输入输出。

Summary（总结）：

There are clear differences between structures and arrays in terms of data structures, data types, member/element types, storage space, and input/output operations. A structure is better suited to represent interrelated data objects with different attributes, while an array is better suited to store and manipulate a large number of data elements of the same type.

结构体和数组在数据结构、数据类型、成员/元素类型、存储空间和输入输出操作等方面存在明显的区别。结构体更适合于表示具有不同属性且相互关联的数据对象，而数组则更适合于存储和操作大量相同类型的数据元素。

8.2 Structure Array and Structure Pointer 结构体数组及结构体指针

8.2节视频

8.2.1 Structure Array（结构体数组）

The structure is a user-defined data type and has the same function as the basic data type. Basic data types can define arrays, and structure types can also define arrays. Arrays are important data structures, especially for writing high-quality and efficient code. Several forms of defining structure arrays are shown below.

结构体是用户自定义数据类型，与基本数据类型作用一致。基本数据类型可以定义数组，结构体类型同样可以定义数组。数组是重要的数据结构，对编写高质量。高效率的代码尤为重要，下面展示定义结构体数组几种形式。

(1) Define a structure array while defining a structure type, wherein that structure array has the following form:

（1）在定义结构体类型同时定义结构体数组，形式为：

```
struct struct name{ member lists } array name
[array length];
```

```
struct 结构体名 { 成员表列 } 数组名
[ 数组长度 ];
```

For example（举例）:

```
struct Student
{
     int num;
     char name[20];
     char sex;
}stu[3];
```

(2) First declare the structure type and then define the structure array, the form is:

（2）先声明结构体类型后定义结构体数组，形式为：

```
struct type array name[array length];
```

```
结构体类型   数组名 [ 数组长度 ];
```

E.g（如）: struct Student Stu[3];

Case 8_2（案例 8_2）

There are three students' information(including student number, name and grade), and the array method is required to display the information of the student with the highest grade.

有三个学生的信息(包括学号、姓名、成绩)，要求使用数组方法，显示成绩最高的学生信息。

Thinking（解题思路）:

First declare the student structure type, then define the structure array and initialize it, mark the structure array subscript with the highest score through the "challenge" algorithm combined with the for loop, and finally output the member of the structure array element corresponding to the subscript, and complete the requirements of the case.

首先声明学生结构体类型，随后定义结构体数组并初始化，通过"打擂台"算法结合for循环，标记成绩最高的结构体数组下标，最后输出下标所对应的结构体数组元素的成员，即可完成案例的要求。

Program Code（程序代码）:

Case8_2.c

```
#include <stdio.h>
struct Student
```

```
3    {
4        int num;
5        char name[20];
6        float score;
7    };
8    int main()
9    {
10       struct Student Stu[3]={{10101, "Zhang An", 78}, {10103, "Xiao Wang", 98.5},
11                              {10106, "Li Li", 86}};
12       int k,i;
13       for(k=0,i=0;i<3;i++)
14       {
15           if(stu[i].score>stu[k].score) k=i;
16       }
17       printf("%d %s %.2f\n",stu[k].num,stu[k].name,stu[k].score);
18       return 0;
19   }
```

Program Output(运行结果):

```
10103 Xiao Wang 98.50
```

Case Analysis(案例解析):

(1) On lines 2 to 7, we declare a structure type called struct Student with three members: num(int), name(char array), and score(float). Note that the semicolon at the end of the structure type declaration cannot be omitted, it is a part of the C syntax rules.

(2) Line 10 defines a one-dimensional structure array Stu of length 3 and initialize the structure array. This array contains three elements: Stu[0], Stu[1], and Stu[2]. All of them are struct Student type, and initialization is similar to array initialization, using {}.

(3) In line 11, we define an integer variable k to mark the subscript of the structure array element with the highest score, and a loop variable i is used to control the change of the subscript.

(4) Line 12-15 is a for loop, and its function is to implement the "challenge" algorithm to find the subscript K corresponding to the structure array element with the highest score. First, initialize k and i to be 0, that is, assume that the Stu[0] student has the highest score. Each array element represents a student's data, and we access its internal score member through the structure variable and the member operator ".". If the score value of Stu[i] is found to be greater than the score value of the current Stu[k], k is updated to i. At the end of the loop, k stores the array element subscript corresponding to the maximum score value.

(5) Line 16 outputs all the information of the structure array element Stu[k]. Note that its members need to be accessed one by one, and the format control character matching the member type is used to output.

(1)在第2~7行，我们声明了一个名为struct Student的结构体类型，其中包括三个成员：num(int型)、name(char数组型)和score(float型)。请注意，结构体类型声明的末尾分号不可以省略，这是C语言语法规则的一部分。

(2)第10行定义了一个长度为3的一维结构体数组stu，并且进行了初始化。这个数组包含三个元素：stu[0]、stu[1]和stu[2]，它们都是struct Student类型，初始化与数组初始化相似，使用{}。

(3)在第11行，我们定义了一个整型变量k来标记拥有最高成绩的结构体数组元素的下标，同时定义了一个循环变量i来控制下标的变化。

(4)第12~15行为一个for循环，功能是实现"打擂台"算法，求拥有最高成绩的结构体数组元素所对应的下标k。首先初始化k和i均为0，即假设stu[0]学生的成绩最高。每个数组元素代表一个学生的数据，我们通过结构体变量和成员运算符"."来访问其内部的score成员。如果发现stu[i]的score值大于当前stu[k]的score值，则更新k为i。循环结束后，k将存放对应最大score值的数组元素下标。

(5)第16行输出结构体数组元素stu[k]的所有信息，注意需要逐一访问其成员，并使用与成员类型匹配的格式控制符进行输出。

8.2.2 Structure Pointer（结构体指针）

A structure can also define a pointer variable in the following form:

```
struct type *pointer variable;
```

结构体同样可以定义指针变量，其定义形式如下：

```
结构体类型 * 指针变量 ;
```

The member of the pointed structure variable can be accessed indirectly through the pointer variable, and the access format is:

```
struct pointer variable-> member
```

通过指针变量可以间接访问所指向的结构体变量的成员，其访问格式为：

```
结构体指针变量 -> 成员
```

This is similar to the pointer variable method used to define basic data types, but now it is replaced by a structure type.

这与以往定义基本数据类型的指针变量方法相似，只不过此时换成了结构体类型。

For example（举例）:

```
struct Student student1;
struct Student *pt=&student1;
```

In the above code, a pointer variable pt pointing to struct Student is defined, and pt points to the student1structure variable. When you use a pointer variable to indirectly access the student1 member, you need to follow the format of pointer variable name-> member name. Then pt-> num is equivalent to student1.num, and also to(* pt).num. Note that the parenthesis(* pt) cannot be omitted here, and it is used to dereference the pointer variable.

For further understanding, we use the pointer method to implement Case 8_1 as follows:

上述代码中，定义了一个指向struct Student类型的指针变量pt，且pt指向student1结构体变量。当使用指针变量间接访问student1成员时，需要遵循指针变量名->成员名的格式。那么pt->num就等价student1.num，同时也等价于(*pt).num。注意，这里的括号(*pt)不能省略，它用于引用指针变量。

为了深入理解，我们将案例8_1用指针方法实现如下：

```
#include <stdio.h>
struct Student
{
    int num;
    char name[20];
    char sex;
};
int main()
{
    struct Student a={10101,"Li Lin",'M'};
    struct Student *p=&a;
    printf("No.:%d\nname:%s\nsex:%c\n",p->num,p->name,p->sex);
    return 0;
}
```

If p points to the variable a, all members in a can be accessed indirectly, and the original a. num, a. name, and a. sex are changed to p-> num, p-> name, and p-> sex.

Rewrite Case 8_2 to use the pointer to reference the array method, and the code is as follows:

p指向了变量a，则可间接访问a中的所有成员，由原来的a.num、a.name、a.sex改为p->num、p->name、p->sex。

将案例8_2改写为使用指针引用数组方法，代码如下：

```
1   #include <stdio.h>
2   struct Student
3   {
4       int num;
5       char name[20];
6       float score;
7   };
8   int main()
9   {
10      struct Student stu[3]={{10101,"Zhang An",78},
11                  {10103,"Xiao Wang",98.5}, {10106,"Li Li",86} };
12      struct Student *p,*q;                    //p 用于访问数组，q 指向当前的成绩最高者
13      for(p=stu,q=stu;p<stu+3;p++)
14      {
15          if(p->score>q->score) q=p;
16      }
17      printf("%d %s %.2f\n",q->num,q->name,q->score);
18      return 0;
19  }
```

Line 11 defines the pointer variables p and q, where p is used to access the array and q is used to point to the array element with the highest current score. In the first expression of the for loop in line 12, p and q are assigned to the initial value stu, that is, they all point to the first element of the structure array. The loop condition is p<stu+3, that is, they do not exceed the range of the array. In line 14, the loop body predicts that if the score of the array element pointed to by p is greater than the score of the array element pointed to by q, the pointer of q is modified. q = p, that is, the address of p is assigned to q, and the assignment between two pointer variables can be visualized as assigning the pointer of p to q. Line 16 accesses all members of the highest-scoring array element indirectly through q.

第11行定义了指针变量p和q，p用于访问数组，q用于指向当前成绩最高的数组元素。第12行中for循环的第1个表达式，将p和q赋予初值stu，即均指向了结构体数组首元素，循环条件为p<stu+3，即不超过数组范围，第14行为循环体预计，若p所指向的数组元素的score大于q所指向的数组元素的score，则修改q的指向，q=p，即将p存放的地址赋值给q，两个指针变量之间赋值，可以形象化称为将p的指向赋给q。第16行通过q间接访问成绩最高的数组元素的所有成员。

8.3 Dynamic Memory Allocation and Linked List Definition 动态内存分配及链表定义

8.3节视频

After understanding the storage allocation of structure variables, we further explore dynamic memory allocation and linked lists. Dynamic memory allocation provides flexible memory management and lays the foundation for building data structures such as linked lists. As a dynamic data structure, linked list plays an important role in programming because of its efficient data organization and operation characteristics. Next, we will deeply analyze the significance and application of the two.

在理解了结构体变量的存储分配后，我们进一步探讨动态内存分配和链表。动态内存分配提供了灵活的内存管理，为构建链表等数据结构奠定了基础。链表作为一种动态数据结构，以其高效的数据组织和操作特性，在编程中发挥着重要作用。接下来，我们将深入剖析这两者的意义与应用。

8.3.1 Dynamic Memory Allocation（动态内存分配）

The computer has the following layout for memory allocation: a program code area for storing binary codes of

计算机对内存分配有如下布局，有程序代码区，用来存放函数的二进制代码，

functions, and a global or static storage area for storing global variables and static local variables, the variables in this area are allocated when the program is compiled, and the storage cycle is long. Corresponding to the area is the stack area, also known as the dynamic storage area, which is used to store function formal parameters, local variables, etc. The variables in these two areas are automatically allocated and released by the system. There is another area we haven't touched, called the heap area, which allows programmers to allocate and free manually. This is the new memory allocation method we need to learn, that is, dynamic memory allocation. Figure 8-1 shows the memory space allocation diagram.

What is the difference between dynamic memory allocation and automatic allocation used by the previous system? The following is a simple comparison, as shown in Table 8-2. The heap area variable is manually applied for and released by the user, and can be applied for and released at any time. The variable has no name, and its address is saved by the pointer variable, and the data is processed by indirect access.

全局或静态存储区，用来存放全局变量以及静态局部变量，这个区域的变量在程序编译时分配，存储周期长。与之对应的是栈区也叫动态存储区，用于存放函数形参、局部变量等，这两个区域的变量由系统自动分配、自动释放。还有一个区域我们没有接触过，叫堆区，这个区域允许程序员手动分配和释放，这就是我们要学习的新的内存分配方式，即动态内存分配方式。图8-1给出了内存空间分配示意图。

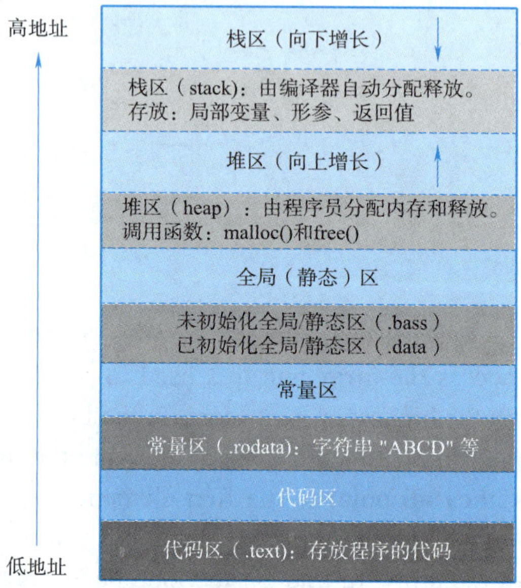

图8-1　内存区域分布示意图

动态内存分配跟以往系统自动分配相比，有什么不同之处呢？下面做个简单的对比，见表8-2。堆区变量是由用户手动申请、手动释放的，且可随时申请随时释放，该变量没有名字，由指针变量保存其地址后，通过间接存取方式处理数据。

表 8-2　栈区和堆区变量区别

栈区变量	堆区变量（动态内存分配）
程序中声明变量名，变量名关联地址，直接存取	不必声明变量名，指针变量存储空间首地址，间接存取
函数结束释放	随时开辟、随时释放

The dynamic allocation of the memory is realized through the library functions provided by the system, mainly includes malloc, calloc, and free. To call the above functions correctly, you need to add the "#include <stdlib.h>" instruction at the beginning of the program. The three functions malloc, calloc, and free are described in turn below.

对内存的动态分配是通过系统提供的库函数来实现的，主要有malloc、calloc、free三个函数，要正确调用上述函数，需要在程序开始加上“#include <stdlib.h>”指令。下面依次介绍malloc、calloc、free三个函数。

1. malloc Function Malloc（malloc 函数）

malloc function prototype:

malloc函数原型：

```
void *malloc(unsigned int size);
```

Function: Allocate a contiguous space of size in memory bytes in the heap.

函数功能：在堆区分配一个内存字节大小为size的连续空间。

If there is enough continuous space, the allocation is successful, and the return value is the first address of the allocated space. If the memory space is insufficient, the allocation fails, and the null pointer NULL is returned.

若有足够大的连续空间，则分配成功，返回值即为所分配空间的首地址，若内存空间不足，则分配失败，返回空指针NULL。

The return value of the malloc function is void *. here, the void does not mean no return value, this is an uncertain pointer type, which can only be determined according to the specific data when it is used. Because there is no variable name in dynamic memory allocation, the space cannot be accessed through the variable name, but only indirectly through the pointer variable, so the pointer variable is often required to receive the return value.

malloc函数的返回值为void*，这里的void不代表无返回值，这是一个不确定的指针类型，只能使用时根据具体数据再确定。由于动态内存分配没有变量名，故不能通过变量名访问空间，只能通过指针变量间接访问，故往往需要指针变量接收返回值。

For example（举例）

```
p=(float *) malloc(4);
```

The function of this statement is to request 4 bytes of continuous storage space(in the heap area) from the system, and convert the first address of this space to float*type(that is, float pointer type), and then assign it to the pointer variable p. To ensure that the allocated memory size exactly matches the float type, it is recommended to use sizeof(float) to get the number of bytes occupied by the float type, so the statement can be rewritten as: p=(float *)malloc(sizeof(float)). In this way, no matter on which platform, as long as the platform supports C language, the above statement can correctly allocate a float type of memory space.

这条语句的作用是向系统申请4字节的连续存储空间（在堆区），并将这块空间的首地址强制类型转换为float*类型（即浮点数指针类型），然后赋给指针变量p。为了确保分配的内存大小与float类型完全匹配，推荐使用sizeof(float)来获取float类型占用的字节数，因此，语句可以改写为：p =(float*)malloc(sizeof(float));。这样，无论在哪个平台上，只要该平台支持C语言，上述语句都能正确地分配一个float类型的内存空间。

2. calloc Function (calloc函数)

Prototype of the calloc function:

calloc函数原型：

```
void * calloc(unsigned n, unsined size);
```

Function: It allocates n continuous spaces with the length of size in the memory heap area. If the allocation is successful, the return value of the function is the first address of the allocated memory area. If the allocation fails, the null null pointer is returned.

函数功能：在内存堆区分配n个长度为size的连续空间，若分配成功，函数返回值为分配内存区域的首地址，若分配失败，则返回NULL空指针。

The calloc function can be used to open up a dynamic storage area for an array. For example, for a one-dimensional array, n can be regarded as the number of array elements, and the length of each element is size.

用calloc函数可以为数组开辟动态存储区域，如一维数组可以将n看成数组元素个数，每个元素长度为size。

For example: p = calloc(50,4), this statement can be interpreted as opening up a temporary allocation area of 50*4 bytes, where 50 is the size of the element, and 4 is the number of each element. It can be seen as a one-dimensional int array with 50 elements, and the first address of the array is assigned to the pointer variable p.

举例：p=calloc(50,4);，这条语句可以解读为开辟了一个50×4字节的临时分配区域，50为元素大小，4为每个元素的个数，可看成有50个元素的一维int数组，并将数组首地址赋给指针变量p。

3. free Function（free 函数）

prototype of free function:

free函数原型：

```
void free(void*p);
```

Syntax description: Here, the void represents that the function has no return value, and void* represents that the formal parameter is a pointer variable. The specific type is uncertain, and it should be determined according to the actual use.

语法说明：这里的void代表函数无返回值，这里的void* 代表形参是一个指针变量，具体类型不确定，要根据实际使用时再确定。

Function: It represents to release the memory area pointed by p, so that this part of memory can be allocated to other dynamic request operations. Dynamically applied space should be released in time after it is used up, otherwise it will cause memory leakage, which will cause the waste of system memory, slow down the running speed of the program and even cause serious consequences such as system crash.

函数功能：释放由p所指向的内存区域，使得这部分内存可以分配给其他动态申请操作。动态申请的空间，用完后要及时释放掉，否则会产生内存泄露，内存泄露会造成系统内存的浪费，导致程序运行速度减慢甚至系统崩溃等严重后果。

The malloc and free functions are mainly used in the linked list we are going to learn. In order to make students better understand these two functions, we will give two specific examples for these two functions.

在我们即将学习的链表中，主要用到malloc和free函数，为了使同学们更好的理解这两个函数，我们针对这两个函数列举两个具体例子。

For example 1（例1）:

```
p=(float*)malloc(sizeof(float));      //开辟4字节的存储空间，并把此地址赋给指针p
free(p);                              //释放动态分配的空间
```

For example 2（例2）:

```
n=(struct Student *)malloc(sizeof(struct Student)); //开辟一个无名的结构体变量空间，令n指向它
free(n);                                             //释放结构体变量空间
```

8.3.2 Linked List Definition（链表的定义）

Dynamic linked list is a flexible data structure, which dynamically allocates node storage space according to data requirements. Because there is no fixed variable name in the dynamically allocated memory, the linked list connects each node through the pointer variable. This structure makes that the linked list has advantages in operations such as insertion and deletion, and is especially suitable for processing data sequences with variable length.

动态链表是一种灵活的数据结构，它根据数据需求动态地分配结点存储空间。由于动态分配的内存没有固定变量名，链表通过指针变量连接各个结点。这种结构使得链表在插入、删除等操作上具有优势，特别适用于处理可变长度的数据序列。

Figure 8-2 gives us an insight into the definition of a linked list. The linked list is identified by the head pointer variable head, which points to the first element of the linked list, that is, the first node. Each linked list element is called a node and contains two parts: the actual data required by the user(the data field) and the address pointing to the next node(the pointer field). These nodes are connected in turn to form a chain structure. For example, one node stores data A and

通过图8-2，我们深入理解链表的定义。链表通过头指针变量head标识，它指向链表的第一个元素，即首结点。每个链表元素称为结点，包含两部分：用户所需的实际数据（数据域），以及指向下一个结点的地址（指针域）。这些结点依次相连，形成链状结构。例如，一个结点存储数据A并指向下一个结点，后者存储数据B

points to the next node, latter stores data B and continues to point to the next node, and so on, until the last node stores data D and points to NULL, and marks the end of the linked list. The arrows in the figure are schematic and do not actually exist in memory. The characteristic of the linked list is that its node address is not continuous, and it is composed of data field and pointer field, which is flexible to adapt to data changes. The head pointer is different from the node, and its type is the same as the node pointer field, which is used to store the address of the node. In particular, the end of the linked list needs to be marked with NULL to provide termination conditions for the subsequent correct processing of the linked list.

Linked list and array, as two distinct data storage structures, have their own unique storage characteristics. Taking the storage of characteristic A, B, C and D as an example, the difference between the two storage structures is shown in Figure 8-3, which is clear at a glance.

Arrays occupy a contiguous space in memory, each element is next to another element, and the addresses are contiguous and ordered. Once an array is created, its size is fixed and cannot be easily changed. Therefore, arrays are particularly suitable for storing a defined number of uniformly-sized data sets.

Linked lists, by contrast, are dynamic and flexible. Each node of the linked list contains data and a pointer pointing to the next node, and the nodes are connected with each other through the pointer and form a data link. This structure makes that the linked list does not need continuous memory space when it stores data, and the addresses of nodes can be discontinuous. More importantly, the linked list can dynamically add or delete nodes to adjust the length of the linked list, it is very suitable for dealing with the scenarios, which has uncertain data size or frequent insertion and deletion operations.

并继续指向下一个结点，依此类推，直到最后一个结点存储数据D并指向NULL，标志着链表的结束。图中的箭头是示意性的，实际并不存在于内存中。链表的特点在于其结点地址不连续，由数据域和指针域组成，灵活适应数据变化。头指针与结点不同，其类型与结点指针域相同，专门用于存储结点的地址。特别要注意，链表的表尾需用NULL来标识结束，为后续正确处理链表提供终止条件。

链表与数组，作为两种截然不同的数据存储结构，各自拥有独特的存储特性。以存储A、B、C、D四个字符为例，二者的存储结构差异如图8-3所示，一目了然。

数组在内存中占据一片连续的空间，每个元素紧挨着另一个元素，地址连续，排列有序。一旦数组被创建，其大小便固定下来，无法轻易改变。因此，数组特别适用于存储数量确定、大小一致的数据集合。

相较之下，链表则呈现出一种动态和灵活的特质。链表的每个结点都包含数据和指向下一个结点的指针，结点之间通过指针相互连接，形成一条数据链。这种结构使得链表在存储数据时无须连续的内存空间，结点的地址可以不连续。更重要的是，链表可以动态地添加或删除结点，从而调整链表的长度，非常适合处理数据规模不确定或需要频繁进行插入、删除操作的场景。

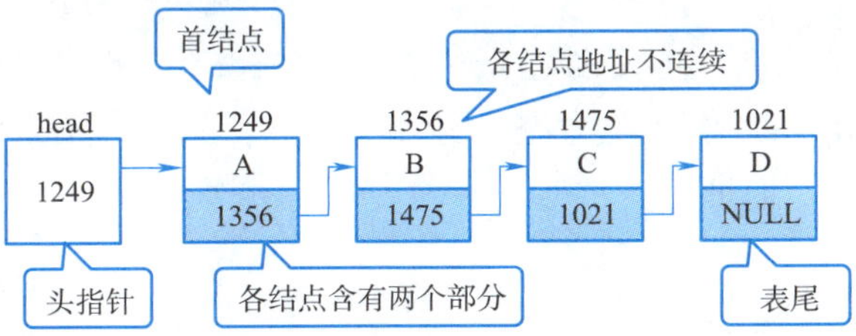

图8-2 一个简单的链表示意图

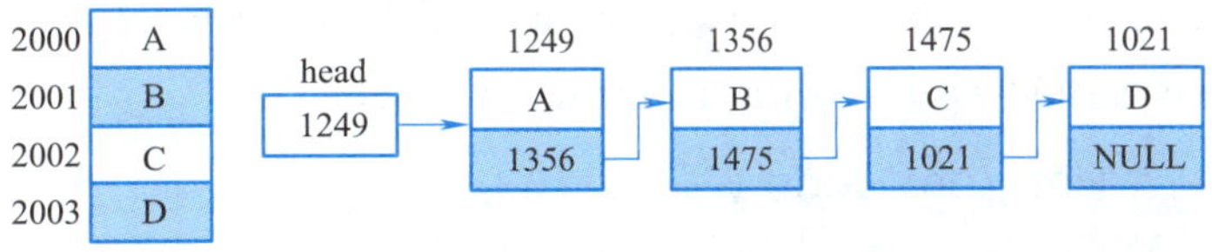

图8-3 数组和链表存储结构示意图

8.4 Creation and Output of Linked List
链表的创建和输出

8.4节视频

We can summarize the basic operations of a linked list into the following: create, output, insert, find, and delete. Each of

我们可将链表的基本操作归纳为以下五个：创建、输出、插入、查找和删除。

these core operations will be described in more detail later. This section focuses on the creation and output of linked lists.

这些核心操作将在后续内容中逐一详细介绍。本节主要探讨链表的创建和输出操作。

8.4.1 Creation of Linked List（链表的创建）

The creation of a linked list is a key process in the execution of a program, which aims to build a complete linked list structure step by step. In this process, we need to loop the following three core steps:

链表的创建是程序执行中的一个关键过程，旨在逐步构建出一个完整的链表结构。这一过程中，我们需要循环执行以下三个核心步骤：

First, open up a new node space to add new members to the linked list.

首先，开辟一个新的结点空间，为链表添加新的成员。

Second, input the required data for the new node to ensure that each node stores the correct information.

其次，为新结点输入所需的数据，确保每个结点存储正确的信息。

Finally, the new node is correctly linked into the existing linked list to ensure the integrity of its logical connection.

最后，将该新结点正确链入已有的链表中，确保其逻辑连接的完整性。

We will analyze the creation process of the linked list in detail through specific cases, and help you understand and master this important operation more intuitively.

我们将通过具体的案例来详细解析链表的创建过程，帮助大家更加直观地理解和掌握这一重要操作。

Case 8_3（案例 8_3）

Write a function to create a dynamic linked list to store student data.

编写函数实现创建一个存放学生数据的动态链表。

Thinking（解题思路）:

First, define a student node structure type. The structure type is different from the previously learned structure type in that it has an additional pointer field to store the next node address. Subsequently, we use malloc to dynamically apply for node space, input student data for the node, and finally link the node to the established linked list. If the linked list is empty, it will be the first element, if it is not empty, it will be linked to the end of the existing linked list.

首先定义一个学生结点结构体类型，该结构体类型跟之前学习的结构体类型不同之处在于多了一个指针域，用来存放下一个结点地址。随后，使用malloc动态申请结点空间，为该结点输入学生数据，最后将该结点链接到已建立的链表中，若链表为空则作为首元素，若非空则链接到已有链表的表尾。

Case Solution（案例实现）:

The linked list creation function is named “creat”, the function parameter can be set to null, and the return value should be the address of the first element of the linked list.

链表创建函数命名为creat，函数形参可以设为空，返回值应为链表首元素地址。

Program Code（程序代码）:

Case8_3.c

```
#include <stdio.h>
#include <stdlib.h>
#define LEN sizeof(struct Student)      //定义符号常量，一个结点类型所占的字节数
struct Student                          //定义学生结点结构体类型
{
    int num;                            //数据域
    float score;                        //数据域
    struct Student *next;               //指针域
};
```

```
10  struct Student *creat()                        //创建链表的函数
11  {
12      struct Student *head=NULL;
13      struct Student *p1=NULL,*p2=NULL;
14      head=p1=p2=(struct Student *)malloc(LEN);
15      scanf("%d%f",&p1->num,&p1->score);
16      while(1)
17      {
18          p1=(struct Student *)malloc(LEN);
19          scanf("%d%f",&p1->num,&p1->score);
20          if(p1->num!=0) {p2->next=p1;p2=p1;}
21          else break;                            //结点的 num 为 0 则结束循环
22      }
23      p2->next=NULL;
24      free(p1);
25      return(head);
26  }
27  int main()
28  {
29      struct Student *pt;                //定义指向结点类型的指针变量
30      pt=creat();                        //接收函数返回值，即保存链表首元素地址
31      return 0;
32  }
```

Program Output（运行结果）:

```
101 89.5 102 90 103 85 0 0
```

Case Analysis（案例解析）:

(1) Line 2 contains the header file that defines the malloc function, which allows us to use the malloc function to dynamically allocate memory.

(2) The next three lines define the symbolic constant LEN, which represents the amount of memory space required by the struct Student type. This constant can be used as an argument to the malloc function.

(3) Lines 4 to 9 define a student structure type, also can be known as a linked list node type, which contain a data field and a pointer field. The data field contains the basic information of the student, and the pointer field is used to link the next node and is a unique member of the dynamic linked list.

(4) Line 12 defines the head pointer head of the linked list. This is because the linked list needs a head pointer to identify its starting position. Note that the type of the head pointer, since it is used to store the address of the first node, it should be struct Student *type.

(5) Line 13 defines two pointer variables, p1 and p2, which will be used to assist in creating the linked list.

(6) Line 14 is used to create the first node of a linked list. It uses the malloc function to dynamically allocate memory, and the argument is the symbolic constant LEN, which is the amount of space required for a struct Student variable. Then,

（1）第2行包含了定义malloc函数的头文件，这使得我们可以使用malloc函数来动态分配内存。

（2）接下来的三行定义了符号常量LEN，它表示struct Student类型所需的内存空间大小。这个常量可以作为malloc函数的参数来使用。

（3）第4~9行定义了一个学生结构体类型，也可称为链表结点类型，包含了数据域和指针域，数据域包含了学生的基本信息，指针域是用于链接下一个结点，属于动态链表特有的成员。

（4）第12行定义了链表的头指针head，这是因为链表需要一个头指针来标识其起始位置，注意头指针的类型，因其作用是存放首结点的地址，故应是struct Student*类型。

（5）第13行定义了两个指针变量p1和p2，它们将用于辅助创建链表。

（6）第14行的作用是创建一个链表的首结点。它使用malloc函数动态分配内存，实参为符号常量LEN，即一个struct Student

the newly allocated memory address is assigned to the head, p1, and p2 pointers, which are pointed to by the three nodes, respectively. Next, it needs to assign values to the num and score fields of the new node through the p1 pointer, where the inputs are 101 and 89.5, as shown in Figure 8-4.

(7) Line 16 starts a loop that keeps adding new nodes to the list. Inside the loop, dynamically request new node memory and point p1 to it. Then, the data of the new node is read in through p1. The following is that the if statement checks the num field of the new node, and if it is not 0, it is linked into the linked list, and if it is 0, the break statement is executed to terminate the loop.

(8) Line 20 uses an if statement to judge the num field of the new node. If it is not 0, we execute the statement p2-> next = p1, so that the tail node of the existing linked list points to the new node. Then, the p2 pointer is updated to point to the current newly linked node, i.e., p2 = p1, and it is in preparation for the next node to be linked. So far, the functions of p1 and p2 are relatively clear. p1 is used to store the node address of each new application, that is, to point to the new node, and p2 points to the tail node, so as to facilitate the link of the new node, as shown in Figure 8-5.

(9) Continuously executing the loop until the num of a certain node is 0, and ending the loop, as shown in Figure 8-6.

变量所需的空间大小。然后，将新分配的内存地址赋值给head、p1和p2指针，三个结点分别指向了该结点。接着，通过p1指针为新结点的num和score字段赋值，这里输入的是101和89.5，如图8-4所示。

（7）第16行开始了一个循环，用于不断将新结点链入链表。在循环内部，动态申请新的结点内存，并将p1指向它。随后，通过p1读入新结点的数据。紧接着的if语句检查新结点的num字段，如果不为0，则将其链入链表，若为0则执行break语句，终止循环。

（8）第20行使用if语句判断新结点的num字段。如果不为0，则执行语句p2->next = p1，使得已有链表的尾结点指向新结点。然后，更新p2指针使其指向当前新链入的结点，即p2=p1，为链入下一个结点做准备。至此p1和p2的作用较为清晰，p1用于存储每次新申请的结点地址，即指向新结点，p2指向尾结点，便于新结点的链入，如图8-5所示。

（9）循环不断地执行，直到遇到某个结点的num为0，结束循环，如图8-6所示。

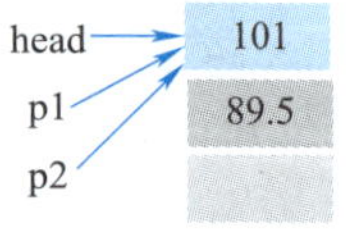

图8-4　创建首结点

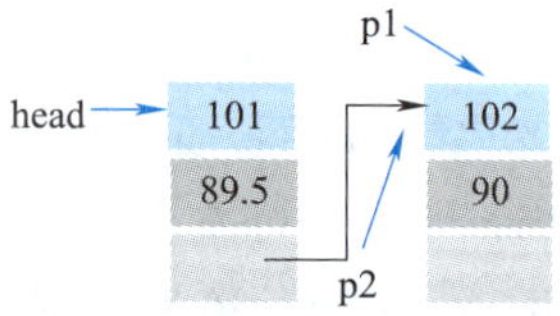

图8-5　链接新结点

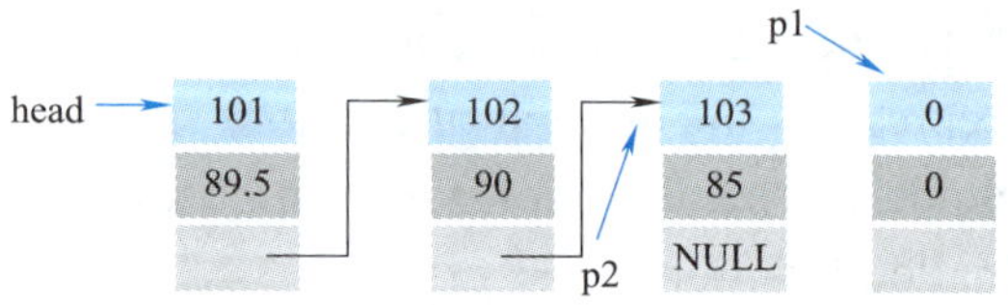

图8-6　链接新结点过程结束

(10)line 23 sets the next field of the tail node to be empty through p2. This step is very important.

(11) In line 24, since the node with num 0 pointed to by p1 is not linked into the linked list, the free function should be used to release it in order to avoid a memory leak.

(12) Lines 27 to 32 are the main function, which calls the creat function to create the linked list. After execution, the creat function returns the head pointer information of the created linked list, namely the address of the first node, which is stored in the variable pt. If this head pointer information is not saved, the created linked list will not be found, resulting in a memory leak.

（10）第23行通过p2置尾结点的next域为空，这一步非常重要。

（11）第24行，由于p1指向的num为0的结点并没有链入链表中，为了避免内存泄露，应使用free函数释放掉。

（12）第27～32行是主函数，它调用creat函数来创建链表。执行后，creat函数返回所创建链表的头指针信息，即首结点的地址，这个地址被保存在变量pt中。如果不保存这个头指针信息，将导致创建的链表无法被找到，从而产生内存泄露。

Case Summary（案例总结）:

(1) This case mainly discusses the one-way linked list, in which each node contains a next pointer to point to the information

（1）本案例主要探讨的是单向链表，其中每个结点包含一个next指针用于指向下

of the next node. In fact, there are several types of linked lists, such as doubly-linked lists and multi-linked lists, which may contain multiple pointer fields. However, this case does not explore these more complex linked list structures in depth, but focuses on laying a solid foundation for the reader, and using one-way linked lists as an example to explain.

一个结点的信息。实际上，链表还有多种类型，如双向链表和多向链表，它们可能包含多个指针域。然而，本案例并不深入探索这些更复杂的链表结构，而是侧重于为读者打下坚实的基础，以单向链表为例进行讲解。

(2) In the one-way linked list, we usually only save the address information of the first node. Because starting from the first node, we can find the next node through the next pointer of each node in turn until the last node. Therefore, it is critical to ensure that the next pointer of the tail node is correctly set to NULL, which marks the end of the linked list.

（2）在单向链表中，我们通常只保存首结点的地址信息。因为从首结点开始，我们可以依次通过每个结点的next指针找到下一个结点，直至尾结点。因此，确保尾结点的next指针被正确设置为NULL是至关重要的，这标志着链表的结束。

(3) The nodes in the linked list are created through dynamic memory allocation, and we only know the memory addresses they occupy, so we need to use the pointers to access these nodes indirectly. When we need to access a specific node, we must have a pointer variable pointing to it. Therefore, it is very important to set and manage the auxiliary pointer variables properly in the linked list operation, and ensure that we can traverse and operate the linked list correctly.

（3）链表中的结点是通过动态内存分配创建的，我们只知道它们所占用的内存地址，因此需要使用指针来间接访问这些结点。当我们需要访问某个特定结点时，必须有一个指针变量指向它。因此，在链表操作中，合理设置和管理辅助的指针变量是至关重要的，以确保我们能够正确地遍历和操作链表。

8.4.2 Output of Linked List（链表的输出）

Output linked list, as its name implies, is to output all the node information in the linked list. Based on the linked list created in Case 8_3, we will write an output function to output all the node information of the linked list.

输出链表，顾名思义，就是要输出链表中的所有结点信息。我们将针对案例8_3中创建的链表基础上，编写一个用于输出链表所有结点信息的输出函数。

Case 8_4（案例 8_4）

On the basis of Case 8_3, an output function is added to output the data domain information of all nodes.

在案例8_3的基础上，增加输出函数，要求输出所有结点的数据域信息。

Thinking（解题思路）:

In Case 8_3, we have created a linked list, and the whole linked list only stores the address information of the first node(head pointer). By passing this first node address, we can use the pointer variable to traverse the entire linked list. It should be noted that the space allocation of a linked list is discontinuous, so you cannot use subscripts to locate elements directly as you do to access arrays, but you should use pointer variables to access them indirectly.

在案例8_3中，我们已经创建了一个链表，整个链表只保存了首结点地址信息（头指针）。通过传递这个首结点地址，我们可以使用指针变量遍历整个链表。需要注意的是，链表的空间分配是不连续的，因此不能像访问数组那样使用下标来直接定位元素，而应使用指针变量间接访问。

Case Solution（案例实现）:

Define a print function used to output the contents of the linked list. It does not need a return value, so the return value is of void type. The function needs to obtain the address of the first node of the linked list to be output from the outside, so the formal parameter should be a pointer to the struct Student type,

定义一个用于输出链表内容的print函数，其不需要返回值，故返回值为void类型，函数需要从外部获取待输出链表的首结点地址，因此形参应为指向struct Student类型的指针，即struct Student*类型。在函

that is, struct Student*type. Inside the function, we will use this pointer to traverse the linked list and output the information for each node. The following code implements the print function.

数内部，我们将使用这个指针来遍历链表，并输出每个结点的信息。下列代码为print函数实现。

Program Code（程序代码）:

Case8_4.c

```
1   void print(struct Student *head)
2   {
3       struct Student *p=NULL;
4       printf("\nNow,These records are:\n");
5       p=head;                                        //令p指向首结点
6       if(head!=NULL)                                 //若不为空
7       do{
8           printf("%d %5.1f\n",p->num,p->score);
9           p=p->next;
10      }while(p!=NULL);                               //循环输出结点信息
11  }
```

Case Analysis（案例解析）:

(1) The first line is the function header, the return value is void, and the formal parameter is struct Student*head, which is used to receive the first address of the external linked list.

(2) Line 3 defines that an auxiliary pointer variable p is suspended at the same time. When defining a pointer variable, it is recommended to suspend it.

(3) In line 5, we assign the value of head to p, so that p points to the first node. Note that the information of head should not be modified in the function, so that the address of the first node is always saved. Because the linked list processed is a one-way linked list, if the head is modified, the information of the first node is lost, which will cause a memory leak. Therefore, an assignment pointer variable p is set to traverse the linked list.

(4) The if statement condition in line 6 controls a do...while statement. The if judgment condition is head!= NULL. If the linked list is not empty, the do...while statement is executed to loop the output node until the tail node. If the linked list is empty, the do...while statement is not executed and the output operation is not performed.

(5) In lines 7~10, the do...while loop executes the loop body statement first and then judges the condition. At this time, if p points to the first node, the information of the first node is output first, and then the statement p=p-> next is executed, it means that p points to the next node. The reason is that p-> next is the next domain of the node pointed by p, which stores the address of the next node. When a value is assigned to p, p also points to the next node.

（1）第1行为函数首部，返回值为void，形参为struct Student *head，用于接收外部链表的首地址。

（2）第3行定义了一个辅助指针变量p同时悬空，定义指针变量时建议悬空处理。

（3）第5行，将head的值赋值给p，令p指向了首结点，注意在函数中不要修改head的信息，令其始终保存首结点地址，因处理的链表为单向链表，若head被修改，则首结点信息丢失，将会产生内存泄露。故设一个赋值指针变量p来遍历链表。

（4）第6行的if语句条件控制的是一条do...while语句，if判断条件为head !=NULL，若成立代表链表不为空，则执行do...while语句循环输出结点直到尾结点为止，若为空则不执行do...while语句，不做输出操作。

（5）第7~10行，do...while循环先执行循环体语句再判断条件，此时p指向了首结点，则先输出首结点信息后，执行了p=p->next语句，含义为令p指向下一个结点，原因是p->next是p所指向结点的next域，其存储了下一个结点的地址，赋值给p后，p也指向了下一个结点。

(6) In line 10, the loop condition is p! = NULL, that is, as long as p is not empty, the loop continues. When p points to the tail node, the loop condition is still true, and the loop body statement is executed. At this time, after the tail node information is output, p = p- > next is executed, and the value of p is NULL. If the condition is not true, the loop ends. At this point, all nodes are output. In fact, the end node NULL flag is often used as a conditional judgment.

(7) The complete code from creation to output is as follows:

（6）第10行的循环条件为p!=NULL，即只要p不为空，则继续循环。当p指向尾结点时，循环条件仍成立，执行循环体语句，此时输出尾结点信息后执行p=p->next，则p值为NULL，条件不成立，循环结束。此时所有结点均输出完毕。实际上尾结点NULL标志常作为条件判断。

（7）完整的从创建到输出的代码如下：

```
#include <stdio.h>
#include <stdlib.h>
#define LEN sizeof(struct Student)        //定义符号常量，一个结点类型所占的字节数
struct Student                            //定义学生结点结构体类型
{
    int num;                              //数据域
    float score;                          //数据域
    struct Student *next;                 //指针域
};
struct Student *creat()                   //创建链表的函数
{
    struct Student *head=NULL;
    struct Student *p1=NULL,*p2=NULL;
    head=p1=p2=(struct Student *)malloc(LEN);
    printf("请输入创建的结点学号和成绩，以0 0作为结束:\n");
    scanf("%d%f",&p1->num,&p1->score);
    while(1)
    {
        p1=(struct Student *)malloc(LEN);
        scanf("%d%f",&p1->num,&p1->score);
        if(p1->num!=0){p2->next=p1;p2=p1;}
        else break;                       //结点的num为0则结束循环
    }
    p2->next=NULL;
    free(p1);
    return(head);
}
void print(struct Student *head)          //输出链表的函数
{
    struct Student *p=NULL;
    printf("\nNow,These records are:\n");
    p=head;                               //令p指向首结点
    if(head!=NULL)                        //若不为空
    do{
        printf("%d %5.1f\n",p->num,p->score);
        p=p->next;
    }while(p!=NULL);                      //循环输出结点信息
}
int main()
{
    struct Student *pt;                   //定义指向结点类型的指针变量
    pt=creat();                           //接收函数返回值，即保存链表首元素地址
    print(pt);                            //调用输出函数
    return 0;
}
```

Program Output（运行结果）:

```
请输入创建的结点学号和成绩，以0 0作为结束：
101  89.5
102  90
103  85
Now,These records are:
101  89.5
102  90.0
103  85.0
```

Case Analysis（案例解析）:

In line 42, we call the creat function to create the linked list and store the address of the first node of the returned linked list in the variable pt. It should be noted that although the execution of the creat function has ended, all the linked list nodes it dynamically requested will not be automatically released. Because the dynamically requested memory space is released only when the free function is explicitly called, the linked list is not automatically released with the end of the function. However, the local variables head, p1, and p2 inside the creat function are released at the end of the function and are not accessible outside the creat function. So, we return the value of head so that pt in the main function can receive it. Then, on line 43, we call the print function and pass pt to it as an argument. Thus, the print function is dealing with the linked list that was just created. It should be noted that the print function only outputs the information of the linked list and does not change the structure of the linked list.

在第42行，我们调用了creat函数来创建链表，并将返回的链表首结点地址存储在变量pt中。需要注意的是，尽管creat函数的执行已经结束，但它所动态申请的所有链表结点并不会自动释放。因为动态申请的内存空间只有在显式调用free函数时才会被释放，故链表不会随着函数的结束而自动释放。但creat函数内部的局部变量head、p1和p2在函数结束后会被释放，并且这些变量在creat函数外部是不可访问的。因此，我们返回了head的值，以便主函数中的pt能够接收它。随后，在第43行，我们调用了print函数，并将pt作为实参传递给它。这样，print函数处理的就是刚刚创建的链表。需要注意的是，print函数仅仅是输出链表的信息，并不会改变链表的结构。

8.5 Insertion, Search and Deletion of Dynamic Linked List 动态链表的插入、查找和删除

8.5节视频

8.5.1 Insertion of Linked List（链表的插入）

The process of inserting a linked list mainly includes two steps: determining the insertion position and linking the node to be inserted into the linked list. In practical applications, there are two commonly used insertion methods, namely, the head insertion method and the tail insertion method. Each of these two methods has its own applicable scenarios, and the appropriate method should be selected to insert the linked list according to the actual needs.

链表的插入过程主要包括两个步骤：确定插入位置以及将待插入的结点链接到链表中。在实际应用中，有两种常用的插入方法，即头插法和尾插法。这两种方法各有其适用场景，根据实际需求选择合适的方法进行链表插入操作。

1. Tail Interpolation（尾插法）

The so-called tail insertion method is to insert a new node into the tail of the linked list each time. During the tail insertion operation, we need to traverse the linked list, find the tail node

所谓的尾插法，就是每次将新结点插入链表的尾部。在进行尾插操作时，我们需要遍历链表，找到当前链表的尾结点，

of the current linked list, and then make the next pointer of the tail node point to the newly created node, so as to add the new node to the end of the linked list. This approach ensures that new nodes are always added to the end of the list, thereby maintaining the order of the list. As shown in Figure 8-7, p1 points to the new node, p2 traverses from the first node to the last node, and then links the new node to the last node through the pointer operation.

然后让尾结点的next指针指向新创建的结点，从而将新结点添加到链表的末尾。这种方法确保了新结点总是被添加到链表的最后，从而维护了链表的顺序。如图8-7所示，p1指向新结点，p2从首结点一直遍历到尾结点，再通过指针操作将新结点链接到尾部即可。

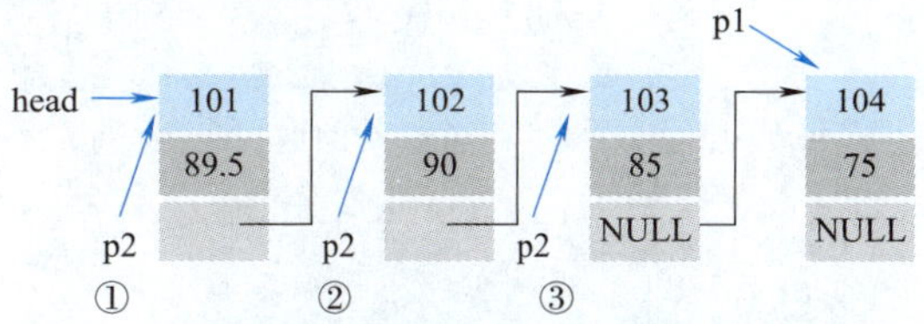

图8-7 尾插法操作示意图

In Figure 8-7, the original linked list has three nodes. Note that the only information we know is the header pointer information. p1 is used to store the dynamically applied new node space, and then the new node is operated through p1 to input the data field for the new node. The student number is 104 and the score is 75. At this time, the new node to be linked is ready. Next, we find the tail node from the first node through the pointer variable p2, let p2 = head, then p2 points to the new node like head, and then let P2 point to the next element step by step through the loop operation until the tail node of the table. The code is as follows:

在图8-7中，原链表有3个结点，注意我们已知的信息只有head头指针信息。使用p1存储动态申请的新结点空间，再通过p1操作新结点，为新结点输入数据域，学号为104、成绩为75，此时待链入的新结点准备完毕。接下来要通过指针变量p2从首结点找到尾结点，令p2=head，则p2与head一样指向了新结点，再通过循环操作令p2逐步指向下一个元素，直到表尾结点为止。代码如下：

```
p1=(struct Student *)malloc(LEN);          // 动态申请新结点，同时令 p1 指向它
scanf("%d%f",&p1->num,&p1->score);         // 通过 p1 间接访问结点，存入数据
p2=head;                                   // 令 p2 指向首结点
while(p2->next!=NULL)                      // 循环控制 p2 指向尾结点
    p2=p2->next;                           // 若非尾结点则向后指一个元素
```

In the above code, the loop condition is p2->next!=NULL. Note that it cannot be written here p2!=NULL. If this is the case, p2 is null at the end of the loop, and the tail node cannot be operated.

在上述代码中，循环条件为p2->next!=NULL，注意这里不能写p2!=NULL，若为该条件，则循环结束时，p2为空，不能操作尾结点。

At this time, p2 points to the tail node, and the link operation can be performed. If the new node is linked in, first, the current tail node stores the address of the new node, that is, the next domain of the table tail node pointed by p2 stores the address of the new node. The code is:

此时p2指向了尾结点，可以进行链接操作，若将新结点链入，首先要令当前的尾结点存放新结点的地址，即令p2指向的表尾结点的next域存放新结点地址，代码为：

```
p2->next=p1;
```

At this time, the link operation has been completed, and the current new node is the new tail node. Be sure to set the tail node to NULL. The code is as follows:

此时已经完成了链接操作，那么当前的新结点即为新的尾结点，务必要令尾结点置NULL，代码如下：

```
p1->next=NULL;
```

The complete function is implemented as follows:

完整的函数实现如下：

```
struct Student *insert_tail(struct Student *head)   // 注意函数首部的写法
```

```
2  {
3      struct Student *p1=NULL,*p2=NULL;
4      p1=(struct Student *)malloc(LEN);
5      printf(" 输入待插入结点的信息:\n:");
6      scanf("%d%f",&p1->num,&p1->score);
7      if(head!=NULL)                              // 判断链表是否为空
8      {                                           // 若非空
9          p2=head;
10         while(p2->next!=NULL)                   // 循环控制 p2 指向尾结点
11         p2=p2->next;                            // 若非尾结点则向后指一个元素
12         p2->next=p1;
13     }
14     else
15     {                                           // 若为空
16         head=p1;
17     }
18     p1->next=NULL;                              // 置表尾为 NULL
19     return head;                                // 返回首结点地址
20 }
```

Pay attention to the writing of the function header. To be on the safe side, we return the linked list header pointer after inserting the new node as the return value, so the return value type is struct Student *, the function name is insert_tail, and the formal parameter is the linked list to be processed, so it is struct Student*head.

注意函数首部的写法，为了保险起见，将插入新结点后的链表头指针作为返回值返回，故返回值类型为struct Student*，函数名见名知意命名为insert_tail，形参需要接待处理的链表，故为struct Student *head。

In addition, various details should be considered when implementing the code. If the linked list is empty, it should be handled separately. At this time, it executes head = p1 to complete the link of the new node. It is worth emphasizing that the new node in line 18 is the current tail node. Note that it should be set to NULL.

另外，在实现代码时要考虑各种细节，若链表为空的情况，要单独处理，此时执行head=p1，即可完成新结点的链入。值得强调的是，第18行的新结点为当前尾结点，注意要置NULL。

2. Head Insertion Method（头插法）

The so-called head insertion method is to insert a new node into the head of the linked list each time, making it a new first node. If the head pointer of the linked list is head, after the new node becomes the first node of the linked list, the pointer of the head should be modified to point to the current new first node.

所谓头插法，即每次将新结点插入链表的头部，使其成为新的首结点。若链表的头指针为head，在新结点成为链表首结点后，要注意修改head的指向，令其指向当前的新首结点。

First, we create a new node and input the corresponding data, and then let the pointer p1 point to the new node. The code is as follows:

首先，我们创建一个新结点并输入相应的数据，然后让指针p1指向这个新结点。代码如下：

```
p1=(struct student *)malloc(LEN);
scanf("%d%f",&p1->num,&p1->score);
```

To insert a new node into the head, you need to perform the following two key operations: one is to make the new node pointed by p1 point to the first node of the original linked list, which can be achieved by executing the p1->next = head;

若要将新结点插入头部，需要执行以下两个关键操作：一是让p1指向的新结点指向原链表的首结点，这可以通过执行p1->next = head;语句来实现；二是修改head的

statement; The second is to modify the value of head to point to the new node, because the new node has become the new first node at this time. Therefore, we assign the value of head to p1, that is, head = p, 1, so that the head points to the new node, and the head insertion operation is completed. The operation diagram is shown in Figure 8-8.The code is as follows:

值，使其指向新结点，因为此时新结点已经成为新的首结点。因此，我们再将head赋值为p1，即head = p1;，这样head就指向了新结点，头插法操作完成。操作示意图如图8-8所示。代码如下：

```
p1->next=head;
head=p1;
```

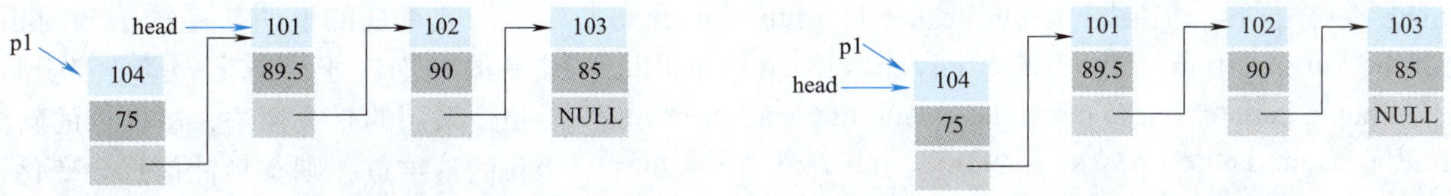

（a）创建新结点并链接到头部　　（b）令head指向新的首结点

图8-8　头插法操作示意图

Think about what happens if you reverse the order of the above two statements? That is to execute head = p1;, then to execute p1->next = head;, and the result after the program execution is shown in Figure 8-9.

There are two phenomena that we don't want to happen:

(1) After the head points to the new node, the new node points to its own circular linked list by mistake. Because head=p1; is executed first, the head points to the new node. Then, p1->next=head; is executed, which results in the next field of the new node pointing to itself, and forms a circular linked list containing only a single node. To avoid this, we should first make the next of the new node point to the original link head node, and then make the head point to the new node.

(2) Another problem is that the original linked list loses the reference of the head pointer, that is, no pointer variable points to the original linked list head node. Due to the loss of the head pointer, the program cannot access the node data in the original linked list through the head pointer. This will lead that the data in the original linked list cannot be properly managed and used, and resulting in a waste of resources. In order to avoid this situation, we must ensure that the update of the head pointer is correct when modifying the linked list structure. And after any operation that may change the head pointer, we must ensure that the head pointer points to the correct first node of the linked list.

The complete header interpolation function is implemented as follows:

请思考：若将上述两个语句颠倒顺序会发生什么问题？即先执行head=p1;，再执行p1->next=head;，程序执行后的效果如图8-9所示。

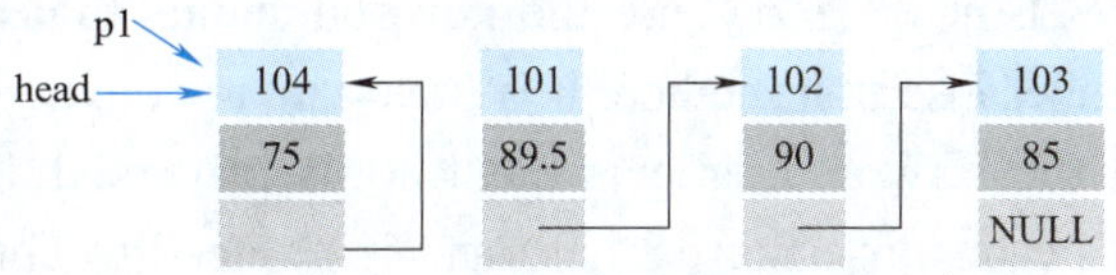

图8-9　程序执行后的效果图

有两个我们不希望出现的现象：

（1）出现head指向新结点后，新结点又错误地指向自己的循环链表情况。因先执行head=p1;，让head指向新结点后，再执行p1->next=head;，导致新结点的next域指向了自身，形成了一个只包含单个结点的循环链表。为了避免这种情况，应先让新结点的next指向原来的链表头结点，然后再让head指向新结点。

（2）另一个问题是原来的链表失去了头指针的引用，即没有任何指针变量指向原来的链表头结点。由于头指针的丢失，程序无法通过头指针来访问原链表中的结点数据。这会导致原有链表中的数据无法被正确管理和使用，从而造成资源浪费。为了避免这种情况，我们在修改链表结构时，必须确保头指针的更新是正确的，并且在任何可能改变头指针的操作之后，都要确保头指针指向的是正确的链表首结点。

完整的头插法函数实现如下：

```
1   struct Student *insert_head(struct Student *head)          //注意函数首部的写法
2   {
3       struct Student *p1=NULL;
4       p1=(struct Student *)malloc(LEN);
5       printf("输入待插入结点的信息:\n:");
6       scanf("%d%f",&p1->num,&p1->score);
7       p1->next=head;
8       head=p1;
9       return head;
10  }
```

The description of the function header is quite similar to that of the tail insertion method. The only difference is that the function name is modified to insert_head, and the head insertion method is used. Therefore, the new node inserted is the new head node, and the head pointer information will change. Be sure to return the modified head pointer information.

In line 3, we define a pointer variable p1 that points to the new node, which will be created later. Next, on line 4, we let p1 point to the new node space obtained through dynamic memory allocation. The sixth line is used to enter the data for the new node.

Lines 7 and 8 are the key statements that implement the header insertion method. It is important to note here that the order of these two statements must not be reversed. The advantage of header insertion is that there is no need to consider whether the linked list is empty or not, because the processing of header insertion is consistent whether the linked list is empty or not.

On the whole, the head insertion method is simpler than the tail insertion method. However, a linked list created using the header insertion method has the elements in the reverse order of the input, and appears in reverse order. Therefore, whether to use the head insertion method or the tail insertion method, we need to make a trade-off according to the specific needs and scenarios.

函数首部的说明与尾插法颇为相似，唯一的区别在于函数名被修改为insert_head，采用的是头插法，故插入的新结点为新的首结点，则头指针信息会变化，务必要返回修改后的头指针信息。

在第3行中，我们定义了一个指针变量p1，用于指向后续将要创建的新结点。紧接着，在第4行，我们让p1指向通过动态内存分配得到的新结点空间。第6行则是用于输入新结点的数据。

第7行和第8行是实现头插法的关键语句。这里需要特别注意，这两条语句的顺序绝对不能颠倒。头插法的好处在于，无须额外考虑链表是否为空的情况，因为无论链表是否为空，头插法的处理形式都是一致的。

从整体上来看，头插法相较于尾插法更为简洁。然而，使用头插法创建的链表，其元素的顺序会与输入的顺序相反，呈现为逆序。因此，在选择使用头插法还是尾插法时，需要根据具体的需求和场景进行权衡。

8.5.2 Search of Linked List（链表的查找）

It is a common operation to find the node of the specified data in the linked list. For example, finding the node with the student number of 103, please note that the address of the linked list is not continuous, and the corresponding node cannot be found directly by using the subscript index method like the array. The linked list needs to use the pointer variable to search one by one from the beginning, so the efficiency of the search is relatively low. The diagram is shown in Figure 8-10.

在链表中查找指定数据的结点是常做的操作，如查找学号为103的结点，请注意链表地址是不连续的，不能像数组那样使用下标索引的方法直接查找到对应结点，链表需要使用指针变量从头开始一个个查找，所以查找的效率比较低。示意图如图8-10所示。

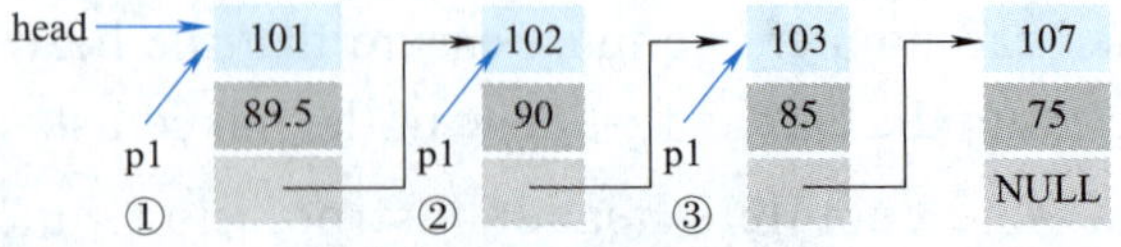

图8-10 查找数据示意图

Set the p1 pointer point to the first node, and use a while loop. If the num value pointed by the current p1 is not 103 and the currently searched node is not the tail node, then it needs to continue to search downward, that is, execute p1 = p1->next. As shown in Figure 8-10. The core code is as follows:

设置p1指针并指向首结点，使用一个while循环，若当前p1指向的num值不为103且当前查找的结点非尾结点，则继续向下查找，即执行p1=p1->next，如图8-10所示。核心代码如下：

```
p1=head;
while(103!=p1->num && p1->next!=NULL)
{
    p1=p1->next;
}
```

At the end of the loop, there are two cases: one is to find the 103 node, and the other is to reach the end node.

循环结束，有两种情况：一种情况是找到了103结点；另一种情况是到了尾结点。

The complete find function for finding a linked list is given below:

下面给出完整的用于查找链表的find函数：

```
1    int find(struct Student *head, int n)
2    {
3        struct Student *p1;
4        p1=head;
5        while(n!=p1->num && p1->next!=NULL)
6            p1=p1->next;
7        if(p1->num==n) return 1;
8        else return 0;
9    }
```

The function receives two parameters: the first parameter is the head pointer information of the linked list, which is used to obtain the linked list information, and the second parameter is the student number data to be searched, which is used as the basis for searching. The return value of the function is an integer. If the matching student number is found, 1 is returned, and the search is successful. If not found, a 0 is returned, and the search fails.

该函数接收两个参数：第一个参数是链表的头指针信息；用于获得链表信息；第二个参数是要查找的学号数据，作为查找的依据。函数的返回值是一个整数，如果找到匹配的学号，则返回1表示查找成功；如果未找到，则返回0表示查找失败。

The third line is the p1 pointer variable declaration. Lines 4 to 6 are the core code part that were just introduced. The student number of the comparison is replaced by the formal parameter variable n, which accepts data from the outside and is more universal. The seventh line is the if statement to judge whether the node with the student number n is found. If the values of p1->num and n are equal, 1 is returned, which means the search is successful. Otherwise, 0 is returned, which means the search fails.

第3行为p1指针变量声明，第4~6行是刚介绍的核心代码部分，比较的学号换成了形参变量n，n从外部接收数据，更具有普适性。第7行为if语句，判断是否找到学号为n的结点，若p1->num和n值相等，则返回1，代表查找成功，否则返回0，代表查找失败。

It should be emphasized that a judgment statement should be added in the 5th line to check if the end of the list is reached, specifically, the condition p1->next!=NULL should be included. If the end is reached, the loop should terminate. Note that if p1!=NULL is used as the loop condition, p1 will be NULL when the loop ends, and attempting to access p1->num in an if statement will result in an error. If p1!=NULL is used, the code can be rewritten as follows:

特别需要强调的是，应在第5行添加一个判断语句来检查是否已到达链表尾部，即需要包含条件p1->next != NULL。如果已到达链表尾部，则应终止循环。值得注意的是，如果使用p1 != NULL作为循环条件，循环结束时p1将为NULL，此时在if语句中尝试访问p1->num将会导致错误。如果使用p1 != NULL作为条件，代码可以改写如下：

```
1    int find(struct Student *head, int n)
2    {
3        struct Student *p1;
4        p1=head;
5        while(n!=p1->num && p1!=NULL)
6            p1=p1->next;
7        if(p1==NULL) return 0;
8        else return 1;
9    }
```

In line 5, the loop condition is modified. There are two cases when the loop ends. One is that n and p1-> num are equal. The other is that p1 is NULL. These information can be used to determine whether the search is successful. As shown in lines 7 and 8, if P1 and NULL are equal, the search fails. If they are not equal, the search succeeds.

第5行中循环条件进行了修改，循环结束有两种情况：一种是n和p1->num相等；另一种是p1为NULL，则可使用该信息来判断是否查找成功，如第7行和第8行所示，若p1和NULL相等，则查找失败，若不相等则查找成功。

From the above analysis, we can see that the core of programming is that its logic is correct and it is easy to understand. Everyone's programming style is different, as long as the code logic is clear, the function is correct, and it is easy to read and maintain, the program can be regarded as high-quality code.

从上述分析中，我们可以看出，编程的核心在于逻辑正确且易于理解。每个人的编程风格各异，只要代码逻辑清晰、功能正确，且易于阅读和维护，便可视作优质代码。

8.5.3 Deletion of Linked List（链表的删除）

Next, we discuss the operation of deleting nodes. If we want to delete the node with student number 103, we must first find 103, assuming that the pointer variable p1 points to the first node, and use the loop to search in turn. On the basis of the search operation in the previous section, we set an additional p2 pointer variable, so that p2 saves the current point of p1, and then p1 continues to search. That is to say, p2 always points to the previous node of p1, also known as the predecessor node, and p1 always points to the next node of p2, also known as the successor node. As shown in Figure 8-11.

接下来探讨删除结点操作，若要删除学号为103的结点，首先要查找到103，设指针变量p1，令p1指向首结点，使用循环依次查找，在上小节的查找操作基础上，多设了一个p2指针变量，让p2保存p1当前的指向后，p1再继续往后查找，即令p2始终指向p1的前一个结点，也称前驱结点，p1始终指向p2的后一个结点，也称后继结点，如图8-11所示。

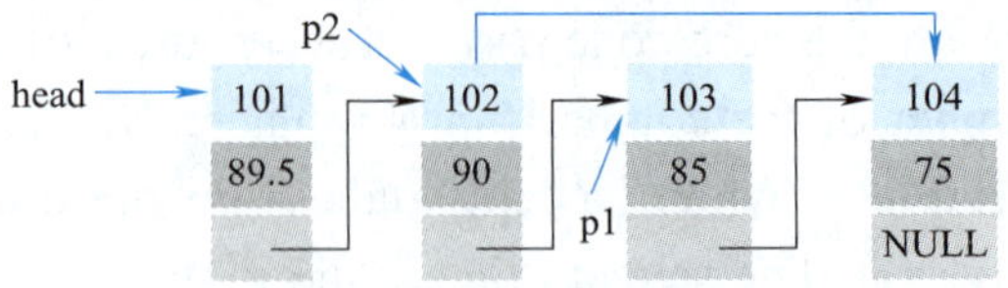

图8-11 删除结点示意图

p1 currently points to the 103 node in the linked list, and p2 points to the previous node of the 103 node, namely the 102 node. If we delete node 103, what we need to do is to adjust the next field of node 102 to point directly to the node after the original node 103, that is, node 104. Since the address of node 104 is stored in the next field of node 103, we can easily get this address through p1->next. In order to do this, we need to use a pointer variable to access and modify the next field of the relevant node. The nodes involved here are mainly nodes 102 and 103, and we use p2 and p1 to point to them respectively. The specific code implementation is: p2->next=p1->next. After this line of code is executed, the next field of node 102 will be modified

p1当前指向了链表中的103结点，而p2则指向了103结点的前一个结点，即102结点。若要删除103结点，我们需要做的是调整102结点的next域，使其直接指向原本103结点之后的结点，即104结点。由于104结点的地址存储在103结点的next域中，我们可以方便地通过p1->next来获取这个地址。为了实现这一操作，我们需要用到指针变量来访问和修改相关结点的next域。这里涉及的结点主要是102和103两

to point to node 104, thus skipping node 103 and achieving to delete node 103 from the linked list. Then the free function can be used to release the memory space of node 103. The space occupied by node 103 will be reclaimed by execute the free(p1) statement.

Two special cases need to be considered: one is that the linked list is empty, and the node cannot be deleted; the other is that the deleted node is the first node, and the head pointer information needs to be modified. In the above case, if the 101 node is deleted, then p1 and head are equal during the searching operation. If the first node needs to be deleted. At this time, you only need to execute head=p1->next.

The complete code for the delete function is as follows:

个结点，我们分别用p2和p1来指向它们。具体的代码实现为：p2->next = p1->next;。这一行代码执行后，102结点的next域将被修改为指向104结点，从而跳过了103结点，实现了将103结点从链表中删除。随后使用free函数来释放103结点的内存空间，其占用的空间将被系统回收，执行free(p1)语句即可。

需要考虑两种特殊的情况：一是链表为空，则不能删除结点；二是删除的结点为首结点，需要修改头指针信息。在上述案例中，若删除101结点，则在查找时p1和head相等，则要删除的是首结点，此时只需要执行head=p1->next即可。

删除函数的完整代码如下：

```
struct Student *del(struct Student *head, int n)
{
    struct Student *p1=NULL,*p2=NULL;
    if(head==NULL)                              //链表为空
    {
        puts("链表为空！");
    }
    else
    {
        p1=head;                                //使p1指向首结点
        while(n!=p1->num && p1->next!=NULL)
        {
            p2=p1;
            p1=p1->next;                        //p1后移一个结点
        }
        if(n==p1->num)                          //如果找到待删除的结点
        {
            if(p1==head)                        //若为首结点
                head=p1->next;                  //把第二个结点地址赋予head
            else
                p2->next=p1->next;              //非首结点
            free(p1);
        }
        else
            puts("找不到要删除的结点！");
    }
    return head;
}
```

Case Analysis（代码解析）:

(1) Pay attention to the writing of the function header. Since it is possible to delete the header node, the header pointer information will be modified, so it is necessary to return the header pointer information of the linked list. There are two formal parameters, struct Student*head is used to receive the pointer information of the linked list to be deleted, and int n is used to receive the student number of the node to be deleted.

（1）注意函数首部的写法，因有可能删除的是首结点，头指针信息会被修改，故需要返回链表头指针信息。形参有两个，struct Student*head用于接收待删除链表头指针信息，int n用于接收待删除结点的学号。

(2) The third line defines the pointer variables p1 and p2 of the auxiliary linked list operation node.

(3) Judging whether the linked list is empty or not in the fourth line, and if the linked list is empty, outputting a line of prompt information that the linked list is empty.

(4) If it is not empty, you need to find the node with student number n in the linked list, as shown in lines 10-15.

(5) The if statement in line 16 is used to judge whether the node with the student number n is found. If the condition is true, it means that the node is found, and the statement after if is executed. If it is not true, a prompt message is output, indicating that there is no data for the node.

(6) Lines 18-21 are an if...else statement, which deals with the statement that deletes the searched node from the linked list, and deals with the deletion of the first node and the non-first node respectively.

(7) Line 22 release the memory space of the deleted node.

(8) Line 27 return the processed linked list header pointer information.

（2）第3行定义了辅助链表操作结点的指针变量p1和p2。

（3）第4行判断链表是否为空，若为空则输出一行链表为空的提示信息。

（4）若非空则需要在链表中查找学号为n的结点，如第10~15行。

（5）第16行的if语句用于判断是否找到学号为n的结点，若条件成立代表查找到该结点，则执行if后的语句，若不成立则输出一条提示信息，表示没有该结点数据。

（6）第18~21行是一条if...else语句，即处理将查找的结点从链表中删除的语句，分别处理删除首结点和非首结点的情况。

（7）第22行释放删除结点的内存空间。

（8）第27行返回处理后的链表头指针信息。

8.6 Comprehensive Cases 综合案例

综合案例 8_6_1　合并两个链表。

案例描述：创建两个链表L1={1, 8, 10}和L2={3, 7}，将两个链表合并为一个链表L1={1, 8, 10, 3, 7}。

解题思路：首先，我们需要定义链表结点的数据类型，包括存储数据的字段和指向下一个结点的指针。接下来，为了创建链表，我们需要编写一个创建链表的函数。最后，为了实现链表的合并操作，我们需要定义一个合并函数。这个函数的主要任务是找到第一个链表（L1）的尾结点，并将其next指针指向第二个链表（L2）的首结点，从而将两个链表连接起来。这样，L1链表和L2链表就合并成了一个新的链表。

案例实现：

Case8_6_1.c

```
#include <stdio.h>
#include <stdlib.h>
#define LEN sizeof(INum)
typedef struct Inum                                   //定义结点结构体类型
{
    int num;
    struct Inum *next;                                //next指针变量，存放下一结点的地址
}INum;
//创建链表函数
INum *creat()
{
    INum *head=NULL;
```

```
    INum *p1=NULL,*p2=NULL;
    head=p1=p2=(INum *)malloc(LEN);
    printf("创建链表，请输入结点数据（0为终止）：\n");
    scanf("%d",&p1->num);
    while(1)                                   //输入数据0终止
    {
        p1=(INum *)malloc(LEN);
        scanf("%d",&p1->num);
        if(p1->num!=0)
        {
            p2->next=p1;
            p2=p1;
        }
        else break;
    }
    p2->next=NULL;
    return(head);
}
//合并两个链表的函数
INum *link(INum *L1,INum *L2)
{
    printf("合并两个链表!\n");
    INum *p1=NULL;
    if(L1==NULL)
    {
        L1=L2;
        return(L1);
    }
    p1=L1;
    while(p1->next!=NULL)
    {
        p1=p1->next;
    }                                          //找到第一个链表的表尾
    p1->next=L2;                               //合并两个链表
    return(L1);
}
void print(INum *head)                         //输出函数
{
    INum *p;
    p=head;
    if(head!=NULL)
    {
        printf("输出链表:\n");
        while(p!=NULL)
        {
            printf("%d ",p->num);
            p=p->next;
        }
        printf("\n");
    }
}
int main()
{
    INum *L1=NULL,*L2=NULL;
    L1=creat();                                //创建第1个链表
    print(L1);                                 //输出第1个链表
    L2=creat();                                //创建第2个链表
```

```
70         print(L2);                                      //输出第2个链表
71         L1=link(L1,L2);                                 //合并链表
72         print(L1);                                      //输出合并后链表
73         return 0;
74     }
```

运行结果：

```
创建链表，请输入结点数据（0为终止）:
1 8 10 0
输出链表:
1 8 10
创建链表，请输入结点数据（0为终止）:
3 7 0
输出链表:
3 7
合并两个链表!
输出链表:
1 8 10 3 7
```

案例解析：

（1）在代码的第3行，我们定义了一个符号常量LEN，其值设定为一个结构体类型INum所占用的字节数。这样的定义有助于我们在后续代码中快速引用该字节数，提高了代码的可读性和可维护性。

（2）第4行定义了一个结构体类型struct Inum，并借助typedef关键字为其重新命名为INum。typedef在C语言中用于为已有的数据类型创建别名，使得代码更为简洁和清晰。特别是在处理复杂数据类型如结构体时，使用typedef能够极大地提升代码的可读性。在本案例中，INum就是struct Inum的别名。

（3）第10~30行是一个名为creat的函数，其主要功能是创建一个链表，并返回链表的头指针信息。这一功能对于构建和操作链表至关重要。

（4）第32~48行定义了一个函数link，用于合并两个链表。该函数接收两个待合并的链表头指针，INum *L1和INum *L2作为参数。在函数内部，首先定义了一个辅助指针变量p1，用于链表操作。若第一个链表L1为空，则直接将L1指向L2的首结点，完成合并；若L1非空，则使用p1遍历L1直到表尾，再将L2的首结点连接到L1的尾结点，从而完成链表的合并。

（5）第35行额外定义了一个辅助链表操作的指针变量p1。

（6）当L1为空时，合并操作变得非常简单，只需将L1指向L2的首结点即可，如第36~40行所示。

（7）若L1非空，则使用p1遍历L1链表，直到表尾，如第41~45行所示，循环结束后，p1指向了L1的尾结点。

（8）第46行，将L2的首结点连接到p1所指向的L1的尾结点，即可完成合并，返回L1，函数结束，如第47行所示。

（9）第49~63行是一个用于输出链表内容的函数，其实现细节在前面的章节已经介绍过，此处不再赘述。

（10）第66行定义了两个指针变量，用于接收creat函数返回的链表头指针信息，即所创建的链表。

（11）第67行调用creat函数创建第一个链表，并将返回的头指针赋给L1。接着，第68行调用输出链表的函数，测试链表是否创建成功。

（12）第69行再次调用creat函数创建第二个链表，并在第70行输出新创建的链表。需要注意的是，创建两个不同的链表需要分别调用两次creat函数，这体现了creat函数的通用性和灵活性，每次调用都能独立地创建一个新的链表。

综合案例 8_6_2 逆序输出链表。

案例描述：本案例旨在创建一个用于存储整数序列数据的链表，并在不改变原链表数据内容的前提下，实现链表的逆序输出。

解题思路：首先定义包含数据域和指针域的结点结构体类型。随后定义创建链表函数，根据输入的整数序列动态生成并连接结点，形成完整链表。最后，设计逆序输出链表函数，遍历链表但反向输出结点数据，确保不改变链表结构和数据的前提下逆序输出结点信息。

案例实现：我们分别使用循环方法和递归方法实现，提升读者的逻辑思维和算法设计能力。

方法一：循环方法。

Case8_6_2_1.c

```
#include <stdio.h>
#include <stdlib.h>
#define LEN sizeof(struct Intnum)
typedef struct Intnum
{
    int num;
    struct Intnum *next;
}INum;
//创建链表的函数
INum *creat()
{
    INum *head=NULL;
    INum *p1=NULL,*p2=NULL;
    head=p1=p2=(INum *)malloc(LEN);
    printf("创建链表，请输入结点数据(0为终止):\n");
    scanf("%d",&p1->num);
    while(p1->num!=0)                              //输入数据0终止
    {
        p1=(INum *)malloc(LEN);
        scanf("%d",&p1->num);
        if(p1->num!=0)
        {
            p2->next=p1;
            p2=p1;
        }
    }
    p2->next=NULL;
    return(head);
}
//逆序输出的函数
void Rprint(INum *head)
{
    if(head!=NULL)                                 //如果链表不为空
    {
        INum *p1=NULL,*p2=NULL;
        p1=p2=head;
        while(p1->next!=NULL)                      //p1指向表尾
        p1=p1->next;
        printf("%d ",p1->num);
        while(p1!=head)
        {
            p2=head;
            while(p2->next!=p1)
```

```
44              {
45                  p2=p2->next;
46              }
47              printf("%d ",p2->num);
48              p1=p2;
49          }
50      }
51  }
52  int main()
53  {
54      INum *head;
55      head=creat();
56      printf("\n逆序输出链表:\n");
57      Rprint(head);
58      return 0;
59  }
```

案例解析：

（1）第4行定义了结构体类型，并使用typedef关键字为其重命名为INum，以增强代码的可读性和可维护性。

（2）第10行定义了创建链表的creat函数，其实现细节与前文一致，此处不再赘述。

（3）第31行定义了用于逆序输出的函数Rprint。由于该函数仅执行输出操作而不改变链表结构，因此无须返回值。函数的形参接收一个链表头指针信息，以便对链表进行遍历和逆序输出。

（4）第33行首先判断链表是否非空。只有当链表非空时，才会执行逆序输出操作；否则，不做任何处理，以避免对空链表进行无效操作。

（5）第35行定义了两个辅助链表操作的指针变量p1和p2。第36行将这两个指针变量初始化为指向链表的首结点，以便后续进行链表遍历操作。

（6）第37、38行是一个while循环，用于将p1指针移动到链表的尾部。循环结束后，p1将指向链表的最后一个结点。第39行输出链表尾部的数据，作为逆序输出的起始点。

（7）第40~49行是一个嵌套的while循环。外层循环的条件是p1不等于链表头指针head，即p1不是链表的第一个结点。内层循环使用p2指针遍历链表，直到找到p1的前驱结点（即p2->next等于p1）为止。此时，p2指向了p1的前一个结点。随后，输出p2指向的结点的数据，实现链表的逆序输出。

（8）在输出完p2指向的结点数据后，将p1向前移动一个元素的位置（即执行p1=p2），以便在下次外层循环中输出下一个结点的前驱结点数据。这样，通过不断循环，可以完成整个链表的逆序输出操作。

（9）在主函数中，首先调用creat函数创建一个链表，并获取其头指针。然后，调用Rprint函数，传入链表头指针作为参数，实现链表的逆序输出。

方法二：递归方法。

Case8_6_2_2.c

```
#include <stdio.h>
#include <stdlib.h>
#define LEN sizeof(struct Intnum)
typedef struct Intnum
{
    int num;
    struct Intnum *next;
}INum;
//创建链表
INum *creat()
```

```
11  {
12      INum *head=NULL;
13      INum *p1=NULL,*p2=NULL;
14      head=p1=p2=(INum *)malloc(LEN);
15      printf("创建链表，请输入结点数据(0为终止):\n");
16      scanf("%d",&p1->num);
17      while(p1->num!=0)                              //输入数据0终止
18      {
19          p1=(INum *)malloc(LEN);
20          scanf("%d",&p1->num);
21          if(p1->num!=0)
22          {
23              p2->next=p1;
24              p2=p1;
25          }
26      }
27      p2->next=NULL;
28      return(head);
29  }
30
31  void Rprint(INum *head)                            //可用递归实现
32  {
33      if(head==NULL)                                 //如果链表为空，则直接结束函数
34          return;
35      else                                           //若链表不为空
36      {
37          Rprint(head->next);                        //递归逆序输出剩下的结点
38          printf("%d ",head->num);
39      }
40  }
41  int main()
42  {
43      INum *head;
44      head=creat();
45      printf("\n逆序输出链表:\n");
46      Rprint(head);
47      return 0;
48  }
```

案例解析：

程序的其他部分与循环方法类似，我们重点分析下递归方法实现的Rprint函数。递归代码在Rprint函数中起到了关键作用，在链表逆序输出中，递归允许我们从链表的末尾开始，逐步向前遍历并打印每个节点的值，从而实现逆序效果。

（1）第33行的if(head==NULL)为递归终止条件，当head为空时，说明已经到达链表末尾或链表本身就是空的。

（2）第37行为递归调用（递归关系），先处理head的下一个结点，即向链表尾部移动。

（3）第38行printf("%d", head->num); 语句的作用是，当递归返回时，输出当前结点的数据，此时是从尾部开始向前输出的。

（4）本案例中递归的工作过程：

①当Rprint函数被首次调用时，它接受链表的头指针head作为参数。首先检查head是否为NULL。如果是，函数直接返回，因为没有更多的结点需要处理。这是递归的终止条件，确保递归不会无限进行下去。

②如果head不为NULL，则递归调用Rprint函数，传入head->next作为参数。这一步是将处理逻辑推向

链表的下一个结点，即更深一层的递归调用。

③递归调用会一直进行，直到到达链表的末尾（即head为NULL），此时递归开始返回。

④当递归返回时，会执行printf语句，输出当前结点的数据。由于递归调用是从链表尾部开始的，因此当递归返回时，输出操作是从尾部向前进行的，从而实现链表的逆序输出。

通过这种方式，递归使得Rprint函数能够无须知道链表的确切长度，就能从尾部开始输出每个结点的数据，直至整个链表被逆序输出完毕。需要注意的是，递归虽然简洁优雅，但也可能导致栈溢出，特别是在处理非常长的链表时。因此，在实际应用中，对于非常长的链表，可能需要考虑使用迭代（循环）的方法或其他优化手段来避免潜在的栈溢出问题。

Summary 小　结

This chapter elaborates the definition of structure type and the declaration of its variables, deeply analyzes the practical application scenarios of structure, and shows the usage of structure pointer and array. At the same time, it also systematically introduces the basic concept of linked list and its basic operation methods, including the creation, output, insertion, search and deletion of linked list, and many complex linked list operations are basically combined or transformed by these basic operations, which lays the foundation for the subsequent in-depth study of advanced data structure processing. Through the study of this chapter, we hope that students can fully grasp the characteristics of the structure type, and skillfully carry out the related operations of the linked list, laying a solid foundation for the follow-up study and practice.

本章详细阐述了结构体类型的定义及其变量的声明方式，深入剖析了结构体的实际应用场景，展示了结构体指针和数组的使用方法。同时，也系统介绍了链表的基本概念及其基本操作方法，基本操作包括链表的创建、输出、插入、查找和删除，很多复杂的链表操作基本由这些基本操作组合或者变换而来，且为后续深入学习高级的数据结构处理奠定基础。通过本章的学习，我们期望同学们能够全面掌握结构体类型的特性，并熟练进行链表的相关操作，为后续的学习和实践奠定坚实基础。

Exercise 习　题

一、单选题

1. 根据下面的定义，能打印出字母M的语句是（　　）。

```
struct Person
{
    char name[9];
    int age;
};
struct Person class[10]={"John",17,"Paul",19,"Mary",18,"Adam",16};
```

A. printf("%c\n", class[3].name);　　B. printf("%c\n", class[3].name[1]);

C. printf("%c\n", class[2].name[1]);　　D. printf("%c\n", class[2].name[0]);

2. 若有以下声明，则对结构体变量stu中成员age的不正确应用方式为（　　）。

```
struct Student
{
```

```
    int num;
    int age;
} stu, *p;
```

A. stu.age　　B. p->age　　C. student.age　　D. (*p).age

3. 若程序中有以下的声明和定义：

```
struct abc
{
    int x;
    char y;
}
struct abc s1, s2;
```

则会发生的情况是（　　）。

A. 编译时报错　　B. 程序将顺序编译、连接执行

C. 能顺序通过编译、连接但不能执行　　D. 能顺序通过编译、但连接出错

4. 设有定义语句：

```
struct
{
    int x;
    int y;
}d[]={{1,3}, {2,7}};
```

则 printf("%d", d[0].y / d[0].x*d[1].x); 的输出结果是（　　）。

A. 0　　B. 1　　C. 3　　D. 6

5. 假设有以下说明语句：

```
typedef struct
{
    int n;
    char ch[8];
} PER;
```

那么下面叙述中正确的是（　　）。

A. PER 是结构体变量名　　B. PER 是结构体类型名

C. typedef struct 是结构体类型　　D. struct 是结构体类型名

二、程序分析题

1. 以下程序的输出结果是____________。

```
struct st
{
    int x;
    int *y;
} *p;
int dt[4]={10, 20, 30, 40};
struct st aa[4]={50, &dt[0], 60, &dt[0], 60, &dt[0], 60, dt[0]};
int main()
{
    p=aa;
    printf("%d", ++(p->x));
    return 0;
}
```

2. 已知如下结构体变量：

```
struct
{
    int day;
} a, *b;
b=&a;
```

可用a.day引用结构体中的成员day，请写出通过指针变量b引用成员a.day的其他两种形式，它们是________和__________。

3. 以下程序的输出结果是__________。

```
struct HAR
{
    int x;
    int y;
    sruct HAR *p;
} h[2];
int main()
{
    h[0].x=1;
    h[0].y=2;
    h[1].x=3;
    h[1].y=4;
    h[0].p=&h[1];
    h[1].p=h;
    printf("%d%d",(h[0].p)->x,(h[1].p)->y);
    return 0;
}
```

4. 已知如下结构体变量，则：

```
struct
{
    int x;
    int y;
} s[2]={{1, 2}, {3, 4}}, *p=s;
```

表达式(p++)->x的值为__________。表达式++p->x的值为__________。

5. 请在横线处填写正确的语句，使程序能实现完整的功能。

```
#include <stdio.h>
static struct Person
{
    char name[20];
    int age;
}p[] ={{"li ming",17},{"wang fang",18},{"zhang chen",19}};

int main()
{
    int i, max, min;
    max=min=p[0].age;
    for(i=0; i<3; i++)                              // 找出年龄最大和最小者
    {
        if(p[i].age>max)
            __________
        else if(p[i].age<min)
            __________
```

```
        printf("Middle:\n");
        for(i=0; i<3; i++)                          //输出年龄居中者的信息
        {
            if(min<p[i].age && p[i].age<max)
            {
                printf("%s %d\n", p[i].name, p[i].age);
                break;
            }
        }
    }
    return 0;
}
```

三、编程题

1. 定义一个结构体，其中包括：职工号、职工名、年龄、工资、地址。按结构体类型定义一个结构体数组，从键盘输入每个结构体元素所需的数据，然后逐个输出这些元素。

2. 定义一个结构体来表示一个员工，包含姓名、工号和工资三个成员。编写一个程序，允许用户输入多个员工的信息，并将其存储在一个结构体数组中。然后，程序应能够按照工资从高到低对员工进行排序，并输出排序后的结果。

3. 定义一个结构体类型来表示一个二维平面上的点，包含x和y两个整型成员。编写一个程序，输入两个点的坐标，计算并输出这两点之间的距离。

4. 设计一个结构体来存储书籍的信息，包括书名、作者、出版社和出版日期。编写一个程序，允许用户输入多本书籍的信息，并将其存储在一个结构体数组中。然后，提供一个查询功能，用户可以根据书名查询书籍的详细信息。

5. 定义一个结构体来表示一个圆形，包含圆心和半径两个成员（圆心可以用一个二维平面上的点来表示）。编写一个程序，允许用户输入多个圆形的信息，然后计算并输出每个圆形的面积和周长。

6. 创建一个结构体来表示一个线段，包含线段的两个端点（每个端点可以用一个二维平面上的点来表示）。编写一个程序，输入两个线段，判断这两个线段是否相交，并输出判断结果。

7. 建立一个链表，每个结点包括学号、姓名、性别和年龄。输入一个年龄，如果链表中的结点所包含的年龄等于此年龄，则将此结点删去。

8. 13个人围成一圈，从第1个人开始顺序报号1、2、3。凡报到“3”者退出圈子，找出最后留在圈子中的人原来的序号。要求使用链表实现。

Chapter 9

File

第 9 章

文　件

Introduction 主要内容

The C programming language is equipped with robust file operation capabilities, including functions for opening, reading, appending, inserting, deleting data, and closing files. Compared to other programming languages, the syntax for file processing in C is relatively intuitive and user-friendly. This chapter offers a detailed explanation of the fundamental principles of file operations, covering basic concepts of files, correct methods for opening and closing files, and techniques for reading and writing files.

C语言具备强大的文件操作能力，涵盖了打开、读取、追加、插入、删除数据以及关闭文件等功能。相较于其他编程语言，C语言在文件处理方面的语法相对直观，易于上手。本章将详细讲解文件操作的核心知识，包括文件的基本概念、如何正确打开与关闭文件，以及文件读写操作的具体方法。

Objectives 学习目标

Through studying, students will acquire a fundamental and comprehensive understanding of the relevant concepts pertaining to files. This will enable them to more easily comprehend the basic procedures involved in files manipulation. Practical exercises will allow students to proficiently grasp the essential principles and techniques for opening, closing, reading, and writing files in C programming language. In the course of their learning, emphasis should be placed on cultivating students' rigorous logical thinking skills to aid them in identifying and resolving common issues that may arise during file operations. By delving into this chapter, students will be able to flexibly utilize the C programming language for file manipulation.

通过学习，学生将对文件的相关概念有一个基础而全面的认知，从而更容易理解文件操作的基本步骤。通过实践操作，让学生掌握C语言对文件的打开、关闭、读入、读出等核心概念与操作。在学习过程中注重训练学生严谨的思维逻辑，帮助学生能够识别并解决在文件操作过程中可能遇到的常见问题。通过本章学习，能够灵活运用C语言进行文件操作。

9.1 File Open and Close 文件的打开与关闭

9.1节视频

Individuals who have experience using computers will be familiar with files, as the operating system is file-based. Written documents, programmed software, images captured with digital cameras, email attachments, and more are all stored in the form of files. When information needs to be accessed from these files, they are utilized. Files play a crucial role in software development and programming activities.

用过计算机的人对文件都不会感到陌生，因为操作系统就是以文件为基本单位的，写好的文档、编好的程序、用数码相机照相的图像、电子邮件的附件等，均是以文件形式保存的信息，需要时就从文件读取信息。文件在软件开发、程序编写中是常被用到的。

9.1.1 What Is a File（什么是文件）

Before delving into the operation of opening and closing files in the C programming language, it is essential to grasp the fundamental concept of files. In simple terms, a file is a collection of data stored on external storage devices such as disks, CD-ROMs, U disks, etc. These external memories possess large capacity and are capable of long-term data storage. In the practice of C programming, working with files is a common occurrence. Source program files(.c), compile-generated object files(.obj), executable files(.exe), and header files containing function declarations and macro definitions all fall within the category of files. The utilization of files provides great convenience. Previously, when a program required input operations, data could only be stored in memory for a short period. Upon restarting the program or releasing memory upon shutdown, the data would be lost and unable to be stored for an extended duration. This was clearly inefficient for processing large amounts of data or tasks requiring repetitive performance. With the use of file storage, necessary data can now be preserved in external memory for prolonged periods. When needed, this data can then be read from the file without frequent interaction with memory: thereby significantly reducing workload.

Furthermore, the file also facilitates the segregation of program and data, allowing for their storage in separate files. This division not only enhances the flexibility of data management but also enables data sharing and improves program maintainability and reusability. Therefore, a thorough understanding of file operations is crucial for mastering the C programming language.

在讲解C语言对文件进行打开与关闭的操作之前，我们首先需要了解文件的基本概念。文件，简单来说，就是存储在外存储器（如磁盘、光盘、U盘等）上的数据集合。这些外存储器具有大容量且可长期保存数据的特性。在C语言编程实践中，我们已多次与文件打交道。例如，源程序文件（.c）、编译生成的目标文件（.obj）、可执行文件（.exe），以及包含函数声明和宏定义的头文件等，都属于文件的范畴。文件的使用为我们带来了诸多便利。以往，当程序需要进行输入操作时，数据只能短暂地存储在内存中，当程序重新运行或者关机时释放内存后，数据就会丢失，不能长期保存。这对于处理大量数据或需要重复执行的任务来说，显然不够高效。而有了文件，我们可以将所需数据长期保存在外存储器中，当需要时再从文件中读取，这样就避免了与内存的频繁交互，大大减少了工作量。

此外，文件还实现了程序与数据的分离。这意味着程序和数据可以分别存放在不同的文件中。这种分离不仅使得数据管理更加灵活，还实现了数据共享，提高了程序的可维护性和可重用性。因此，掌握文件操作对于C语言编程来说至关重要。

9.1.2 Classification of Files（文件的分类）

So how are files classified? Files are classified according to their content into two categories: program files and data files.

那么文件又是如何进行分类的呢？依据内容的不同，文件可被划分为程序文件

Program files mainly contain the source code of a program, while data files are used to store data, such as original input or output results required during program execution. From an organizational perspective, files can be divided into text files and binary files. Text files store ASCII codes of one character per byte, while binary files maintain consistency with the binary representation in memory. For example, storing the number 13297 requires four bytes in memory for a binary file, whereas it would require five bytes in a text file to store each digit as a single character with its corresponding ASCII value. Specifically, the '1' character stores its corresponding ASCII value 49, which is the binary representation of 49. The second byte stores the ASCII value of the '3' character 51, and so on, the remaining three characters are also stored separately in their ASCII binary form. Figure 9-1 shows the storage format of the number 13297 in two files.

和数据文件两大类。程序文件主要包含程序的源代码，而数据文件，顾名思义，则是用于存储数据的，如程序执行时所需的原始数据或输出的结果等。而从文件的组织形式来看，文件又可分为文本文件和二进制文件。文本文件的特点是每个字节存储一个字符的ASCII码；而二进制文件则与内存中的二进制表示方法保持一致。例如，若要存储数字13297，在内存中会使用4个字节来存放。同样地，在二进制文件中，该数字的存储形式与内存中的表示方法一致。然而，在文本文件中，每个数字会被单独视作一个字符来存储，每个字符占用一个字节，因此存储数字13297需要5个字节，分别用来存放这5个数字字符的ASCII码。具体来说，'1'字符会存储其对应的ASCII码值49，这个字节即为49的二进制表示；第二个字节存储'3'字符的ASCII码值51，依此类推，剩余的三个字符也将以它们ASCII码的二进制形式分别存放。数字13297在两种文件中的存储形式如图9-1所示。

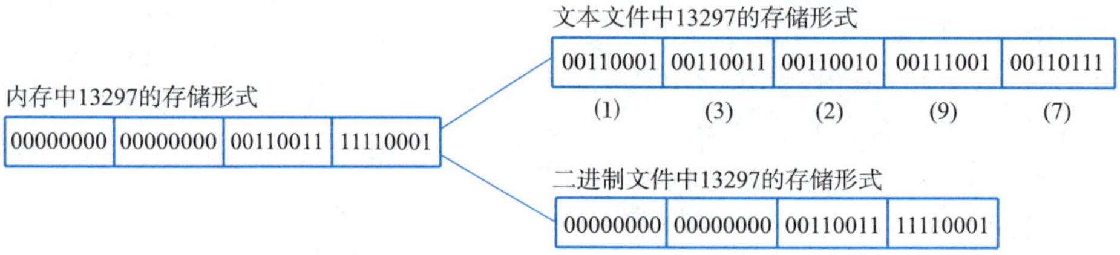

图9-1 13297在两种文件中的存储形式

Let's examine the advantages and disadvantages of these two approaches. The text file format is more convenient for reading and outputting because its content is presented in human-readable characters. However, when reading a text file from memory, it needs to be converted to a binary format, which can result in relatively slow read speeds. Additionally, due to the need for text files to store information in a more redundant way, they typically require more storage space.

In contrast, the binary file approach has a significant advantage in terms of saving storage space because it stores data directly as an in-memory representation. This consistency makes reading and writing binary files faster and easier for long-term data storage. However, binary files cannot be output directly as characters, which makes them less convenient to read and output than text files. If you want to output binary file content into readable character form, you must first convert it to ASCII characters, which adds an additional conversion process.

下面我们来分析这两种方式的优缺点。文本文件方式在阅读和输出方面更为便捷，因为其内容以人类可读的字符形式呈现。然而，当从内存读取文本文件时，需要将其转换成二进制格式，这可能导致读取速度相对较慢。此外，由于文本文件通常需要以更冗余的方式存储信息，因此它们通常占用更多的存储空间。

相比之下，二进制文件方式在节省存储空间方面具有显著优势，因为它直接以内存中的表示形式存储数据。这种一致性使得二进制文件的读写速度更快，并且更便于数据的长期保存。然而，二进制文件不能直接以字符形式输出，这导致它们在阅读和输出方面不如文本文件方便。如果需要将二进制文件内容输出为可读的字符形式，必须先将其转换成ASCII字符，这增加了额外的转换过程。

In conclusion, both text file mode and binary file mode have their own advantages and limitations. When deciding which method to use, it is important to carefully consider the specific application scenario and requirements.

综上所述，文本文件方式和二进制文件方式各有其优势和局限性。在选择使用哪种方式时，需要根据具体的应用场景和需求进行权衡。

9.1.3 File Name（文件名）

Each file is assigned a unique file name to distinguish it from other files. The file name comprises three parts: the file path, the main file name, and the file suffix. In Figure 9-2, for example, "d:\cc\temp"represents the file path indicating its location on an external storage device, specifically pointing to the temp subfolder in the cc folder on drive d. "file1" serves as the main file name and follows identifier naming rules typically set by the user. As for "dat", it is a file extension used to indicate the nature or type of a file, such as ".dat" representing a data file. Various applications utilize different suffixes. For instance, Word documents commonly end with ".docx", slide documents with ".pptx", and C language source files with ".c". These suffixes not only aid in identifying the type of a given file but also determine how it is opened and edited within specific software platforms.

每个文件都拥有一个独特的文件名，这是用以区分不同文件的标识。文件名主要由三部分构成：文件路径、主文件名和文件扩展名。在图9-2所示的例子中，“d:\cc\temp”即为文件路径，它指示了文件在外部存储设备中的具体位置，具体指向d盘下cc文件夹中的temp子文件夹。而“file1”则是主文件名，它遵循标识符的命名规则，通常由用户自行设定。至于“dat”，它是文件扩展名，用于标明文件性质或类型，如“.dat”表示数据文件。不同的应用通常会有不同的文件扩展名。比如Word文档通常以“.docx”作为扩展名，幻灯片文档则是“.pptx”，而C语言的源文件扩展名为“.c”。这些后缀不仅有助于我们识别文件的类型，也决定了文件在特定软件中的打开和编辑方式。

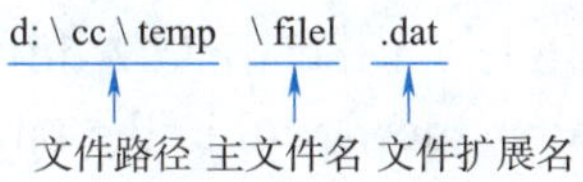

图9-2 文件名示例

In the C programming language, special attention must be given to the representation of file names or paths, particularly when dealing with escape characters. When a file name or path contains special characters such as backslashes(\), it is not possible to use a single backslash directly because backslashes are considered escape characters in C strings. To accurately represent an actual backslash, double backslashes(\\) should be used instead.

在C语言中，文件名或路径的表示需要特别注意转义字符的使用。尤其是当文件名或路径中包含像反斜杠（\）这样的特殊字符时，不能直接使用单个反斜杠，因为反斜杠在C语言字符串中是一个转义字符。为了正确表示一个实际的反斜杠，应该使用双反斜杠（\\）来替代。

For example, if there is a file located at “D:\temp\data.txt”, then the corresponding path in C programming language be expressed as follows:

例如，假设我们有一个文件位于“D:\temp\data.txt”，那么在C语言中表示这个路径时，应该写成如下形式：

```
const char *filepath="D:\\temp\\data.txt";
```

In the code above, each pair of double backslashes(\) is interpreted by the compiler as a separate backslash character. This is because in C programming language strings, a double backslash is an escape sequence used to represent an ordinary backslash character.

在上述代码中，每对双反斜杠（\）都会被编译器解释为一个单独的反斜杠字符。这是因为在C语言字符串中，双反斜杠是一个转义序列，用于表示一个普通的反斜杠字符。

9.1.4 File Buffer（文件缓冲区）

During the process of reading and writing files, data is not

文件在读写的过程中，并不是直接将

directly transferred from the peripheral to the program's memory area. Instead, it need pass through a file buffer. Files are stored in external memory, such as storage disk files, which have relatively slow read and write speeds. In contrast, memory has faster read and write speeds. External memory has a larger capacity for long-term storage, while memory has a smaller capacity and cannot be used for long-term storage. The speed difference between the two necessitates the use of a file buffer to improve efficiency in reading and writing operations. Programs and data are initially stored in external memory and are only called into the memory when needed or ready to be executed. To bridge this gap, input buffers are created in the memory to facilitate data transfer from the disk, while output buffers assist with writing data back to the disk. It is important to note that the input and output buffer is a memory area specifically allocated for data transfer.

As shown in Figure 9-3, data is retrieved from the disk and initially stored in a buffer. Subsequently, the computer accesses the data from the buffer before reading it from the disk upon completion of buffering. This approach reduces the frequency of read and write operations on the disk. Additionally, due to faster processing speed within the buffer compared to direct disk operations, its implementation significantly enhances overall computer performance.

Buffer flushing is a critical process that involves the complete transfer of data from a buffer to the peripherals or memory regions associated with it. When the buffer reaches its maximum capacity, this refresh process occurs automatically. In the case of the output buffer, once it is full, the data in the buffer is written to the target peripheral file at once. As for the input buffer, when it is full, the accumulated data is read into the specified program data area in memory for further processing by the program.

There are also several specific situations that can cause a buffer flush. For example, the buffer is typically flushed when a return character is encountered during input or output. Additionally, upon closing the file, the buffer is flushed to ensure all data has been written or read. Lastly, by utilizing the "fflush" function in the standard library, programmers can actively trigger a buffer refresh operation.

文件直接从外设流入内存中的程序数据区，中间需要经过一个文件缓冲区。文件存在外部存储器，如存储磁盘中的文件，外存储器的读写速度会比较慢，而内存的读写速度比较快，前者容量大，可长期保存，后者容量相对小，不能长期保存。两者之间存在速度差，为了提高读写的效率，文件缓冲区的存在变得尤为重要。程序和数据均是存储在外部存储器中的，只有要执行或者需要时才调入内存中，为了解决数据差异，在内存中开辟了输入输出缓冲区，输入缓冲区是从磁盘中读数据，输出缓冲区是辅助写入磁盘。注意输入输出缓冲区是在内存中开辟的一个区域。

如图9-3所示，我们从磁盘里取数据，先把读出的数据放在缓冲区，计算机再从缓冲区中取数据，等缓冲区的数据取完后再去磁盘中读取，这样就可以减少磁盘的读写次数，再加上计算机对缓冲区的操作快于对磁盘的操作，故应用缓冲区可大幅度提高计算机的运行速度。

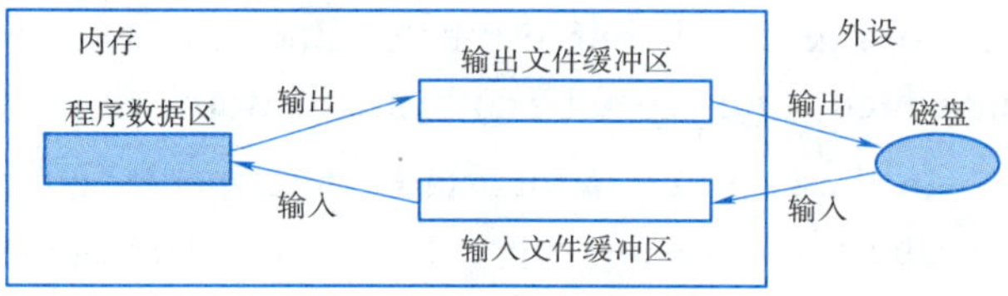

图9-3　文件的读写过程

缓冲区刷新是一个至关重要的过程，它涉及将缓冲区中的数据完全传输到与之相关联的外设或内存区域。当缓冲区达到其容量上限时，这一刷新过程会自动进行。对于输出缓冲区而言，一旦其填满，缓冲区内的数据就会一次性被写入目标外设文件。而对于输入缓冲区，当其满时，则会将累积的数据一次性读取到内存中的指定程序数据区，以供程序进一步处理。

还有几种特定情形也会导致缓冲区刷新。例如，在输入或输出过程中遇到回车符时，缓冲区通常会被刷新。此外，当文件被关闭时，为了确保所有数据都已写入或读取完毕，缓冲区也会被刷新。最后，通过使用标准库中的fflush函数，程序员可以主动触发缓冲区的刷新操作。

9.1.5 File Open and Close Operations（文件的打开与关闭操作）

We have previously introduced the relevant concepts of file operations. Now, how can we write a program to effectively manipulate files?

File operations are typically divided into three steps: opening the file, reading and writing to the file, and closing the file. In the C programming language, files are stored as streams and are usually located on external media(such as disks). They are only brought into memory when they are being used. Therefore, it is necessary to open and close the file when performing operations on it.

The file pointer variable is required to operate the above three steps, and the definition form is as follows:

```
FILE *pointer variable;
```

This pointer variable is utilized to store the initial address of a file. The FILE type is defined in stdio.h, and inclusion of this header file allows for direct usage.

上文我们已经介绍完了文件的相关概念，那么如何编写程序操作文件呢？

文件操作的分成三个步骤：打开文件、文件读写以及关闭文件。C语言是流式的文件存储，文件通常是驻留在外部介质（如磁盘等）上的，在使用时才调入内存中来。因此在对文件进行操作时需要进行打开和关闭。

操作上述三个步骤需要文件指针变量，定义形式：

```
FILE * 指针变量名；
```

这个指针变量用来存放某个文件的首地址。FILE类型定义在stdio.h中，包含这个头文件就可以直接使用。

1. Open File（打开文件）

The C programming language file is opened using the fopen function, which is located in the stdio.h header file. The fopen function is utilized to open a file in a specified manner, and its prototype typically appears as follows:

C语言文件的打开操作通过函数fopen实现，其所在的头文件是stdio.h。fopen函数用于以指定的方式打开文件，其原型通常为：

```
FILE *fopen(const char *filename, const char *mode);
```

Where “filename” is a string representing the name(including the path) of the file to be opened, and “mode” is also a string indicating the mode in which the file is to be opened. For instance, "r" denotes opening a file in read-only mode, while "w" signifies opening a file in write-only mode. Please refer to Table 9-1 for a list of common methods for opening a file.

其中，filename是一个字符串，表示要打开的文件的名称（包括路径）；mode也是一个字符串，表示以何种模式打开文件。例如，"r"表示以只读方式打开文件，"w"表示以只写方式打开文件。文件常用的打开方式见表9-1。

表 9-1 常用文件打开方式及其含义

使用方式	含义	使用方式	含义
"r"	打开文本文件，只读文件须已存在	"r+"	打开文本文件，读写文件须已存在
"w"	创建文本文件，只写（有则删，无则建）	"w+"	创建文本文件，先写后读，写前删除全部数据
"a"	打开文本文件，追加文件须已存在	"a+"	打开文本文件，读、追加文件须已存在
"rb"	打开二进制文件，只读文件须已存在	"rb+"	打开二进制文件，读写文件须已存在
"wb"	创建二进制文件，只写（有则删，无则建）	"wb+"	创建二进制文件，先写后读，建立新文件
"ab"	打开二进制文件，追加文件须已存在	"ab+"	打开二进制文件，读、追加文件须已存在

After successfully opening the file, the fopen function returns a pointer of type FILE, which can be utilized for subsequent read or write operations. In case of a failed opening, NULL is returned. Therefore, it is typically necessary to check the return value of fopen to ensure that the file was successfully opened before proceeding with any further operations.

Table 9-1 presents a list of common file opening methods. The first three rows in this table pertain to text files, with "r" indicating read-only, "w" indicating write-only, and "a" indicating append. It is important to note that for read and append operations, the files must already exist, whereas for write operations, the original file's data will be cleared. Additionally, the "+" symbol denotes both read and write operations. When adding "b", it signifies binary file operations: "rb" for read-only, "wb" for write-only, and "ab" for append. Furthermore, "rb+", "wb+", and "ab+" represent combined read and write operations on binary files.

成功打开文件后，fopen会返回一个指向FILE类型的指针，该指针可以用于后续的读取或写入操作。如果打开失败，则返回NULL。因此，在调用fopen后，通常需要检查其返回值以确保文件已成功打开。

表9-1中列出了几种常见的文件打开方式，这个表前三行操作的是文本文件，其中"r"代表只读，"w"表示只写，"a"代表追加，注意读和追加需要文件先存在，"w"会先清空原文件数据再写。加上“+”符号代表读写操作；加上"b"之后，操作的就是二进制文件，"rb"、"wb"、"ab"这三个分别代表只读、只写和追加，"rb+"、"wb+"、"ab+"这三个是对二进制文件的读写操作方式。

A concrete example is given below:下面给出一个具体的例子：

```
FILE *fp;
fp=fopen("c:\\n\\b\\test.dat", "r");
```

We declare the file pointer variable "fp" and attempt to open the file located at the specified path(c:\n\b\test.dat) in read-only mode("r") by invoking the "fopen" function. The "fopen" function requires the file name and opening method as arguments, and the backslash “\” in the path should be represented as “\\”. If the file is successfully opened, fopen will return a non-empty FILE pointer which will be assigned to fp; if the operation fails, NULL will be returned, requiring error handling.

There is a significant issue with the example provided above: if the file fails to open, the compiler does not directly report an error, leading to subsequent operations on the file being invalid. Therefore, it is essential to ensure that the file is successfully opened. The revised code is as follows:

我们定义文件指针变量fp，并通过调用fopen函数尝试以只读模式（"r"）打开指定路径（c:\n\b\test.dat）的文件。fopen需传递文件名和打开方式作为参数，且路径中的反斜杠“\”需用“\\”表示。若文件成功打开，fopen返回非空FILE指针并赋值给fp；若打开失败，则返回NULL，需进行错误处理。

上述例子中存在一个关键问题：如果文件打开失败，编译器并不会直接报错，这导致后续对文件的操作都将是无效的。因此，确保文件是否成功打开是非常必要的。以下是修改后的代码：

```
FILE *fp;
fp =fopen("c:\\n\\b\\test.dat", "r");
if(fp==NULL)
{
    printf("File open error!\n");
    exit(0);                    //标准库函数，使用需要包含stdlib.h，作用是退出程序
}
```

2. Close File（关闭文件）

After utilizing a file, it is essential to promptly close the file in order to prevent misuse or data loss. The function fclose is used to close the file, and its prototype is as follows:

```
int fclose(File *fp);
```

在使用完一个文件后，为了防止被误用或是数据丢失，我们应当及时关闭文件。文件的关闭使用函数fclose，函数原型是：

```
int fclose (FILE *fp);
```

Function: Close the file pointed by fp. The return value is 0 when it is closed normally, and the non-0 value is returned when an error occurs.

函数功能：关闭fp所指向的文件，正常关闭返回值为0，出错时返回非0值。

In the following code, fp is utilized to manipulate the test.dat file. Upon completion of the operation, it is imperative to call fclose(fp) function in order to close the file and release the associated resources. It is essential to always remember to call fclose function after using the file in order to prevent resource leakage.

下面的代码中，fp被用来操作test.dat文件。完成操作后，应调用fclose(fp)来关闭文件并释放资源。确保在使用文件后总是调用fclose，以避免资源泄露。

```
FILE *fp;
fp=fopen("c:\\n\\b\\test.dat", "r");
if(fp==NULL)
{
    //文件打开失败，进行错误处理
    printf("文件打开失败，请检查文件路径和权限。\n");
    //可以在这里选择退出程序或进行其他错误处理
    …
}
//如果文件成功打开，则可以继续进行文件读取等操作
fclose(fp);                              //文件使用完毕后，记得关闭文件
```

Identification and Resolution of Common Problems 常见问题识别及解决方案

(1) When a file is opened using "r" mode, the file must exist beforehand, and only reading data from the file is supported.

(2) When a file is opened in "w" mode, only writing content to the file is permitted. If the file does not exist, the system will automatically create a new file with the specified file name. If the file already exists, the original contents will be overwritten and the file will be rebuilt. When an existing file is opened in write or read/write mode, the original content will be deleted.

(3) In order to add new information to an existing file, you should open the file in "a" mode. It is important to note that when using append mode, the file must already exist; otherwise, it will result in an error.

（1）当使用"r"模式打开一个文件时，该文件必须事先存在，且仅支持从该文件中读取数据。

（2）使用"w"模式打开文件时，仅允许向该文件写入内容。如果文件不存在，系统会自动以指定的文件名创建新文件；若文件已存在，则会覆盖原有内容并重建文件。以写或读写模式打开已存在的文件时原有内容将被清除。

（3）若要向已存在的文件追加新的信息，应使用"a"模式打开文件。但请注意，在使用追加模式时，文件必须事先存在，否则将会导致错误。

9.2 File Read and Write Operations 文件的读写操作

9.2节视频

Once the file is opened, it allows for read and write operations to facilitate data inflow and outflow. The process of reading and writing involves utilizing buffers through which

文件打开后，我们可进行读写操作，实现数据的流入与流出。读写过程涉及缓冲区，其中的数据流动称为数据流。本节

the flow of data is referred to as a data stream. This section is dedicated to the discussion of reading and writing files, providing examples to enhance comprehension.

重点讲解文件的读写操作，并通过实例加深理解。

9.2.1 File Read and Write Functions（文件读写函数）

The functions “fputc” and “fgetc” are utilized in the C programming language for file manipulation. They are employed to write a character to a file and read a character from a file, respectively. These two functions are widely used in file I/O(input/output) programming, particularly when dealing with text files.

fputc和fgetc是C语言中用于文件操作的函数，分别用于向文件写入一个字符和从文件读取一个字符。这两个函数在文件I/O（输入/输出）编程中非常常用，特别是在处理文本文件时。

1. fputc Function（fputc函数）

The “fputc” function is utilized for the purpose of writing a character to a specified file. The function prototype is as follows:

fputc函数用于将一个字符写入指定的文件中。其函数原型如下：

```
int fputc(int ch, FILE *stream);
```

Parameter meaning: The first parameter, “ch” indicates the character to be written to the file. When passed to the function, this argument is converted to an unsigned char and written to the file. The second parameter, “stream” is a file pointer to the target file, which must be a valid file pointer and must already be open in write mode(such as "w" or "a") or read/write mode(such as "r+").

参数含义：其中第1个参数ch表示要写入文件的字符。这个参数在传递给函数时会被转换为unsigned char，然后写入文件。第2个参数stream为指向目标文件的文件指针，它必须是一个有效的文件指针，并且该文件必须已经以写入模式（如"w"或"a"）或读写模式（如"r+"）打开。

Return value: If the function is successful, it will return the character that was written. In case of an error, EOF(end of file, typically defined as -1) will be returned.

返回值：如果函数执行成功，它返回写入的字符。如果发生错误，则返回EOF（文件结束符，通常被定义为-1）。

Example（使用示例）：

```
FILE *fp=fopen("example.txt", "w");
if(fp != NULL)
{
    char ch='A';
    fputc(ch, fp);                    //将字符'A'写入文件
    fclose(fp);                       //关闭文件
}
```

2. fgetc Function（fgetc）函数

The “fgetc” function is utilized for reading a character from the specified file. The function prototype is as follows:

fgetc函数用于从指定的文件中读取一个字符。其函数原型如下：

```
int fgetc(FILE *stream);
```

Function argument: “stream” is a file pointer to the source file, and it must be a valid file pointer. Additionally, the file must already be open in read mode(e.g., "r") or read/write mode(e.g., "r+" or "a+").

函数参数：stream为指向源文件的文件指针，它必须是一个有效的文件指针，并且该文件必须已经以读取模式（如"r"）或读写模式（如"r+"或"a+"）打开。

Return value: If the function succeeds, it returns the read character. If the end of file(EOF) is reached or an error occurs, the EOF(end of file character, usually defined as -1) is returned.

返回值：如果函数执行成功，它返回读取的字符。如果读取到文件结束（EOF）或发生错误，则返回EOF（文件结束符，通常被定义为-1）。

For example（示例）：

```
FILE *fp=fopen("example.txt", "r");
if(fp != NULL)
{
    int ch=fgetc(fp);                              //从文件读取一个字符
    if(ch != EOF)
    {
        printf("Read character: %c\n", ch);
    }
    fclose(fp);                                    //关闭文件
}
```

By utilizing the functions “fputc” and “fgetc”, text content can be written and read character by character, providing flexible capabilities for text processing. These two functions are commonly used in conjunction when working with files, with one function used for writing data and the other for reading data. They enable developers to perform precise manipulations on files at a character level, which is particularly advantageous when working with text files that consist of sequences of characters.

使用fputc和fgetc可以逐字符地写入和读取文本内容，从而实现灵活的文本处理功能。这两个函数在处理文件时通常一起使用，一个用于写入数据，另一个用于读取数据。它们允许开发者以字符为单位对文件进行精确的操作，这在处理文本文件时特别有用，因为文本文件本质上是由字符序列组成的。

9.2.2 Application Examples（应用举例）

The following will consist of a series of selected cases, providing in-depth analysis of text file read and write operations, aimed at enhancing understanding and mastery of the document processing process. The purpose of these cases is to assist readers in gaining a more intuitive comprehension of text file reading and writing methods, enabling them to flexibly apply this knowledge in practical applications to enhance the efficiency and accuracy of document processing.

下面将通过一系列精选案例，深入剖析文本文件的读写操作，以增进对文件处理流程的理解与掌握。这些案例旨在帮助读者更直观地了解文本文件的读入读出方法，从而在实际应用中能够灵活运用，提升文件处理的效率和准确性。

Case 9_1（案例 9_1）

Type characters through the keyboard and save them one by one to the disk file until '#'.

通过键盘输入字符，并将其逐个保存到磁盘文件中，直到输入'#'为止。

Thinking（解题思路）：

This example illustrates the process of opening a file in write-only mode, reading characters from the keyboard, and saving them to the file one by one. The input of '#' signifies the termination of character input and saving. Initially, the file needs to be opened in preparation for writing operations. Subsequently, the getchar function is utilized to read a character

本案例演示了如何使用只写模式打开文件，从键盘读取字符，并将其逐个保存到文件中。当输入'#'时，停止字符的输入和保存。首先，我们需要打开文件以准备写入操作。随后，利用getchar函数从键

from the keyboard and determine if it is '#'. If not, the fputc function is employed to write the character to the file. This sequence continues until '#' is encountered, at which point program execution ceases. Finally, upon completion of writing all characters, the fclose function is used to close the file and release related resources.

盘读取一个字符，并检查该字符是否为'#'。如果不是，我们就使用fputc函数将该字符写入文件中。这个过程将持续进行，直到读取到'#'为止，此时我们将停止程序，最后在完成所有字符的写入后，使用fclose函数关闭文件，释放相关资源。

Program Code（程序代码）：

Case9_1.c

```
1    #include <stdio.h>
2    #include <stdlib.h>
3    int main()
4    {
5        FILE *fp;                                  // 定义文件指针变量
6        char ch, *filename="out.txt";
7        if((fp=fopen(filename,"w"))==NULL)         // 打开文件同时判断是否失败
       {
8            printf("Cannot open file\n");
9            exit(0);                               // 若打开失败退出程序
10       }
11       printf("Please input string:");
12       ch=getchar();                              // 输入一个字符给 ch
13       while(ch!='#')
14       {
15           fputc(ch,fp);                          // 向 fp 指向的文件输入 ch
16           putchar(ch);                           // 将 ch 输出到显示器上查看
17           ch=getchar();                          // 输入一个字符给 ch
18       }
19       fclose(fp);                                // 关闭文件
20       return 0;
21   }
```

Program Output（运行结果）：

The program output is shoun in Figure 9-4.

运行结果如图9-4所示。

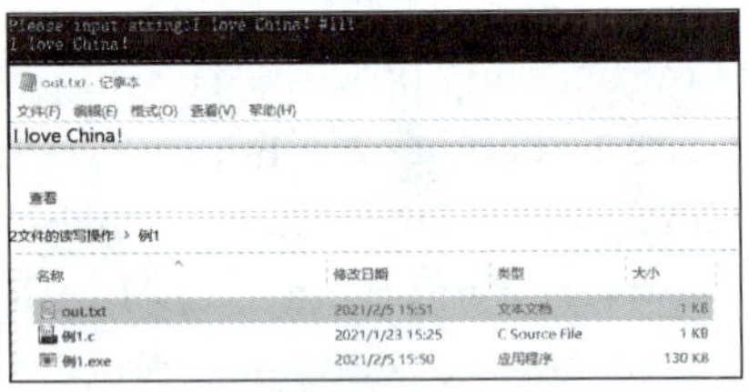

图9-4 程序运行结果（含文件内容显示）

Case Analysis（案例解析）：

(1) Line 5 defines the file pointer variable “fp”, and the file type is File.

(2) Line 6 defines a character variable “ch”, and a pointer variable “filename”, which points to a constant string “out. txt”.

(3) If statement from line 7 to line 11. First, the fopen(filename, "w") function is called to open the file named “out.txt” in write-only mode. If the file does not exist, a new file named out.txt will be created. If the file already exists, it will be cleared. The return value of "fopen" function is assigned to the file pointer "fp" and checked for null. If "fp" is null, the if statement’s condition is met and the code inside the if statement block is executed to report an error and terminate program execution.

（1）第5行定义为文件指针变量fp，文件类型为File。

（2）第6行定义了一个字符型变量ch，以及一个指针变量filename，该变量指向了一个常量字符串"out.txt"。

（3）在第7~11行为一条if语句。首先调用fopen(filen-ame, "w")函数，以只写模式打开名为“out.txt”的文件，如果该文件不存在，系统将创建一个名为out.txt的文件，若文件已存在，系统将清空该文件。fopen的返回值赋给了文件指针fp，并检查它是否为空。如果fp为空if语句的条件成立，随后将执行if语句块内的代码，进行错误报告并终止程序执行。

(4)The line 13, the "getchar" function is used to read a character from the keyboard and assign it to the variable "ch". Lines 14-19 contain a while loop that calls the fputc function to write the character ch to the file pointed by fp if ch is not equal to '#'. At the same time, "getchar" function is used to display the character ch on the screen for visual inspection of the result. Finally, another character is read in using "getchar" function, and this process continues in a loop until the termination symbol '#' is encountered, at which point the loop terminates.

（4）第13行使用getchar函数从键盘读入一个字符赋值给ch变量。第14~19行是一个while循环，如果ch不等于'#'则调用fputc函数向fp指向的文件写入字符ch，同时用putchar函数将字符ch显示到屏幕上，用于直观查看结果。最后再次用getchar函数读入一个字符，不断循环直到遇到结束符'#'，此时终止循环。

Lines 13-19 can be simplified and rewritten as:

第13~19行，可以简化改写为：

```
while((ch=getchar())!='#')
{
    fputc(ch,fp);                          //向fp指向的文件输入ch
    putchar(ch);                           //将ch输出到显示器上查看
}
```

(5) The "fclose" function call on line 20 is a crucial step in properly closing the opened file. This ensures that the file is closed after the completion of file operations, thereby preventing misuse and data loss. It is important to always remember to close the file as soon as it is finished.

（5）第20行的fclose函数调用是关闭已打开文件的关键步骤，确保文件操作完成后正确关闭文件，防止误用和数据丢失。务必记得在文件使用完毕后及时调用。

Case 9_2（案例 9_2）

Read the contents of the text file "out. Txt" in Case 9_1 and display them on the operation interface.

读取案例9_1中文本文件"out.txt"的内容，并显示在运行界面上。

Thinking（解题思路）:

After opening the file and testing whether it opened successfully, we utilize the "fgetc" function to sequentially read characters from the file. Following each read, we check for the end of file flag EOF. If EOF is not encountered, it indicates that the end of the file has not been reached. Simultaneously, we employ the "putchar" function to display the read characters on screen in real time, allowing users to observe the file content as it is being processed. This process continues until the "fgetc" function returns EOF, signifying that the entire file has been completely read.

打开文件并测试是否成功打开后，我们使用fgetc函数从文件中逐个读取字符。在每次读取后，检查所读取的字符是否为文件结束标志EOF。如果读取的字符不是EOF，那么这表示尚未到达文件的末尾。同时利用putchar函数将读取到的字符输出到屏幕上，以便用户能够实时查看文件内容。这个过程将循环进行，直到fgetc函数返回EOF，标志着整个文件已经被完全读取。

Program Code（程序代码）:

Case9_2.c

```
#include <stdio.h>
#include <stdlib.h>
int main()
{
    FILE *fp;
    char ch,*filename="out.txt";
    fp=fopen(filename,"r");                    //以只读方式打开文件
    if(fp==NULL)
```

```
9       {
10          printf("Cannot open file\n");
11          exit(0);                              //若打开失败则退出程序
12      }
13      while((ch=fgetc(fp))!=EOF)                //从文件读取 ch，若不到达文件末尾则循环
14      putchar(ch);                              //输出 ch
15      fclose(fp);                               //关闭文件
16      return 0;
17  }
```

Program Output（运行结果）:

The program output is shoun in Figure 9-5.

Case Analysis（案例解析）:

(1) The program executes line 7(fp=fopen(filename, "r");), which calls the “fopen” function to attempt to open a file named “out.txt” in read-only mode. The variable “filename” is a character pointer that points to the constant string “out.txt”. The parameter "r" indicates that the file will be opened in read-only mode, requiring the file to exist beforehand. If the file is successfully opened, the “fopen” function returns a pointer to the file assigned to the variable "fp". If opening fails, NULL is returned.

(2) Lines 8 to 12 are used to check whether the file was opened successfully. If the value of "fp" is NULL, the file failed to open, and the program will execute exit(0) to terminate immediately.

(3) Line 13-14 is a while loop, and its condition is(ch = fgetc(fp))! = EOF. First, the “fgetc” function is executed to read characters one by one from the “out.txt” file pointed to by “fp”. The characters are then assigned to the variable “ch”, and a check is performed to see if ch is equal to the end-of-file flag EOF. If it is not, this indicates that the end of the file has not been reached, and “putchar(ch)” will be executed in a loop to output the read character “ch” to the display. The loop ends when the end of the file is reached, i.e., when ch equals EOF.

(4) After using the file, it is important to ensure data integrity and security by using “fclose(fp)” function to close the file, in order to prevent data loss and misoperation.

运行结果如图9-5所示。

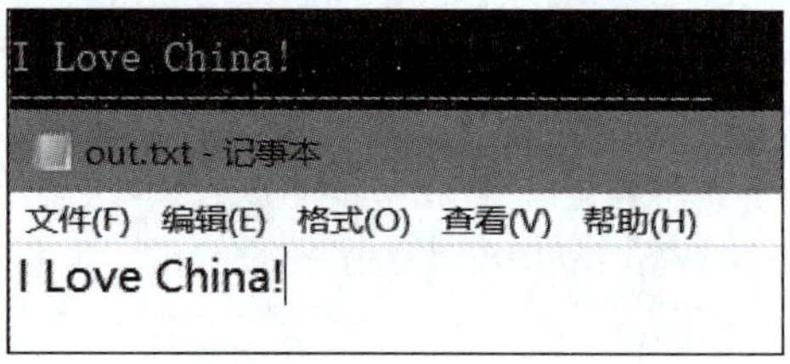

图9-5　程序运行结果（含文字内容显示）

（1）在执行第7行fp=fopen(filename, "r"); 时，程序调用了fopen函数尝试以只读模式打开一个名为“out.txt”的文件。filename是一个字符指针变量，它指向的是常量字符串“out.txt”。"r"参数表示文件将以只读模式被打开，这要求文件必须事先存在。如果文件成功打开，fopen函数将返回一个文件指针，该指针被赋值给变量fp；如果打开失败，则返回NULL。

（2）第8~12行用于检查文件是否成功打开。如果fp的值为NULL，表示文件打开失败，此时程序将执行exit(0)立即终止。

（3）第13~14行是一个while循环，其条件为(ch = fgetc(fp)) != EOF。首先执行fgetc函数，从fp指向的“out.txt”文件中逐个读取字符，并将其赋值给变量ch，随后检查ch是否等于文件结束标志EOF。如果不等，说明尚未到达文件末尾，此时将执行循环体内的putchar(ch);语句，将读取到的字符ch输出到显示器。当读取到文件末尾，即ch等于EOF时，循环将结束。

（4）在使用完文件后，为了防止数据丢失和误操作，应使用fclose(fp)来关闭文件，这是确保数据完整性和安全性的重要步骤。

Case 9_3（案例 9_3）

Realize the file copy function, that is, copy the contents of a text file to another text file.

实现文件复制功能，即将一个文本文件的内容复制到另一个文本文件中。

Thinking（解题思路）:

In order to copy the contents of file 1 to file 2, it is essential to first confirm the existence of file 1. Then, proceed to open file 1 in read-only mode and create or open file 2 in write-only mode. Subsequently, characters are read one by one from file 1 using the fgetc function within a loop, with each character checked for the end-of-file flag EOF. As long as the character being read is not EOF(i.e., has not reached the end of the file), utilize the fputc function to write the character to file 2. This process continues until EOF is encountered, signifying completion of the file copy operation. Through this meticulous character-by-character copying approach, assurance is provided that all content from file 1 is accurately replicated in file 2. Finally, remember to close both files using the fclose function.

为了实现将文件1的内容复制到文件2的功能，我们首先需要确保文件1已经存在。然后，以只读模式打开文件1，并以只写模式创建或打开文件2。接下来，我们将通过循环调用fgetc函数从文件1中逐个读取字符，检查每个字符是否为文件结束标志EOF。只要读取的字符不是EOF，即表示还未到达文件末尾，我们就使用fputc函数将该字符写入文件2。这一过程将持续进行，直到读取到EOF为止，此时文件复制操作即告完成。通过这种逐字符的复制方式，我们可以确保文件1的所有内容都被精确地复制到文件2中。最后不要忘记使用fclose函数关闭两个文件。

Program Code（程序代码）:

Case9_3.c

```
1    #include <stdio.h>
2    #include <stdlib.h>
3    int main()
4    {
5        FILE *in, *out;
6        char ch,infile[10],outfile[10];
7        scanf("%s",infile);
8        scanf("%s",outfile);
9        if((in=fopen(infile, "r"))== NULL)
10       {
11           printf("Cannot open infile.\n");
12           exit(0);
13       }
14       if((out=fopen(outfile, "w"))== NULL)
15       {
16           printf("Cannot open outfile.\n");
17           exit(0);
18       }
19       while((ch=fgetc(in))!=EOF)
20           fputc(ch, out);
21       fclose(in);
22       fclose(out);
23       return 0;
24   }
```

Program Output（运行结果）:

The program output is shoun in Figure 9-6.

运行结果如图9-6所示。

Case Analysis（案例解析）:

(1) Line 5 defines two file pointer variables, “in” and “out”. When processing a file, it is necessary to first define a file pointer variable in order to subsequently manipulate the file.

(2) Line 6 defines a variable “ch”, to store a single character

（1）第5行定义了两个文件指针变量：in和out。在处理文件时，首先需要定义文件指针变量，以便后续操作文件。

（2）第6行定义了一个变量ch，用于存储从文件中读取的单个字符。同时，定义了

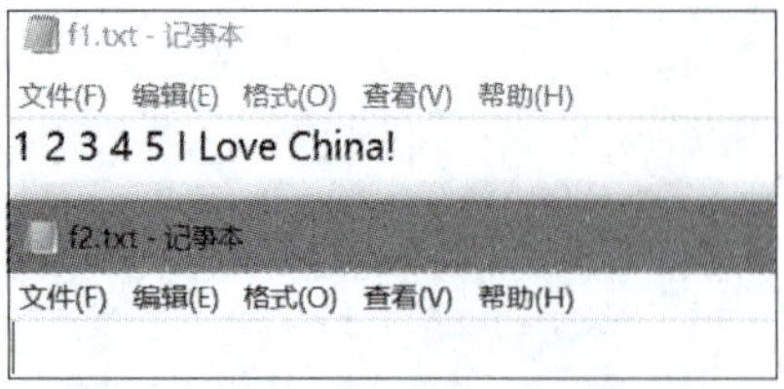

（a）程序运行前

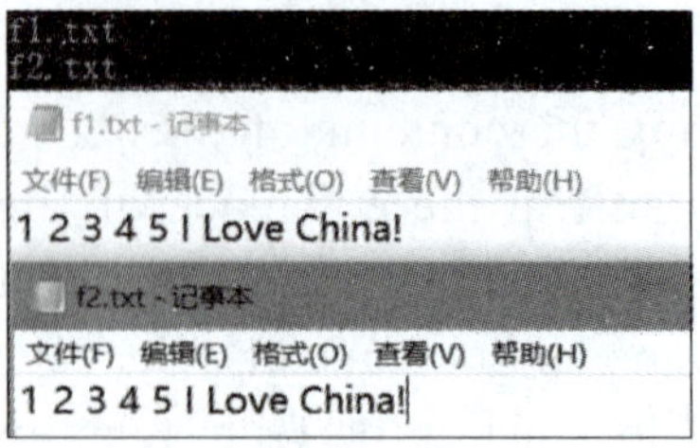

（b）程序运行后

图9-6 程序运行结果（含文件内容显示）

read from the file. Simultaneously, two one-dimensional arrays of char type “infile” and “outfile” are defined. These arrays will be utilized to store the two file names entered by the user from the keyboard.

(3)Lines 7 and 8 prompt the user to input two strings as file names, which are then stored in the “infile” and “outfile” character arrays for reading and writing purposes, respectively.

(4)Lines 9-13 attempt to open the file specified by “infile” in read-only mode and verify whether the file was successfully opened.

(5)Lines 14-18 attempt to open or create the file specified by “outfile” in write-only mode and verify whether the file was successfully opened for writing.

(6)Lines 19-26 contain a while loop that continuously reads a character from the “infile” file into the variable “ch” and checks if “ch” is the end-of-file flag EOF. As long as the end of the file is not reached, the “ch” is written to the “outfile” file. The loop ends when the end-of-file flag EOF is encountered, at which point the file copy function is complete.

(7)After completing the file operation, it is important to ensure that the files are closed one by one, as demonstrated in lines 27 and 28, in order to properly release resources.

两个char型的一维数组infile和outfile，它们将用于存储用户从键盘输入的两个文件名。

（3）第7行和第8行通过键盘输入获取了两个字符串，分别作为要读取和写入的文件名，并将它们存储在infile和outfile字符数组中。

（4）第9~13行尝试以只读方式打开infile指定的文件，并检查文件是否成功打开。

（5）第14~18行尝试以只写方式打开或创建outfile指定的文件，并检查文件是否成功打开以供写入。

（6）第19~26行包含一个while循环，该循环不断从infile文件中读取一个字符到变量ch中，并检查ch是否为文件结束标志EOF。只要没有到达文件末尾，就将ch写入到outfile文件中。当遇到文件结束标志EOF时，循环结束，此时文件复制功能已完成。

（7）在完成文件操作后，务必逐个关闭文件，如第27行和第28行所示，以确保资源得到正确释放。

9.3 Comprehensive Cases 综合案例

综合案例 9_3_1 文件小写字符转大写字符。

案例描述：将文件file1中的所有字符内容转换为大写，并将转换后的内容保存到file2文件中。

解题思路：本案例通过两个函数f1和f2实现模块化编程。

函数f1：从键盘读取字符串并写入file1.txt文件。

（1）以只写模式打开file1.txt文件，并利用if语句验证文件是否成功打开。若打开失败，则输出错误信息。

（2）使用fgets函数（而非gets，因为gets已被废弃，存在安全风险）从键盘安全地读取一个字符串，并通过fputs函数将该字符串写入file1.txt文件中。

（3）关闭文件以确保数据完整性和资源释放。

函数f2：从file1.txt读取字符串，将小写字母转换为大写，并保存到file2.txt中。

（1）同时以只读模式打开file1.txt和以只写模式打开file2.txt。利用if语句检查两个文件是否均成功打开。如果任一文件打开失败，则输出错误信息。

（2）使用fgets函数从file1.txt中逐行读取字符串。对于读取到的每一行，通过遍历字符串，将其中的小写字母转换为大写字母。随后，使用fputs函数将转换后的字符串写入file2.txt文件中。

（3）关闭所有打开的文件，以确保资源得到妥善释放。

程序代码：

Case9_3_1.c

```c
#include <stdio.h>
#include <stdlib.h>
void f1()                                         //从键盘读入一行字符写入 file1.txt 文件
{
    FILE *fp;
    char str[100];
    if((fp=fopen("file1.txt","w"))==NULL)
    {
        printf("文件打开失败！\n");
        exit(0);
    }
    printf("请输入一个字符串:\n");
    gets(str);
    fputs(str,fp);                                //将 str 数组中字符串写入文件中
    fclose(fp);
}
void f2()                                  //从 file1.txt 文件读数据转换成大写存入 file2.txt
{
    FILE *fp1,*fp2;
    char str[100];
    int i=0;
    if((fp1=fopen("file1.txt","r"))==NULL)
    {
        printf("文件打开失败！\n");
        exit(0);
    }
    if((fp2=fopen("file2.txt","w"))==NULL)
    {
        printf("文件打开失败！\n");
        exit(0);
    }
    fgets(str,99,fp1);//fgetc
    while(str[i]!='\0')
    {
        if(str[i]>='a'&&str[i]<='z')
        {
            str[i]=str[i]-32;          //小写转换成大写
        }
        i++;
    }
    fputs(str,fp2);
    puts(str);
    fclose(fp1);
    fclose(fp2);
}
int main()
{   //主函数功能是调用两个函数
    f1();
```

```
49        f2();
50    }
```

运行结果：

运行结果如图9-7所示。

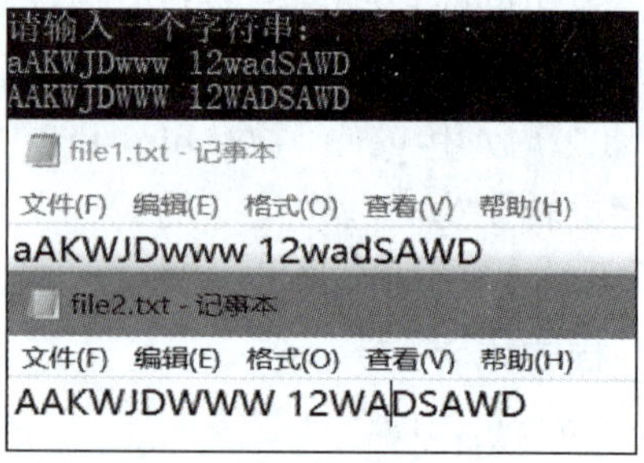

图9-7　程序运行结果（含文件内容显示）

案例解析：

（1）第3行是f1函数的定义，该函数不需要返回值和参数。

（2）f1函数内部执行以下关键步骤：定义文件指针变量，打开文件，从键盘读入字符串并写入文件，最后关闭文件。

（3）第5行定义了一个文件指针变量fp，第6行定义了一个char型一维数组str，用于存储从键盘输入的字符串数据。

（4）第7~11行尝试以只写模式打开file1.txt文件，并检查是否成功打开。如果打开失败，则退出f1函数。注意，这里的exit(0);可以替换为return;，以仅退出f1函数而非整个程序。

（5）第13行的gets函数用于从键盘读取字符串，并将其存储在str数组中。

（6）第14行使用fputs函数将str数组中的字符串写入fp指向的file1.txt文件。注意区分fputs和fputc函数：fputs写入一行字符串，以'\0'为结束标志；fputc每次写入一个字符。

（7）第15行关闭fp指向的file1.txt文件，结束f1函数。

（8）第17行是f2函数的定义，该函数同样不需要返回值和参数。

（9）第19行定义了两个文件指针变量，分别用于指向file1.txt和file2.txt文件。

（10）第20行定义了一个str字符串数组，用于存储从file1.txt读取的一行字符串。

（11）第22~26行尝试以只读模式打开file1.txt文件，并检查是否成功打开。

（12）第27~31行尝试以只写模式打开file2.txt文件，并检查是否成功打开。

（13）第32行描述了fgets函数的作用和参数。fgets从指定文件中读取一行字符串，最多读取99个字符（因为str数组长度为100，需要为'\0'结束符预留空间）。

（14）第33~40行描述了一个while循环，用于遍历str数组中的每个字符。如果字符是小写字母（ASCII码在'a'和'z'之间），则将其转换为大写字母（通过减去32）。循环继续直到遇到字符串结束符'\0'。

（15）第41行使用fputs函数将处理后的str数组中的字符串写入fp2指向的file2.txt文件。

（16）第42行调用puts函数将str数组中的字符串输出到屏幕上。

（17）第43、44行分别关闭file1.txt和file2.txt文件，结束f2函数。

案例拓展：字符串文件读写函数

fgets和fputs是C语言中用于文件I/O（输入/输出）的标准库函数。它们分别用于从文件中读取一行文本和将字符串写入文件。

（1）fgets函数。fgets函数用于从指定的文件流中读取一行数据，并将其存储在提供的字符数组中。函数的原型如下：

```
char *fgets(char *str, int n, FILE *stream);
```

str：指向一个字符数组的指针，用于存储读取的字符串。

n：要读取的最大字符数（包括空字符'\0'）。通常这个数值应该是目标数组的大小。

stream：指向FILE对象的指针，指定了读取操作的文件流。

fgets会读取直到遇到换行符（'\n'）、EOF（文件结束标志）或者读取了n-1个字符为止。读取的字符串将以空字符（'\0'）结尾，并存储在str指向的数组中。

说明：由于gets函数被认为是不安全的，因为它不检查目标缓冲区的大小，很容易导致缓冲区溢出。因此，常常建议使用fgets函数来代替gets。如综合案例9_3_1中的第13行gets(str);可以改为：fgets(str, sizeof(str), stdin)。

（2）fputs函数。fputs函数用于将一个字符串写入到指定的文件流中。函数的原型如下：

```
int fputs(const char *str, FILE *stream);
```

str：指向要写入的字符串的指针。

stream：指向FILE对象的指针，指定了写入操作的文件流。

fputs函数会将str指向的字符串写入到stream指定的文件流中，直到遇到字符串的结束符'\0'。函数返回非负值表示成功，EOF表示出错。

下面是一个简单的示例，展示了如何使用fgets从文件中读取一行并使用fputs将其写入另一个文件中：

```
#include <stdio.h>
int main()
{
    FILE *inputFile=fopen("input.txt", "r");
    FILE *outputFile=fopen("output.txt", "w");
    if(inputFile==NULL || outputFile==NULL) {
        cout("Error opening file");
        return 1;                          //结束函数
    }
    char line[1024];                       //假设一行不超过1023个字符
    while(fgets(line,sizeof(line),inputFile)!=NULL)
    {
        fputs(line, outputFile);
    }
    fclose(inputFile);
    fclose(outputFile);
    return 0;
}
```

在这个示例中，程序首先打开名为input.txt的文件进行读取，并打开（或创建）名为output.txt的文件进行写入。然后，它使用fgets从inputFile中读取一行文本，并使用fputs将该行写入到outputFile中。这个过程会一直持续到inputFile中没有更多的行为止。最后，程序关闭两个文件并退出。

综合案例 9_3_2 员工信息建档与查找。

案例描述：设计一个程序，用以执行以下任务：

（1）将员工信息（职工号、姓名、年龄和工资）保存到磁盘文件中。

（2）向文件中新增员工信息。

（3）在屏幕上展示文件中存储的所有员工信息。

（4）根据输入的职工号，检索并显示对应员工的信息；若无对应员工，则给出提示。

解题思路：我们将员工信息定义为结构体，利用文件I/O函数（如fwrite和fread）以二进制形式读写这些结构体数据。首先，定义员工结构体类型STAFF，包含职工号num、姓名name、年龄age、工资pay等成员。随后将案例要求的4个功能，写成4个函数。

实现输入数据的input函数：首先，以只写模式打开一个名为f1.txt的二进制文本文件。接着，在for循环中，利用scanf函数从键盘逐个读取员工数据，并将这些数据分别赋值给结构体中的各个成员变量。然后，使用fwrite函数，以数据块的形式将这些结构体数据写入到f1.txt这个二进制文件中。通过重复这个过程n次，就可以将n个员工的数据完整地保存到f1.txt文件中。

实现追加员工信息的add函数：以追加工作方式打开二进制文件f1.txt。写入数据的方式与input函数相同。

实现输出所有员工信息的display函数：以只读模式打开二进制文本文件f1.txt。使用fread函数以数据块的方式读取文件内容，该函数在成功读取数据时会返回读取到的数据项个数，若读取失败或到达文件末尾则返回0。在while循环中，调用fread函数来读取数据项，并将读取到的数据存储在一个结构体变量中。每当fread函数的返回值非零时，即表示成功读取了一个数据项，就输出该结构体中的数据信息。一旦fread函数返回0，就意味着文件已经读取完毕，此时将结束循环。

实现查找员工信息的find函数：按照display函数读取数据块的方式遍历f1.txt文件中的所有数据项，同时检查每一个数据项中的员工编号，以确认是否存在与给定编号相匹配的记录。一旦发现匹配的员工编号，将输出该结构体数据项中包含的详细信息。如果在遍历完整个文件后仍未找到对应的员工数据，则向用户反馈“查无此人”的消息。

程序代码：

Case9_3_2.c

```
#include <stdio.h>
#include <stdlib.h>
#include <string.h>
typedef struct
{
    int num;
    char name[20];
    int age;
    float pay;
}STAFF;
void input(int n)                    //将n个员工的数据写入磁盘f1.txt文件中保存
{
    FILE *fp;
    STAFF st;
    int i;
    if((fp=fopen("f1.txt","wb"))==NULL)
    {
        printf("文件打开失败！\n");
        exit(0);
    }
    for(i=1;i<=n;i++)
    {
        printf("从键盘读入第%d个员工的数据(职工号、姓名、年龄、工资):\n",i);
        scanf("%d%s%d%f",&st.num,st.name,&st.age,&st.pay);
        if(fwrite(&st,sizeof(STAFF),1,fp)!=1)
            printf("写入文件失败！\n");
    }
    fclose(fp);                      //关闭文件
}
void display()                       //将文件内容输出到显示器上查看
{
    FILE *fp;
    STAFF st;
    if((fp=fopen("f1.txt","rb"))==NULL)
    {
        printf("文件打开失败！\n");
        exit(0);
    }
    printf("\n输出所有职工信息:\n");
    printf("职工号 姓名   年龄    工资\n");
    while(fread(&st,sizeof(STAFF),1,fp))
    {
        printf("%-7d%-7s%-7d%-8.2f\n",st.num,st.name,st.age,st.pay);
    }
    fclose(fp);                      //关闭文件
}
void add(int n)                      //增加新员工信息到文件的末尾
{
```

```
    int i;
    FILE *fp;
    STAFF st;
    if((fp=fopen("f1.txt","ab"))==NULL)
    {
        printf(" 文件打开失败！ \n");
        exit(0);
    }
    for(i=1;i<=n;i++)
    {
        printf(" 从键盘读入第 %d 个增加的新员工的数据 ( 职工号、姓名、年龄、工资 ):\n",i);
        scanf("%d%s%d%f",&st.num,st.name,&st.age,&st.pay);
        if(fwrite(&st,sizeof(STAFF),1,fp)!=1)
            printf(" 写入文件失败！ \n");
    }
    fclose(fp);                                 // 关闭文件
}
void find(int n)                                // 根据职工号查找员工
{
    FILE *fp;
    STAFF st;
    int flag=1;
    if((fp=fopen("f1.txt","rb"))==NULL)
    {
        printf(" 文件打开失败！ \n");
        exit(0);
    }
    while(fread(&st,sizeof(STAFF),1,fp))
    {
        if(st.num==n)
        {
            flag=0;
            printf("\n 查找的职工号 %d 的信息为 :\n",n);
            printf(" 职工号 :%d\n 姓名 :%s\n 年龄:%d\n 工资:%.2f\n",st.num,st.name,st.age,st.pay);
        }
    }
    if(flag) printf("\n 查无此人！ \n");
    fclose(fp);                                 // 关闭文件
}
int main()
{
    int m,n,num;
    printf(" 请输入要建档的员工信息数:\n");
    scanf("%d",&m);
    input(m);
    display();
    printf("\n 请输入要追加的员工信息数:\n");
    scanf("%d",&n);
    add(n);
    display();
    while(1)
    {
        printf("\n 请输入要查找的职工号:\n");
        scanf("%d",&num);
        find(num);
    }
    return 0;
}
```

运行结果:

```
请输入要建档的员工信息数:
2
```

```
从键盘读入第 1 个员工的数据（职工号、姓名、年龄、工资）：
101 张三 20 3400
从键盘读入第 2 个员工的数据（职工号、姓名、年龄、工资）：
102 李四 40 8000

输出所有职工信息：
职工号   姓名   年龄   工资
101      张三    20    3400.00
102      李四    40    8000.00

请输入要追加的员工信息数：
1
从键盘读入第 1 个增加的新员工的数据（职工号、姓名、年龄、工资）：
103 王五 34 5800

输出所有职工信息：
职工号   姓名    年龄    工资
101      张三    20      3400.00
102      李四    40      8000.00
103      王五    34      5800.00

请输入要查找的职工号：
102

查找的职工号 102 的信息为：
职工号：102
姓名：李四
年龄：40
工资：8000.00

请输入要查找的职工号：
111

查无此人！
```

案例解析：第4~10行是一个结构体的声明。定义了员工结构体，其中包含职工号num、姓名name、年龄age和工资pay等成员。为了方便使用，我们使用了typedef将这个结构体类型重命名为STAFF，从而创建了一个无名的结构体类型。

input 函数解读

（1）第11行为input函数首部。不接收返回值，它接收一个整型变量n作为参数。该函数的主要作用是创建f1.txt文件，并将n个员工的数据写入该文件并保存在磁盘上。

（2）第13~14行定义指针变量和结构体变量。操作文件需要先声明一个文件指针变量fp，为了存储读入的员工信息，还需定义一个STAFF类型的结构体变量st。

（3）第16~20行以二进制写模式打开f1.txt文件，并判断是否打开成功。注意本案例处理的是二进制文件。

（4）第24行使用scanf函数从键盘读取数据，并将其存储到结构体变量st的各个成员中。注意，访问结构体成员时，例如访问name成员应使用st.name。由于name是字符数组，数组名即代表其地址，所以st.name前不需要加取地址符&。而对于其他变量，需要在前面加上&符号。此外，格式字符串中的格式控制符必须与成员的数量和类型相匹配。

（5）第25、26行将st以数据块形式写入f1.txt文件。fwrite函数以数据块的形式将数据写入二进制文件。

①fwrite函数有4个参数：第1个参数指定要写入文件的数据块的起始地址；第2个参数指定数据块的字节长度；第3个参数指定要写入的数据项数量；第4个参数指定目标文件。

②fwrite(&st, sizeof(STAFF), 1, fp);语句在fp指向的文件中写入一个数据项，其起始地址为&st，字节长度为sizeof(STAFF)。如果fwrite函数成功写入，将返回1。如果返回值不等于1，则表示写入失败，此时将执

行if语句内的代码，输出“写入文件失败”。

③通过for循环可以将n个员工的信息写入文件。

（6）第28行关闭已打开的文件。

display 函数解读

（1）第30行为display函数的定义开始，该函数没有返回值，也不需要传递任何参数。它的主要功能是将文件中存储的所有员工信息输出到屏幕上。

（2）第32行声明了一个文件指针变量fp，用于在后续的操作中指向文件。第33行则定义了一个STAFF类型的结构体变量st，用于在程序中暂存员工的信息。

（3）第34~38行的代码以二进制只读模式（“rb”）打开名为f1.txt的文件，并检查文件是否成功打开。需要注意的是，以只读二进制模式打开文件时，该文件必须已经存在。

（4）第41~44行是一个while循环，它的功能是从文件中读取员工的信息，并将其显示在屏幕上。fread函数以数据块的形式读取数据，这里的数据块类型是员工结构体STAFF。fread函数与fwrite函数在功能上是对称的。

①fread(&st, sizeof(STAFF), 1, fp)的含义是从文件指针变量fp所指向的文件中，读取一个大小为sizeof(STAFF)的数据项，并将其存储在变量st的地址（&st）中。

②fread函数在成功读取数据后会返回读取的数据项数量，否则返回0。在while循环中，如果fread函数成功读取了数据（返回值为1），则循环条件成立，执行循环体内的语句，输出员工结构体的各个成员数据。

③输出格式中，%-7d表示输出结果占据7列且左对齐，%-8.2f表示输出结果保留两位小数，占据8列且左对齐，“-”代表左对齐。

④如果fread函数返回0，表示文件已经读取完毕，此时循环条件不成立，循环结束。

（5）第45行代码负责关闭已打开的文件。

add 函数解读

（1）第47行是add函数的开始，该函数没有返回值，它接收一个整型参数n，这个参数表示需要追加到文件中的员工信息的数量。

（2）第49行声明了循环变量i，第50行声明了文件指针变量fp，第51行则声明了结构体类型变量st。

（3）第52~56行的代码以二进制追加模式（"ab"）打开f1.txt文件，并检查文件是否成功打开。追加员工信息的操作与之前的input函数中的操作类似，但关键区别在于文件的打开方式为追加模式（"ab"），如果该文件已存在，文件指针将会放在文件的结尾，新的内容将会被写入到已有内容之后；如果该文件不存在，系统会创建新文件进行写入，此时，文件指针将会放在文件的开头。

（4）第57~64行的代码负责从键盘读取员工信息，并将其依次写入到f1.txt文件中。这一过程与input函数中的操作一致，因此不再详细叙述。

（5）第65行代码负责关闭已打开的文件。

find 函数解读

（1）第67行为find函数的定义起始行，它没有返回值，接收一个形参，即需要查找的员工职工号。每位员工都有一个唯一的职工号，这成为我们查找员工的重要依据。

（2）第71行声明了一个名为flag的标志变量，并将其初始值设为1。在查找过程中，若找到目标员工，则将flag置为0。若查找结束后flag仍为1，则表示员工名单中无此人。

（3）第72~76行，程序以二进制只读模式（"rb"）尝试打开f1.txt文件，并验证文件是否成功打开。

（4）第77~86行，程序从文件中逐个读取员工信息，并判断其职工号是否与待查找的职工号n相符。若匹配，则说明已找到该员工，将flag设为0，并输出员工信息。具体流程为：

①使用fread(&st, sizeof(STAFF), 1, fp)从文件中读取员工信息并存储在st结构体中。在while循环内，若

fread成功读取数据并返回1，则执行循环体内的操作。

②若读取到的员工信息中的职工号st.num与n相等，表示查找成功，此时将flag设为0，并输出st中的所有员工数据。

③若读取到文件末尾，fread将返回0，此时while循环条件不成立，循环终止。

（5）若查找结束后flag仍为1，则在第87行输出“查无此人！”的信息，表示员工名单中无此人。

（6）最后，在第88行关闭已打开的文件。

main 函数解读

（1）在第92行，我们定义了三个必要的变量：m、n和num。这些变量都需要通过键盘输入来获取值。其中，m代表将要建立的员工数量，n代表要追加到文件中的员工数量，而num则代表需要查找的特定员工的工号。

（2）第95行中，我们调用了input函数来创建m个员工的信息，并将这些信息写入f1.txt文件中。

（3）紧接着，在第96行，我们调用了display函数来展示f1.txt文件中所包含的所有员工信息。

（4）到了第99行，我们执行了add函数，这个函数的作用是将n个新员工的信息追加到f1.txt文件的末尾。

（5）为了验证追加操作是否成功，我们在第100行再次调用了display函数，以展示更新后的f1.txt文件中的所有员工信息。

（6）最后，在第101~106行之间，我们设置了一个循环，用于不断接收用户输入的职工号，并调用find函数进行查找。需要注意的是，这里的while(1)循环仅用于测试目的，在实际系统中应该设定一个明确的退出条件，以避免出现无限循环的情况。

案例总结：若打开由本案例生成的员工文件则会发现部分信息会显示为乱码。这是因为fwrite和fread函数是用于对二进制文件进行成块的读写操作。当以二进制方式处理文件时，除了字符型数据以ASCII码显示、可以在记事本中正常查看外，其他数值类型都是按照其在内存中的二进制格式进行存储的。如果尝试以ASCII码的形式去解释这些二进制数据，就会出现乱码的情况。为了确保数据的正确读写，并提高效率，建议使用fwrite函数进行写入操作后，相应地使用fread函数进行读取。这样可以避免程序出错，并确保数据的一致性。

Summary
小 结

This chapter primarily introduces the fundamental concepts and classifications of files, and elaborates on the distinctions between text files and binary files, as well as their respective advantages and disadvantages. The emphasis is placed on the proper methods for opening and closing files. When utilizing the fopen function, it is crucial to ensure the accuracy of the file path and opening method. After using a file, it is important to promptly close it in order to prevent resource leakage. For read operations, the file must already exist; otherwise, it will not open. For write operations, if the file does not exist, the system will automatically create a new file. If the file already exists, the system will clear its content before writing to it. Finally, through a comprehensive case study, we demonstrate the process of reading from and writing to a file with an aim to deepen understanding and strengthen practical application abilities.

本章节主要介绍了文件的基本概念和分类，详细阐述了文本文件与二进制文件之间的区别及其各自的优缺点。重点在于讲解文件的打开与关闭操作。在使用fopen函数时，务必确保文件路径和打开方式的准确性。使用完文件后，切记及时关闭以避免资源泄露。对于读取操作，文件必须事先存在，否则将无法打开；而对于写入操作，如果文件不存在，系统将自动创建一个新文件，若文件已存在，则会先清空内容再进行写入。最后，通过综合案例，我们演示了文件读写的操作流程，旨在加深理解并强化实际应用能力。

Exercise
习　　题

一、单选题

1. 若在使用fopen函数时发生错误，则函数的返回值是（　　）。

A. 0　　B. 1　　C. 地址值　　D. EOF

2. 若要用fopen函数打开一个不存在的文本文件，该文件要既能读也能写，则文件打开方式字符串应是（　　）。

A. "a+"　　B. "w+"　　C. "r+"　　D. "w"

3. fgetc函数的作用是从指定文件读入一个字符，下列文件打开方式中，错误的是（　　）。

A. 读　　B. 读写　　C. 追加　　D. 读或读写

4. 若要使用fopen函数打开C盘上子目录ABC下名为abc.txt的文本文件进行读、写操作，下列符合此要求的函数调用是（　　）。

A. fopen("C:\ABC\abc.txt","r")　　B. fopen("C:\\ABC\\abc.txt","r+")

C. fopen("C:\\ABC\\abc.txt","w")　　D. fopen("C:\ABC\abc.txt","w+")

5. 当前目录下存在文件file1.txt文件，执行函数fopen("file1.txt","w+")的功能是（　　）。

A. 打开 file1.txt 文件，清除全部内容并写入操作

B. 打开 file1.txt 文件，只能写入新的内容

C. 打开 file1.txt 文件，只能读取原有内容

D. 打开 file1.txt 文件，清除全部内容并先写后读操作

6. 有以下程序：

```
#include <stdio.h>
void write(char *f, char *s)
{
    FILE *fp;
    fp = fopen(f, "w");
    fputs(s, fp);
    fclose(fp);
}
int main()
{
    write("1.txt", "ABCDE");
    write("1.txt", "abc");
    return 0;
}
```

上述程序运行后，文件 1.txt 中的内容是（　　）。

A. abc　　B. abcde　　C. ABCde　　D. ABCDEabc

7. 以下程序出错的原因是（　　）。

```
#include <stdio.h>
int main()
{
    FILE *fp;
    char c;
    fp=fopen("text.txt", "w");
    c=fgetc(stdin);
    while(c!='#')
    {
```

```
        fputc(c, fp);
        c=fgetc(stdin);
    }
    return 0;
}
```

A. 函数 fopen 调用形式错误　　B. 文件指针 stdin 没有定义

C. 函数 fgetc 调用形式错误　　D. 输入文件没有关闭

8. 假设文件test存在且内容为空，以下程序执行后，文件test中的内容是（　　）。

```
#include <stdio.h>
#include <string.h>
void fun(char *fname, char *st)
{
    FILE *myf;
    myf=fopen(fname, "a");
    for(int i=0; i<strlen(st); i++)
    fputc(st[i], myf);
    fclose(myf);
    return;
}
int main()
{
    fun("test.txt", "hello, ");
    fun("test.txt", "new world");
    return 0;
}
```

A. hello, new world　　B. new woeldhello

C. new world　　D. hello,rld

9. C语言中，fclose函数在表示关闭不成功时返回（　　）。

A. 0　　B. -1　　C. EOF　　D. 非零值

10. 下列关于文件操作的叙述中，正确的是（　　）。

A. C 语言中的文本文件以 ASCII 码形式存储数据

B. C 语言中对二进制位的访问速度比文本文件慢

C. C 语言中随机读写方式同样适用于文本文件

D. C 语言中顺序读写方式不适用于二进制文件

二、程序分析题

1. 有如下程序代码，从键盘输入123456789#后，请写出输出结果。

```
#include <stdio.h>
#include <stdlib.h>
int main()
{
    FILE *fp;
    int i=0;
    char ch,*filename="file1.txt";
    if((fp=fopen(filename,"w"))==NULL)
    {
    printf("Cannot open file\n");
    exit(0);
    }
    printf("Please input string:");
    ch=getchar();
    while(ch!='#')
    {
```

```
        i++;
        fputc(ch,fp);
        putchar(ch);
        if(i%3==0) putchar('#');
        ch=getchar();
    }
    fclose(fp);
    return 0;
}
```

2. 有如下程序代码，第一题从键盘输入了123456789#后创建了一个file1.txt文件（本题运行的是同一个file1.txt文件），求得到的结果。

```
#include <stdio.h>
#include <stdlib.h>
int main()
{
    FILE *fp;
    char ch;
    long num=0L;
    if((fp=fopen("file1.txt","r"))==NULL)
    {
        printf("Open error\n");
        exit(0);
    }
    while((ch=fgetc(fp))!=EOF)
    {
        num++;
    }
    printf("num=%ld\n",num);
    fclose(fp);
}
```

3. 下列程序是读取文件内容并输出，请找出程序中的一处错误并改正。

```
#include <stdio.h>
#include <stdlib.h>
int main()
{
    FILE *fp;
    char ch,*filename="out.txt";
    if((fp=fopen(filename,"r"))==NULL)
    {
        printf("Cannot open file\n");
        exit(0);
    }
    printf("\n");
    while((ch=fgets(fp))!=EOF)
    putchar(ch);
    fclose(fp);
    return 0;
}
```

4. 有如下程序代码，从键盘输入了“#H我爱ello中国Wo好r和l学d人”后，则文件file1.txt里的存储内容是什么？

```
#include <stdio.h>
#include <stdlib.h>
int main()
{
    FILE *fp;
    char str[100];
```

```
    if((fp=fopen("file1.txt","w"))==NULL)
    {
        printf(" 文件打开失败！ \n");
        exit(0);
    }
    printf(" 请输入一个字符串:\n");
    gets(str);
    int i=0;
    while(str[i]!='\0')
    {
        if((str[i]>='a'&&str[i]<='z')||(str[i]>='a'&&str[i]<='Z'))
        {
            fputc(str[i],fp);
        }
        i++;
    }
    fclose(fp);
    return 0;
}
```

5. 下列程序的功能是将旧文件内容复制到新文件中，请找出程序中的错误并改正。

```
#include <stdio.h>
#include <stdlib.h>
int main()
{
     FILE *in, *out;
     char ch,infile[10],outfile[10];
     scanf("%s",infile);
     scanf("%s",outfile);
     if((in=fopen(infile, "r"))==NULL)
     {
         printf("Cannot open infile.\n");
         exit(0);
     }
     if((out=fopen(outfile, "w"))==NULL)
     {
         printf("Cannot open outfile.\n");
         exit(0);
     }
     while((ch=fgetc(in))!=NULL)
         fputc(ch, in);
     return 0;
}
```

三、编程题

1. 设计一段程序，将文件2中的内容追加到文件1中去。

2. 有两个磁盘文件A和B，各存放一行字母，今要求把这两个文件中的信息合并（按字母顺序排列），输出到一个新文件C中去。

3. 有5个学生，每个学生有3门课程的成绩，从键盘输入学生数据（包括学号、姓名、三门课的成绩），计算出平均成绩，将原有数据和计算出平均分数存放在磁盘文件stud中，最后输出文件stud里的内容。